High-energy Combustion Agents of Organic Borohydrides
Synthesis, Characterization and Applications

High-energy Combustion Agents of Organic Borohydrides
Synthesis, Characterization and Applications

By

XiaoHong Zhang
Xi'an Modern Chemistry Research Institute, China
Email: zhangxh204@sina.cn

WeiQiang Pang
Xi'an Modern Chemistry Research Institute, China
Email: nwpu_pwq@163.com

Luigi T. DeLuca
Politecnico di Milano, Italy
Email: luigi.t.deluca@gmail.com

and

Yu Zhao
Xi'an Modern Chemistry Research Institute, China
Email: 609092981@qq.com

ROYAL SOCIETY OF CHEMISTRY

Print ISBN: 978-1-83916-713-3
PDF ISBN: 978-1-83767-001-7
EPUB ISBN: 978-1-83767-002-4

A catalogue record for this book is available from the British Library

The Royal Society of Chemistry is a charity, registered in England and Wales, Number 207890, and a company incorporated in England by Royal Charter (Registered No. RC000524), registered office: Burlington House, Piccadilly, London W1J 0BA, UK, Telephone: +44 (0) 20 7437 8656.

Visit our website at www.rsc.org/books

Printed in the United Kingdom by CPI Group (UK) Ltd, Croydon, CR0 4YY, UK

Foreword

In recent years, with the development of science and technology, solid propellants and explosives with high combustion performance have drawn widespread attention by worldwide researchers, and the combustion agents to be burned in the corresponding high-energy formulations have undergone rapid development. Metal fuels, due to their very high gravimetric heat release and combustion performance improvement, are among the most promising ingredients for solid propellants and explosives, attracting much attention from researchers all over the world.

High-Energy Combustion Agents of Organic Borohydrides: Synthesis, Characterization and Applications is a professional book designed to systematically introduce the latest research progress in the areas of high-energy combustion agents, based on organic borohydrides, and solid propellant application performance. This publication covers the most recent achievements of the advanced borohydride combustion agents, experimentally studies the synthesis and characterization of borohydrides, analyzes the interactions between an organic borohydride (tetraethylammonium dodecahydrododecaborate, BHN) and the propellant main ingredients, discusses the properties of BHN/nAl (nano-sized aluminum powder) composites, and examines the effect of borohydrides on the properties of HTPB-based fuel-rich solid propellants. At the end, the prospective development of high-energy combustion agents in solid propellants and explosives are summarized, and the foreseeable development trends of military applications are presented.

This volume constitutes a collective scientific research achievement by the team of authors through referring to and summarizing recent domestic as well as international research findings in the relative fields. Thus, it has great value both as a theoretical research accomplishment, thanks to the wide scope of its compilation, and as a reference document, thanks to the large amount of reported experimental data.

High-energy Combustion Agents of Organic Borohydrides: Synthesis, Characterization and Applications
By XiaoHong Zhang, WeiQiang Pang, Luigi T. DeLuca and Yu Zhao

Published by the Royal Society of Chemistry, www.rsc.org

The publication of this book is an important contribution to international technical literature and will certainly play important role in the theoretical as well as technological development of high-energy organic combustion agents for solid propellants and explosives.

Vladimir Zarko
Voevodsky Institute of Chemical Kinetics and Combustion,
Novosibirsk 630090, Russia

Preface

In recent years, developing high-energy combustion agents used for solid propellants and explosives has become a hot topic worldwide. The main reasons for utilizing innovative high-energy combustion agents in various chemical propulsion devices are their high heat release and high combustion efficiency. Borohydrides have been paid attention by aviation and rocket researchers because of their high heat of combustion. Although innovative combustion agents have the very attractive feature of producing high energy or high burning rate, their combustion efficiency and industrial application present a great challenge to engineers and scientists. Various techniques have recently been developed to overcome these intrinsic difficulties. For example, many fundamental research investigations have also been conducted to explore detailed physicochemical processes associated with combustion and lab-scale application of innovative combustion agents. In order to present a systematic reference to readers, it is very suitable to publish a book on these innovative combustion agents.

The organization of this book was initiated in early 2021, and the outline was first reviewed and discussed by several co-authors and press editors, who provided invaluable suggestions that were greatly appreciated. This book is organized as a milestone of advanced research on high-energy combustion agents as used in chemical rocket propulsion technologies. The most recent achievements are summarized in seven chapters as follows. Chapter 1 includes ionic hydroborate, carborane and its derivatives, metal carborane, borane energetic ionic liquids (salts), synthesis progress of carborane derivatives, and applications progress of carborane derivatives and hydroborate salts. Chapter 2 experimentally studies the synthesis and characterization of typical high-energy combustion agents of organic borohydrides ($B_{10}H_{10}[(C_2H_5)_4N]_2$, BHN), bis(diaikyl-5-aminotetrazolium) dodecaborates, cationic polynitrogen heterocyclium decahydrodecaborates (CPHD),

High-energy Combustion Agents of Organic Borohydrides: Synthesis, Characterization and Applications
By XiaoHong Zhang, WeiQiang Pang, Luigi T. DeLuca and Yu Zhao

Published by the Royal Society of Chemistry, www.rsc.org

dihydrazine-1,2,4,5-tetrazine decahydrodecaborate, energetic caged-carborane derivatives by means of IR, SEM, DSC, TG–DTG, *etc.*, techniques from the particle size and particle size distribution, thermal decomposition and heat of combustion aspects. Chapter 3 analyzes the interactions between $B_{10}H_{10}[(C_2H_5)_4N]_2$ and energetic components and inert materials in solid propellants ingredients by means of differential scanning calorimetry (DSC) and vacuum stability test (VST) methods, examines the stability of $B_{10}H_{10}[(C_2H_5)_4N]_2$ under impact load such as single explosion shock and electric detonation impact stimuli, and analyzes the reaction performance of $B_{12}H_{12}[(C_2H_5)_4N]_2$ particles as well. Chapter 4 discusses the properties of $B_{10}H_{10}[(C_2H_5)_4N]_2$/nAl (nano aluminum powder) composites and the effects of $B_{10}H_{10}[(C_2H_5)_4N]_2$ on the properties of HTPB-based fuel-rich solid rocket propellants. Chapter 5 covers the effects of hydroborate metal additives, such as hydroborate iron salt (BH–Fe), hydroborate copper salt (BH–Cu), hydroborate lead salt (BH–Pb), *etc.* on the properties of composites and fuel-rich propellants from the SEM, surface and interface properties, thermal decomposition, and combustion diagnosis aspects. Chapter 6 analyzes the combustion of NG-based propellants with carborane methyl acrylate, investigates the boron cage compounds on combustion of liquid–solid high-energy fuels, and summarizes the recent advances of camouflaged carboranes and polyhedral boranes. Chapter 7 presents the prospective development of innovative combustion agents, including hydrogen storage composites, metal hydrides, metal borohydride compounds, ternary metal composites, and metallic-based energetic composites (MECs), which is a milestone of advanced research on high-energy combustion agents used in chemical rocket propulsion technologies.

Several international researchers have made significant contributions to the theory and experiments on high-energy organic combustion agents. To improve the quality of this book, many international experts were involved as external reviewers of the chapters in addition to the editors. It is our great pleasure to thank the following international experts for their substantial help in raising the quality of the book. Without their efforts, the publication of this volume would have been impossible. Our special thanks go to Professor Luigi T. DeLuca at Space Propulsion Laboratory (SPLab), Politecnico di Milano, Milan, Italy for his preface contribution to this book. We give our warmest thanks to Professors V. E. Zarko and Oleg Glotov from Voevodsky Institute of Chemical Kinetics and Combustion, Russia, and Professor Alon Gany from Faculty of Aerospace Engineering, Technion-Israel Institute of Technology, Haifa, Israel for their invaluable suggestions. Furthermore, our special thanks go to Professors FengQi Zhao, XueZhong Fan, BoZhou Wang, HuiXiang Xu, YunNa Xue, and Drs Fan Jiang and WeiQiang Wang from Xi'an Modern Chemistry Research Institute as well as other colleagues for their great support and help in the scientific experiments. The authors would also like to thank their colleagues Mr KeYong Wang, Ms WenJing Zhou, XiaoNing Ren, and Juan Zhao of the Analysis and Measurements Center, Xi'an Modern Chemistry Research Institute for carrying out SEM, particle size distribution, thermal decomposition, hazardous properties and flame structure analyses. The authors would also like to thank their colleagues for some useful suggestions regarding the English language, statistical data collection errors and further helpful advice. Finally, we especially thank the editorial staff for this book from the Royal Society of Chemistry, Merlin Fox and Amina Headley for their patience, kind support, and assistance in the completion of this publication. Without their work, this book would not have been organized and published. The authors sincerely hope that the research efforts described in this book will continue to exhibit fast growth and lead to maturation of advanced technologies, which will improve the quality of chemical propulsion for better defense and space exploration. We also sincerely wish that this international effort will help all readers gain a better understanding of the puzzling intricacies and appealing secrets of high-energy organic combustion agents as well as of the perplexing difficulties, but also fascinating horizons, of space propulsion.

Each chapter was edited carefully according to the production procedures employed by novel energetic materials. Thus, this book can be considered as supplemental reading for researchers, students, teachers, and engineers in universities and industries dealing with energetic materials production, characterization, and applications.

XiaoHong Zhang, WeiQiang Pang, Luigi T. DeLuca, and Yu Zhao
Xi'an, China and Milan, Italy

Author Biographies

Prof. Dr **XiaoHong Zhang**, doctor supervisor, is engaged in solid propellants, explosives, combustion, and explosion. He has served as the chief editor of academic journals, such as *Chinese Journal of Explosives and Propellants*, *FirePhysChem*, *etc*. He has published 100 peer-reviewed papers, 100 patents, and 10 books and book chapters.

Prof. Dr **WeiQiang Pang** is Master Supervisor, and Visiting Scholar at Space Propulsion Laboratory (SPLab), Politecnico di Milano, Italy. He is engaged in innovative composite energetic materials, combustion, simulation and application of solid propellants. He has served as the (young) editorial board member of *Energy Storage and Saving*, *Chinese Journal of Explosives and Propellants*, *Journal of Solid Rocket Technology*, *Equipment Environmental Engineering*, *etc*. He has also been special issue editor of *Crystals*, *FirePhysChem*, *Equipment Environmental Engineering*, *Journal of Weapon Equipment Engineering*, *etc*. He has published nearly 200 peer-reviewed papers, 50 patents, and 11 books.

Prof. Dr **Luigi T. DeLuca** was a Professor of Aerospace Propulsion at the Polytechnic University of Milan, Italy, where he founded its Space Propulsion Laboratory (SPLab). Now retired, Professor DeLuca has authored 18 books and more than 300 scientific publications. He is a Fellow of the Korean Government Brain Pool Program and a Visiting Professor at the Nanjing University of Science & Technology, China.

Associate Prof. **Yu Zhao**, is engaged in innovative composite energetic materials, combustion, and application in solid propellants. He has published 50 peer-reviewed papers, 20 patents, and 2 books.

High-energy Combustion Agents of Organic Borohydrides: Synthesis, Characterization and Applications
By XiaoHong Zhang, WeiQiang Pang, Luigi T. DeLuca and Yu Zhao

Published by the Royal Society of Chemistry, www.rsc.org

Contents

High-energy Combustion Agents of Organic Borohydrides: Synthesis, Characterization and Applications
By XiaoHong Zhang, WeiQiang Pang, Luigi T. DeLuca and Yu Zhao

Published by the Royal Society of Chemistry, www.rsc.org

CHAPTER 1

Comprehensive Survey of Combustion Agents

1.1 Introduction

A high burning rate propellant can make a solid rocket motor produce large thrust in a short time, which meets the requirements of rocket motor for strategic and tactical missiles. At present, the application of a burning rate catalyst to improve the burning rate of solid propellant is one of the main research directions in this field.[1,2] Compared with the previous inert catalyst, the new energetic burning rate catalyst not only improves the burning rate of propellant, but also enhances the energy of the propellant, which has become the focus of current research.

Borohydride compounds have been paid attention to by aviation and rocket workers because of their high combustion calorific value. It is an effective and practical way to obtain the high burning rate of solid propellant by using a borohydride burning rate regulator.[3,4] Boron and hydrogen compounds mainly include open cage borane, closed cage type ionic hydroborate, carborane and metal carborane. At present, there are many kinds of borohydride compounds which can regulate the burning rate of solid propellant, among which carborane derivatives with closed cage structure and ionic hydroborates are more studied. Metal carborane has been developed rapidly since it was reported in the 1960s, and its application in adjusting the burning rate of solid rocket propellant has also been reported. New borane energetic ionic liquids have good application prospects, and have been listed as the research plan of high energy rocket propellant abroad, but the research in this field is still in its infancy. In this chapter, the research progress of combustion agents, including ionic hydroborates, carborane and its derivatives, metal carboranes, borane energetic ionic liquids were summarized, the synthesis

High-energy Combustion Agents of Organic Borohydrides: Synthesis, Characterization and Applications
By XiaoHong Zhang, WeiQiang Pang, Luigi T. DeLuca and Yu Zhao

Published by the Royal Society of Chemistry, www.rsc.org

and application progress of carborane derivatives, hydroborate salts, carborane derivatives at home and abroad in recent years are overviewed, and the development of new high-efficiency energetic borane based burning rate regulators is prospected.

1.2 Combustion Agents

1.2.1 Ionic Hydroborates

Borohydride has been paid attention to by aviation and rocket workers because of its high combustion calorific value. It has been proved in practice that simple borohydride compounds have high chemical activity, high toxicity, incomplete combustion and high cost due to the open structure of bridged hydrogen (Figure 1.1), so it is difficult for them to be directly used as high-energy fuel. After that, a series of new stable borohydride compounds with closed cage structure were developed, which can be compatible with propellant components, in particular they can make composite propellant and double base propellant obtain good ballistic performance with high burning rate or even ultra-high burning rate (Figure 1.2).

Ionic hydroborates mainly refer to the salts of decahydrodecaboric acid and dodecahydrododecaboric acid, which have stable properties and low toxicity. The preparation of ionic hydroborates is mainly based on the pyrolysis of tetraalkylammonium borohydride,[6] followed by a simple double decomposition reaction or ion exchange process.[7] Polyhydric polyborates containing different cations (such as Li^+, K^+, Cs^+) can be prepared. The pyrolysis reaction is as follows:

$$10Et_4NBH_4 \rightarrow (Et_4N_2)B_{10}H_{10} + 8Et_3N + 8C_2H_6 + 11H_2 \quad (1.1)$$

Spielvogel *et al.*[8] provided a new method for the synthesis of hydroborates by pyrolysis of tetraalkylammonium borohydride under controlled conditions. A variety of ammonium hydroborates such as $B_9H_9^-$, $B_{10}H_{10}^{2-}$, $B_{11}H_{11}^-$, $B_{12}H_{12}^{2-}$, *etc.* can be obtained by choosing different pyrolysis temperatures. Compared with the Dunks process method,[9] the yield of this

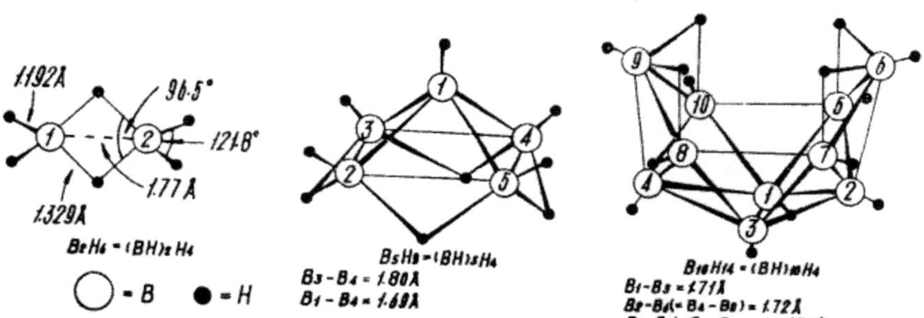

Figure 1.1 Simple borohydride with open structure. Reproduced from ref. 5.

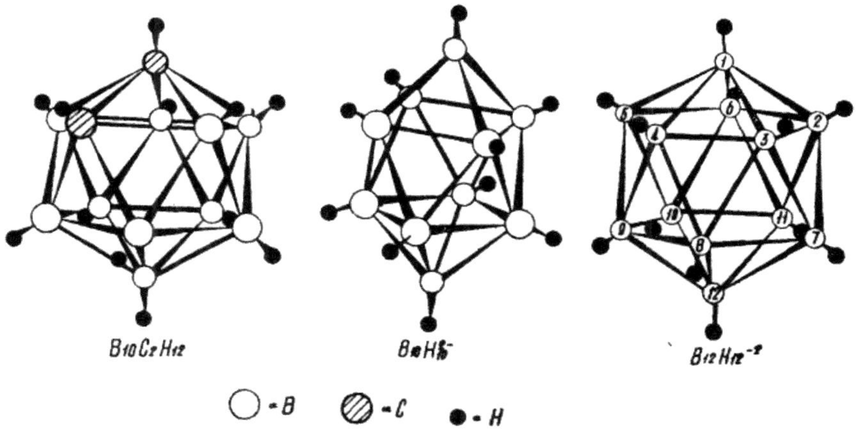

Figure 1.2 Borohydride with closed cage structure. Reproduced from ref. 5.

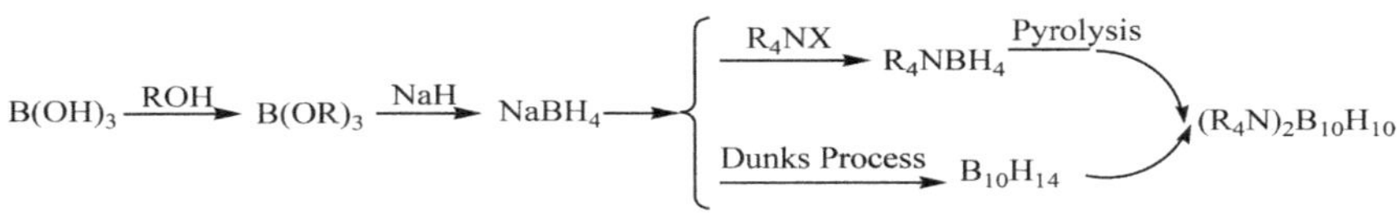

Figure 1.3 Two methods of preparation of $(R_4N)_2B_{10}H_{10}$. Reproduced from ref. 10.

method is significantly increased from 30% to more than 80%. The preparation process of tetraalkyl ammonium decahydrodecaborate is shown in Figure 1.3.

Ivanov *et al.*[11] summarized the preparation methods of dodecahydrododecaborates (such as ethylborane method and hydroborate thermolysis). On this basis, an economic and efficient preparation process of dodecahydrododecaborates was developed. The yield of $Na_2B_{12}H_{12}$ was about 80%. At the same time, because of the absence of the ethyl borane in the whole process, the process is safe. The process flow is shown in Figure 1.4.

Furthermore, for the application of ionic hydroborates in propellants, it has been widely reported that decahydroborates and their coprecipitates with nitrates, such as $K_2B_{10}H_{10}$, $Cs_2B_{10}H_{10}$, $(NH_4)_2B_{10}H_{10}$, $Cs_2B_{10}H_{10}$, $CsNO_3$ can increase the burning rate of composite and double base propellants significantly. For example, the coprecipitation of $Cs_2B_{10}H_{10}$ and KNO_3 was used as oxidant in HTPB propellant, and the burning rate of 300 mm s^{-1} can be obtained.[5] The mass amount of decahydrodecaborate in the formulation is large, about 15–26%, because a large amount of heat energy released by the rupture of the boron cage in the hydroborate is needed to accelerate the combustion of the propellant so as to achieve the purpose of adjusting the burning rate of the propellant. Although the amount of decahydrodecaborate in the formula is large, it has some disadvantages over carborane in terms of simple synthesis process and low toxicity.

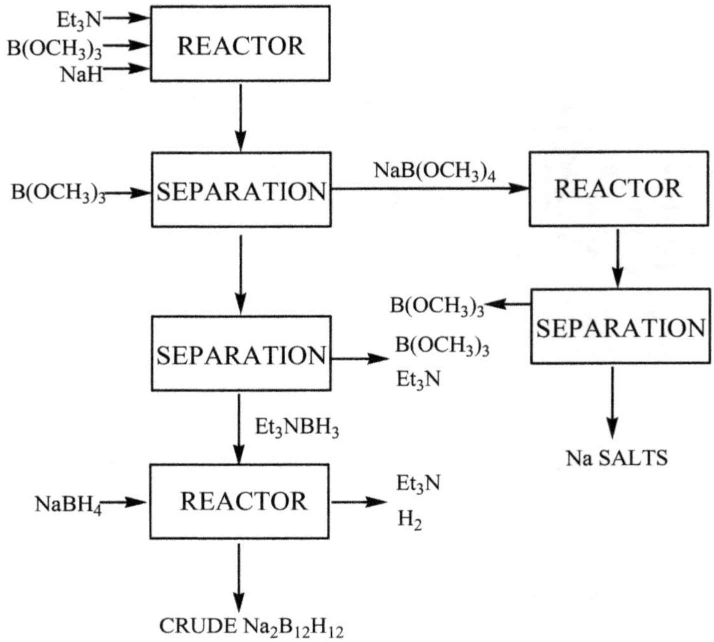

Figure 1.4 Synthesis of $Na_2B_{12}H_{12}$ from NaH and $B(OCH_3)_3$. Reproduced from ref. 10.

As an additive, dodecahydrododecaborate has been little researched on the burning rate of solid propellant. For instance, Chen Futai *et al.*[12] studied the effect of $[N(C_2H_5)_4]_2B_{12}H_{12}$ on the combustion properties of nitrate ester plasticized polyether (NEPE) propellant (the main components are cyclotrimethylene trinitramine (RDX), cyclotetramethylene tetranitramine (HMX), ammonium perchlorate (AP)), and the results indicate that this material has no obvious catalytic effect on the thermal decomposition of AP, but accelerates the thermal decomposition of nitramine, thus increasing the burning rate of NEPE propellant, and the "plateau" phenomenon appears over the high pressure range. When the amount of catalyst combined with $[N(C_2H_5)_4]_2B_{12}H_{12}$ and carbon is 1.5–2%, the burning rate of NEPE propellant at 7 MPa is 35.6% higher than that of lead citrate catalyst.

With the deepening of research, we halogenated the above two hydroborates (including fluorination, chlorination and bromination). For example, Castel *et al.*[13] studied the new fluorination process of $B_{10}H_{10}{}^{2-}$ and $B_{12}H_{12}{}^{2-}$ hydroborates reacting directly with F_2 in different acidic media, and prepared a series of hydroborates with a different number of fluorine atoms. Peryshkov *et al.*[14] summarized the preparation process of dodecafluorododecaborate, and found that the previous preparation was completed in acidic medium. On this basis, a process for preparing dodecafluorododecaborate was developed, in which acetonitrile was used as the reaction medium, and a certain proportion of mixed gas of F_2 and N_2 was continuously introduced into the system to react with dodecahydrododecaborates. At the same time, it was found that hydrofluoric acid and other protonic acids could terminate the fluorination reaction. In another example, Drozdova *et al.*[15] reacted $B_{10}H_{10}^{2-}$ with HCI, HBr, and HI in dichloroethane to form monohalogenated, dihalogenated and trihalogenated products. The results indicate that the proportion of product depends on the amount of hydrogen halide, reaction temperature, and reaction time.

The halogenated ions have more outstanding thermal and chemical stability, and can be used as electrolytes and as catalyst components in various reactions to improve the catalytic activity of metal cations. However, the effectiveness of halogenated ions in adjusting the burning rate of solid propellants is rarely reported, However, recent patents published in the United States have shown that energetic ionic salts prepared from partially fluorinated and perfluorinated boron cluster anions can be used as high-energy density materials to replace boron powder, aluminum powder, and other components in propellant formulations,[16] but specific data are not provided.

1.2.2 Carborane and Its Derivatives

Carborane and its derivatives have stable properties and are easily chemically modified,[17] which can meet the needs of applications in many fields.[18,19] These derivatives include isopropyl carborane, *n*-butyl carborane and *n*-hexyl carborane (NHC). They are liquid at room temperature, which is conducive to their uniform distribution in propellant, so as to ensure the burning rate stability of solid propellants. At present, NHC (containing mass fraction of boron 47.29%, boiling point 1178 °C at 64.5 Pa, density 0.901 $g\,cm^{-3}$, freezing point less than −50 °C) is most widely used. For example, the abdominal snake rocket launcher developed by General Dynamics for the army uses NHC 13.1% in mass fraction in the composition as the burning rate regulator to make the propellant reach a high burning rate of 144 $mm\,s^{-1}$ (13.73 MPa).[20]

NHC is a colorless and transparent liquid with high boiling point. It is stable in air, and has nonspontaneous combustion and low toxicity. The first method developed was the pyrolytic method of ethylborane, which was developed by Callery Chemical Company in the United States.[21] At present, a large-scale production unit of NHC has been established in the United States. However, it is necessary to develop other safe and efficient preparation processes due to the dangerous use of highly toxic ethylborane and its gas-phase cracking reaction. For example, Hill and Johnson developed the ammonium hydroborate pyrolysis method. The method successfully prepared *n*-hexane carboborane by the steps of decomposition of tetraalkylammonium borohydride, using sodium borohydride instead of highly toxic acetoborane and decaborane as starting materials.[22]

Boron containing compounds (especially boron hydrogen compounds) have been paid much attention by researchers for their ability to obtain higher burning rate of propellants, such as $B_{10}H_{10}^{2-}$, which can improve the thermal decomposition of HMX and increase the burning rate of solid propellants to 10–100 m s^{-1}.[23] Since the 1970s, a lot of research on carborane derivatives has been carried out in several countries, such as the United States, Japan, *etc.* The *n*-hexylcarborane can not only greatly increase the burning rate of solid propellant (more than 100 mm s^{-1}), but also improve the mechanical properties of propellants, and reduce the impact drop by a small margin.[24] For instance, the burning rate of ammonium perchlorate/triethylene glycol dinitrate (AP/TEGDN) propellant containing 48% AP in mass fraction is increased to 101.6 mm s^{-1} (14 MPa) by 12% in mass fraction addition of 11-isopropenyl-12-propyl carborane (IPCA).[25] After adding 0.2% of 1,2-*m*-2,3-epoxypropyl carborane and 3.5% 1-(2,3-epoxypropyl) carborane to the propellant containing 68% AP, the burning rate of the propellant is increased by 18 mm s^{-1}. Furthermore, when the oxidizing groups (such as ClO_4^-) were introduced to boron compounds (Table 1.1), it was found that the higher the number of boron cage groups, the closer the distance between them, and the stronger the catalytic effect of the compound on propellant combustion, which may be due to the formation of intra-molecular catalysis between the boron cage and ClO_4^-, enhancing the catalytic effect of this compound.[26]

In order to prepare ultra-high burning rate regulator, Hill *et al.*[27] introduced a nitro group into isopropenyl carborane to obtain 1-nitro-2-(1-carboranyl) propylene, which is an excellent burning rate regulator for double base propellant and modified double base propellants. Recently, Lin *et al.*[28] introduced a nitro group and an azide group into carborane derivatives when modifying 1-bromomethyl carborane, and obtained two new types of energetic caged carboranes as burning rate regulators of solid propellants, which are expected to make up for the energy loss of a propellant system due to large addition.

N-Hexylcarborane has been widely used in high burning rate propellants due to its excellent comprehensive properties. However, due to its migration, it has a negative effect on the properties of propellants. Therefore, a lot of research work has been carried out to solve this problem. At present, there are two main ways to overcome the mobility of borohydride burning rate regulators:

(1) The solid borohydride with oxidizing group is synthesized, which can be used instead of part of ammonium perchlorate (AP), so as to reduce the influence of the increase of solid mass fraction in the formula on the technological and mechanical properties of propellant. At present, the main synthetic materials are methyl carboboralkyl dicarboxylate methyl perchlorate (MTCMP) (Scheme 1.1) and methyl carboboralkyl

Table 1.1 Effect of the boron cage group and ClO_4^- group on the combustion catalysis of AP-based propellant. Reproduced from ref. 1.

Sample	Number of boron cage group	Number of ClO_4^-	Burning rate (mm s^{-1}) (14 MPa)	Degree of catalysis
NHC	1	0	144.78	Common
MTCMP	2	1	203.2–228.6	Very strong
CDFMP	1	1	177.80	Strong

bis-(ferrocenyl) methyl perchlorate (CDFMP) (Scheme 1.2), which can be used in HTPB propellant with high burning rate to obtain high energy and high burning rate. For instance, Shan *et al.*[1] summarized the effects of compounds containing a boron cage group and a ClO_4^- group on the burning rate of AP propellant. Under the same conditions, NHC, MTCMP, and CDFMP can make the burning rate of AP propellant reach 144.78 mm s^{-1}, 203.2 mm s^{-1}, and 177.8 mm s^{-1}, respectively.

(2) Carborane derivatives with active functional groups were synthesized and connected to the polymer chain of a binder by chemical reaction, or various polymers containing carborane groups were synthesized to make them have adhesive function. These carborane derivatives mainly contain a carbon double bond or epoxy group,[29] hydroxyl group,[30] and other functional groups. They can be polymerized with other double bond monomers (such as butadiene and acrylic acid) to synthesize copolymers with adhesive function or directly used as curing agents. Typical carborane derivatives include 1,2-bis(2,3-epoxypropyl) (Scheme 1.3) carborane and carboranyl acrylate (CMA) (Scheme 1.4).[31] For example, the burning rate of a carboxybutane propellant with *n*-butyl ferrocene as catalyst (7%) is 30.48 mm s^{-1}. When 1,2-bis(2,3-epoxy propyl) carborane (0.2%) was used instead of the common epoxy curing agent, the burning rate increased to 40.64 mm s^{-1}.

1.2.3 Metal Carboranes

How to form metal borane complexes by out of cage bonding, how to insert metal atoms into boron cages, the number of embedded metal atoms and the

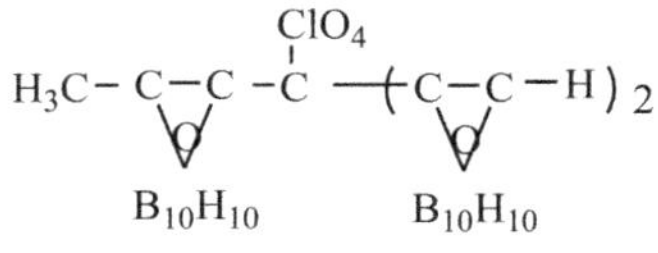

Scheme 1.1

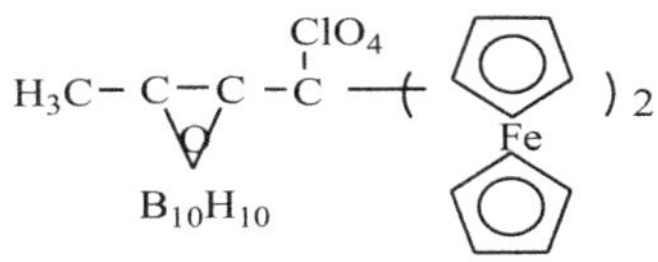

Scheme 1.2

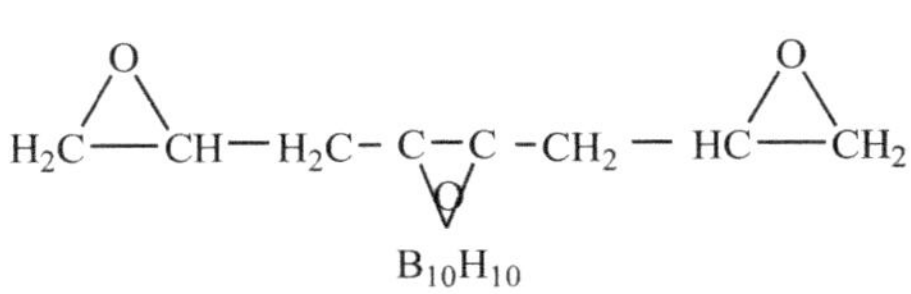

Scheme 1.3

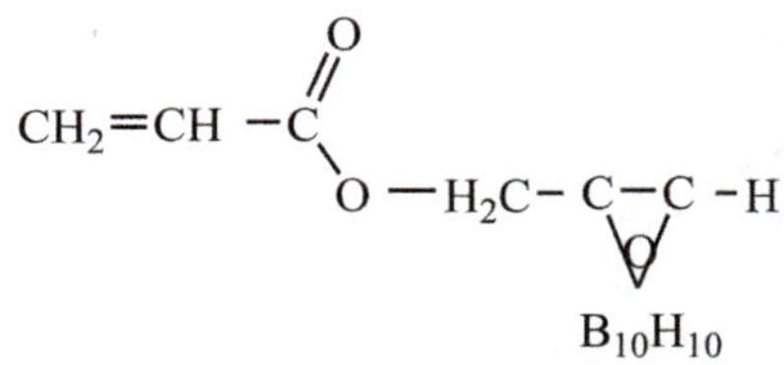

Scheme 1.4

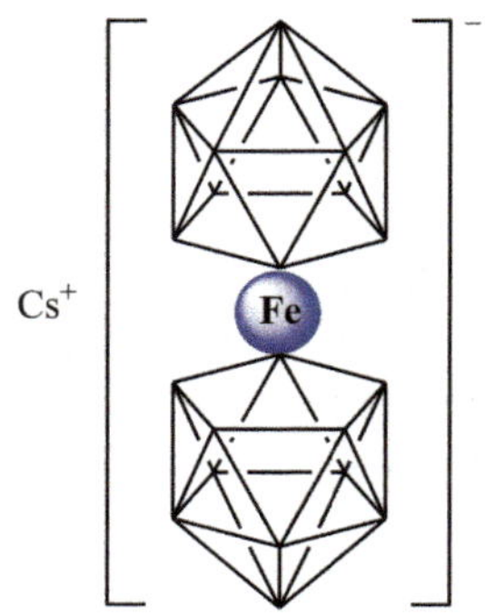

Figure 1.5 Structure of Cs[$(C_2B_9H_{11})_2$Fe]. Reproduced from ref. 10.

structure and properties of products have been the focus of research in this field.[32,33] At present, many new carborane metal complexes have been developed, but the research on their application is still in the theoretical stage.

Because carborane, ferrocene,[34,35] and other transition metals (such as iron, copper, cobalt, *etc.*) are efficient burning rate catalysts for propellants, some metal carborane burning rate catalysts based on these two kinds of materials, mainly dicarborane transition metal complexes (BDCMS) have been prepared. The structure is shown in Figure 1.5.

Some transition metal carboranes (such as Cu based, Ni based, and Fe based) were synthesized[36] and their properties as burning rate catalysts were studied. The results show that the iron-based carborane has ideal burning rate regulation performance. Nechai *et al.*[37,38] studied the catalytic mechanism of iron-based BDCMS on the combustion of propellant components. The results show that the thermal degradation temperature of BDCMS has an obvious effect on the combustion of catalytic components. Through thermal analysis of BDCMS, it is found that BDCMS has obvious degradation above 300 °C. In addition, when the mass fraction of BDCMS in nitrocellulose propellant is low (*e.g.* 1–2%), it has little effect on the burning rate of the propellant. However, when the mass fraction reaches 5%, the burning rate of the propellant can be greatly increased by 48.7 mm s^{-1}. Furthermore, Sivaevl *et al.*[39,40] summarized the synthesis methods and chemical properties of cobalt based, nickel based and iron-based BDCMS and their derivatives, and pointed out that they can be used in solvent extraction of radionuclides, catalysts, and biomedical fields.

1.2.4 Borane Energetic Ionic Liquids (Salts)

At present, after more than ten years of development, there are more than 100 kinds of ionic liquids, which have been widely used as solvents and electrolytes, and are also actively developing in other fields.

It was reported that ionic liquids can be used as new propellants, explosives and ammunition due to their unique properties, and the corresponding characterized performance of energetic ionic liquids as high-precision demonstration propellants was carried out by the US Air Force Research Laboratory (AFRL) in 2009. Li *et al.*[41] reviewed the research history, molecular design and properties of energetic ionic compounds, and pointed out that the research of energetic ionic compounds expands the research fields suitable for high-energy insensitive propellants, new high-energy materials, explosives and propellants.

Due to the adjustability of ionic liquids, some ideal structures can be designed by selecting suitable anions and cations, and some ionic liquids (salts) can be prepared as high energetic catalysts for the burning rate enhancement of modified propellants. At present, a series of energetic ion salts have been prepared by the combination of nitrogen containing heterocyclic cations and $B_{10}H_{10}{}^{2-}$ or $B_{12}H_{12}{}^{2-}$ and their properties in propellant and explosives. For the preparation of this kind of energetic ionic salts, there are two methods: double decomposition reaction and acid–base neutralization. For example, Nieuwenhuyzen *et al.*[42] combined imidazole cations with boron cluster anions through double decomposition reaction to prepare a series of low melting point ionic liquids. The boron cluster anions introduction are $[Co(C_2B_9H_{11})_2]^-$, $[C_2B_9H_{12}]^-$, $[B_{10}Cl_{10}]^{2-}$, and $[B_{12}Cl_{12}]^{2-}$. At the same time, it is found that the ionic salt with anion $[B_{12}Cl_{12}]^{2-}$ has higher thermal decomposition temperature.

The common nitrogen-containing heterocyclic compounds are azoles and azines. Keitz[43] and Weigele *et al.*[44] reported some new nitrogen-containing heterocyclic compounds. In order to prepare energetic materials with unique properties, nitrogen rich compounds such as triazole, tetrazole, and tetrazine can be combined with a boron cage group to prepare energetic ionic salts. For example, Shackelford *et al.*[45,46] prepared a series of heterocyclic carborane ionic salts by combining a variety of unsaturated nitrogen-containing high-energy heterocyclic cations with $[B_{12}F_{12}]^{2-}$, $[B_{12}H_{12}]^{2-}$, and $[CB_{11}H_{12}]^-$ in a new process. These materials have the characteristics of high energy and high density, and they can be used as high-energy density propellants and fuels. The main nitrogen-containing heterocycles used are triazole compounds (Schemes 1.5 and 1.6).

In recent decades, although borane chemistry has developed rapidly and made some achievements in biomedicine, new materials, molecular devices and other emerging fields, there are relatively few reports on the research of

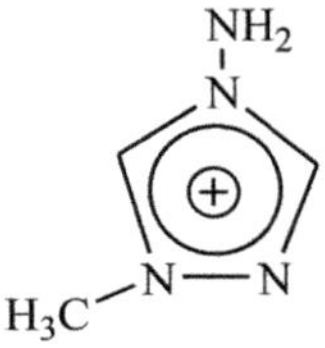

Scheme 1.5

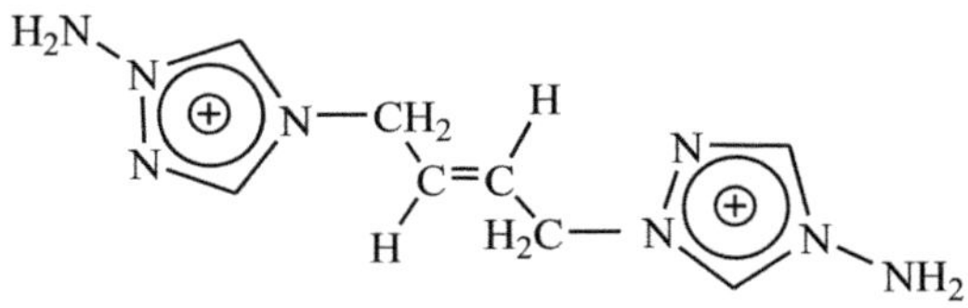

Scheme 1.6

borohydride as a burning rate regulator of solid propellants. Therefore, how to apply the current research results to this field to achieve a new breakthrough in burning rate improvement of solid propellant can be an important research direction. In addition, the development of a safe, efficient and low-cost preparation process is the premise of its large-scale application, which also needs to be paid attention to.

1.3 Synthesis Progress of Carborane Derivatives

Carborane is a general term for a class of compounds formed by carbon atoms replacing boron atoms at the apex of borane compounds. Since the first synthesis of carborane in the 20th century, it has been of great interest to the United States, Russia and other space powers. It was found that carborane derivatives can effectively improve the burning rate of propellant and improve the process performance. Carborane derivatives, for their high potential, good combustion performance, and good compatibility with composite materials, have a wide application background in propellants.[47]

1.3.1 Structure and Properties of Carborane Derivatives

For borane, different positions or different numbers of boron vertices are replaced by carbon, forming many carborane derivatives. According to the configuration difference, carborane derivatives can be divided into closed type (*closo*-), open type (*nido*-, nested structured), and *arachno* type (*arachno*-)[48] (Scheme 1.7). The closed structure of carbodecaborane ($C_2B_{10}H_{12}$) is one of the hot topics in current research. The number of valence electrons on the top of boron in the nested decaborane is 24 except in the formation of a B–H bond. The two carbon atoms just provide two additional electrons to fill the

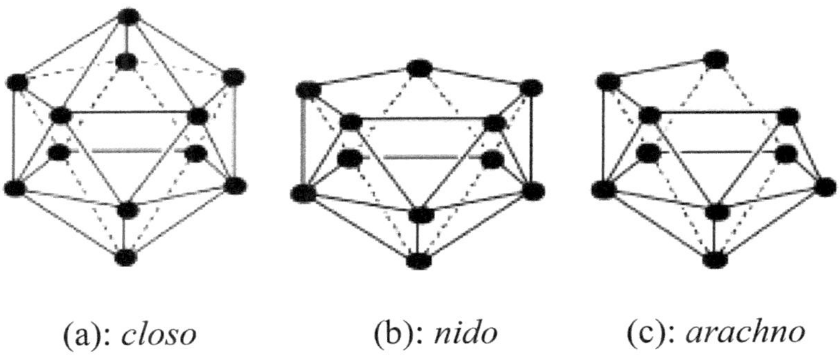

Scheme 1.7 Schematic presentation of *closo*-, *nido*-, and *arachno*-type polyhedral boranes. Reproduced from ref. 33.

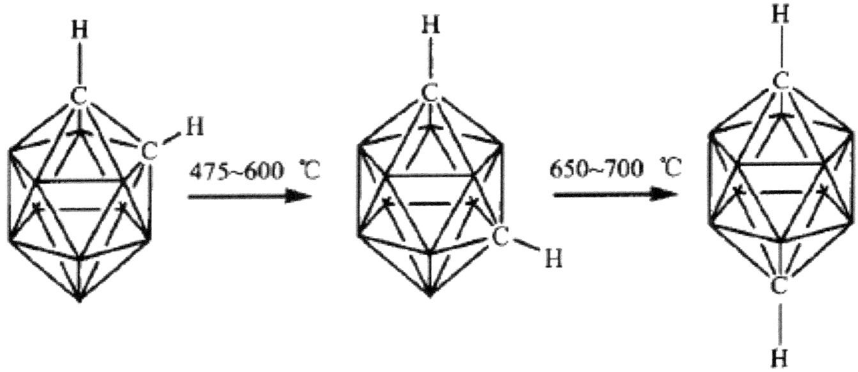

Figure 1.6 Three isomers of carborane and their thermal rearrangement (unmarked vertex is BH). Reproduced from ref. 47.

13 molecular orbitals required by the regular icosahedron, forming a closed carboborane with 12 vertices. Each vertex atom of the carborane cage is six coordinated, and there are three center two-electron chemical bonds. This belongs to a highly electron deficient structure, which will cause the shortening of adjacent bonds and further improve the stability of the compound.[49] The C atoms in the cage structure can be at positions 1 and 2, 1 and 7, and 1 and 12, respectively, forming three carborane isomers called *ortho*, *meta*, and *para*.[50] When the *o*-carborane is heated to 475 °C, it would be thermally rearranged into *m*-carborane. When heated to 650 °C, it can be further thermally rearranged into *p*-carborane, so *o*-carborane (melting point 254 °C) has received the most attention of researchers. The three carborane isomers and their thermal rearrangement reactions are shown in Figure 1.6.

Among the borane compounds, the closed form $[B_nH_n]^{2-}$ ($n = 6–12$) has attracted wide attention, among which $[B_{10}H_{10}]^{2-}$ (decadecadecaborate) and $[B_{12}H_{12}]^{2-}$ (dodecadecaborate) are the most representative. Both are triangular polyhedrons with a high stability $[B_{10}H_{10}]^{2-}$ skeleton that has a double

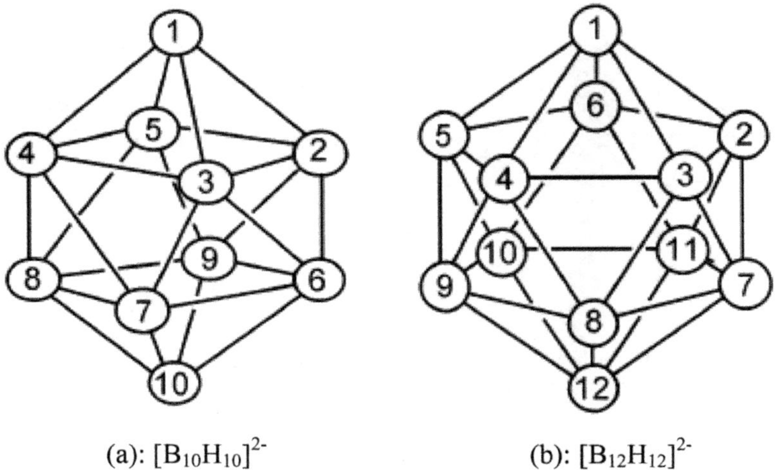

Figure 1.7 Structures and numbering schemes of $[B_{10}H_{10}]^{2-}$ and $[B_{12}H_{12}]^{2-}$. Reproduced from ref. 33.

capped tetragonal anti-prism structure, and $[B_{12}H_{12}]^{2-}$ is a regular icosahedral configuration (Figure 1.7).

1.3.2 Synthesis Progress of Carborane Derivatives

1.3.2.1 Synthesis of Carborane Derivatives

o-Carborane is generally prepared by the addition of substituted alkynes and decaborane. The intermediate products can be divided into two type methods according to whether they are separated or not (Figure 1.8): a one-step method is to synthesize *ortho* carborane directly from decaborane, Lewis base and alkyne. The two-step method is to synthesize *ortho* carborane from the decaborane complex and alkyne after reaction of decaborane with the Lewis base. Wherein, R = H, alkyl, aryl, haloalkyl, alkoxy, *etc*; L = acetonitrile, DMF, DMA, *etc.* Since the strong alkali alcohol solution or amine solution disintegrates the boron cage, substituted alkynes containing amino groups and hydroxyl groups cannot be used as reactants in the alkyne addition method.

Chen *et al.*[51] dissolved the acetonitrile complex of decaborane with a molar ratio of 1 : 1 and 1,4-diacyloxy-2-butylene in toluene, refluxed at 100 °C for 6 h, and purified to obtain 1,2-diacyloxymethyl carborane, with a yield and purity of 81.2% and 96.2%, respectively. Xue *et al.*[52] reacted 3.6 g of decaborane tetrahydrothiophene complex, 15 mL of toluene solution, and 1.9 g of terminal octyne at 90–120 °C for 1–6 h; after post-treatment, *n*-hexylcarborane was obtained with a yield of 40.2%.

Beihang completed the synthesis of *n*-hexylcarborane (NHC) by using the improved solvent method.[53] The proposed technical route is shown in Figure 1.9.

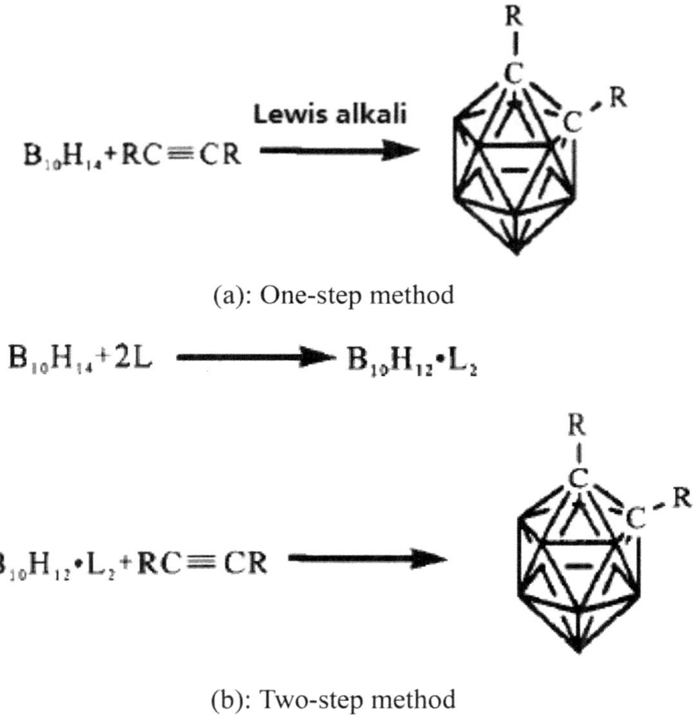

Figure 1.8 Synthesis method of carborane. Reproduced from ref. 47.

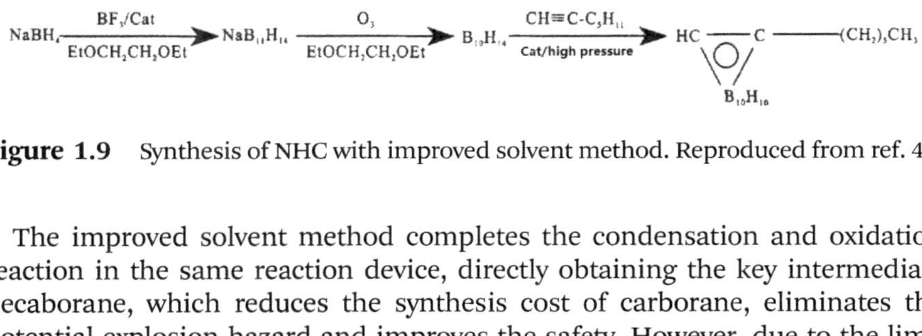

Figure 1.9 Synthesis of NHC with improved solvent method. Reproduced from ref. 47.

The improved solvent method completes the condensation and oxidation reaction in the same reaction device, directly obtaining the key intermediate decaborane, which reduces the synthesis cost of carborane, eliminates the potential explosion hazard and improves the safety. However, due to the limitation of the solvent method itself, a large amount of boron elements are converted into useless boric acid in the synthesis process, and the atomic utilization rate is low. As a result, the synthesis cost is still as high as 10 000–15 000 Yuan per kg, which makes it difficult to meet the demands of large-scale preparation of carborane derivatives. It can be seen that in order to meet the production demand of carborane derivatives for solid propellants, it is necessary to first solve the problem of how to realize the safe, efficient, low-cost and large-scale production of the upstream material decaborane, and then carry out the preparation of functional carborane derivatives.

1.3.2.2 Modification of Carborane

The modification of carborane mainly has the following four modes:

(1) The weakly acidic C-terminal H is attacked by reagents such as alkyl lithium and phenyl lithium, and then mono-substituted or disubstituted alkyl, acyl and aldehyde groups are introduced at the C-terminal by substitution, substitution and coupling.[54,55] Cheng *et al.*[56] added 20.0 mmol of *n*-butyl lithium and 10.0 mmol of *o*-carborane to the anhydrous toluene/ether mixed solution under stirring at 0 °C. The mixture was warmed to room temperature and stirred for 30 min, then a toluene/ether solution of ethylene oxide was added at 0 °C. After stirring at room temperature for 24 h, the crude product disubstituted hydroxyethyl *o*-carborane was obtained with a yield of 92%. Dozzo *et al.*[57] synthesized mono-substituted carborane formyl derivatives with *o*-carborane, methyl formate and *n*-butyl lithium as raw materials, opened up a high-yield preparation route of *o*-carborane aldehyde derivatives, and also converted these aldehyde derivatives into carborane aminoesters through reductive amination.

(2) The backbone electrons of carborane are highly delocalized and it can be electrophilically substituted like planar aromatic hydrocarbons to realize the conversion from B–H to B–X, such as the metallization, alkylation, halogenation, and hydrogenation of boron vertices.[58–60] The order of electrophilic substitution is: [B (9, 12)]>[B (8, 10)]>[B (4, 5, 7, 11)].

    Zhu *et al.*[54] refluxed 0.6 mol of *o*-carborane, 0.6 mol of *p*-cyanobenzyl bromide and 0.6 mol of anhydrous aluminum trichloride in toluene solution for 5 h, extracted with dilute hydrochloric acid, dried and purified to obtain a coupling product containing a B–C bond and a B–Cl bond. The highly regioselective B (8,9,10,12)-tetrafluoro substituted carborane derivatives were obtained by using a F^+ reagent in a palladium catalyst system.[61] Compared with the traditional SbF_5 fluorination system, this method does not require special reaction vessels and solvents, and avoids the generation of HF, which is safer and more efficient.

(3) When carborane reacts with strong and non-bulky nucleophiles, the skeleton boron atoms can be removed, and the boron cage would be degraded to form a *nido*-$C_2B_9H_{12}^-$ anion, as shown in Figure 1.10(a), and then it can form ligand with metal.[62]

(4) *o*-Carborane can also undergo dehydrogenation reaction to form two active intermediates: 1,2-dehydrogenated *o*-carborane or 1,3-dehydrogenated *o*-carborane, as shown in Figure 1.10(b), which may undergo a pericyclic reaction with alkyne, alkyne, aromatic compound or heteroaromatic compounds, alkyne, aromatic compound or heteroaromatic compounds.[63]

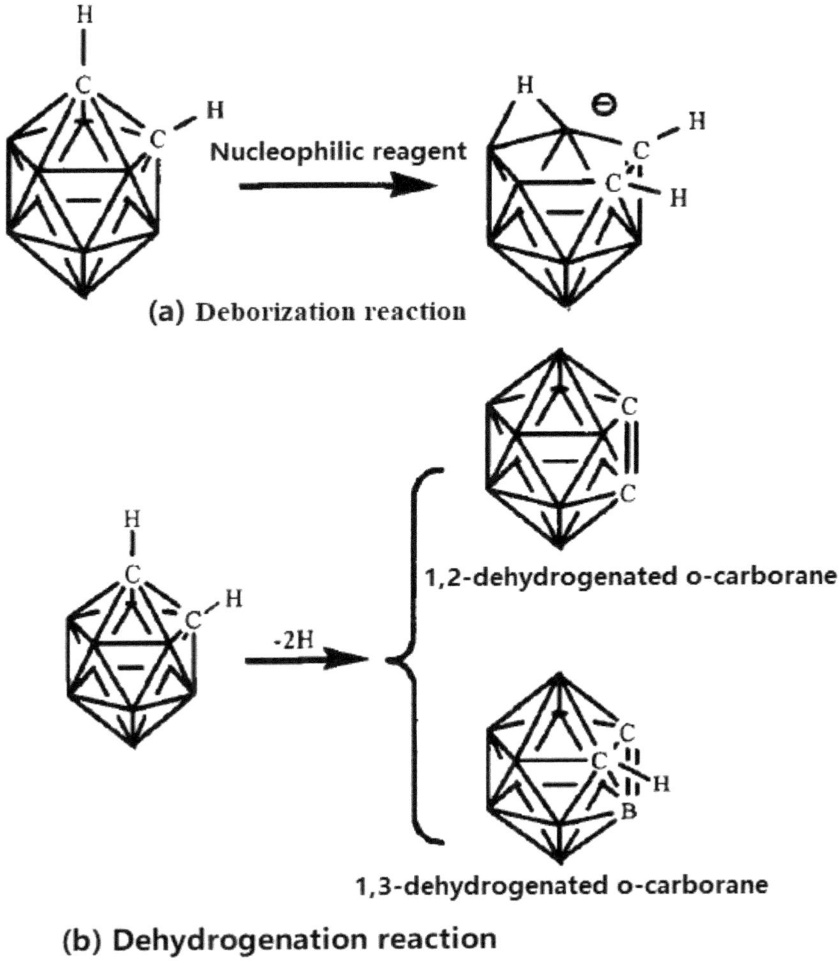

Figure 1.10 Dehydrogen reaction and deboron reaction of carborane. Reproduced from ref. 47.

1.3.3 Application Progress of Carborane Derivatives

1.3.3.1 Burning Rate Modifiers

Due to the advantages of stable properties and good compatibility with propellant components, many kinds of carborane derivatives have been used as propellant and fuel components in tactical missile weapons in recent decades. The mass fraction of *n*-hexylcarborane (NHC) in the butyl hydroxyl formula used in the "Agkistrodon" rocket launcher produced by General Motors of the United States is 13.1%, and the burning rate is increased to 144 mm s^{-1}, which is about 1–1.2 times higher than the initial speed of the previous generation of equipment.[10] Sayles[64] tested the gap formula containing 4.7% carboborate alkyl methacrylate, and the burning rate was 259.1 mm s^{-1} at 13.8 MPa.

The research on carborane derivatives with single functional groups have proved that they have good compatibility with each component of propellant and an obvious burning rate regulation effect in China. Under certain compositions conditions, some derivatives can improve the pressure exponent of the propellant and have the potential to be applied to fuel-rich solid propellant.[65] Shan *et al.*[66] found that carborane methyl acrylate (CMP) can catalyze the high-temperature decomposition of AP and can increase the burning rate of AP/nitroglycerin (NG) propellant by 83–155%. When the addition of CMP is 8% in mass fraction, the burning rate of propellant reaches 107.53 mm s^{-1} at 12 MPa. When the addition of CMP is 12% in mass fraction, the burning rate of propellant is 106.38 mm s^{-1} under 8 MPa. Lin *et al.*[67] synthesized 1-nitromethyl carborane, with a decomposition peak of 168.7 °C and the decomposition heat release is 245.6 J g^{-1}. As an energetic burning rate catalyst, it can compensate for the energy loss of propellant caused by large addition.

In addition to adjusting the burning rate, carborane derivatives also have other functions: carborane derivatives which are liquid at room temperature can be added to the formula as plasticizers to improve the process performance and mechanical properties, such as *n*-hexylcarborane and isopropyl carborane. Derivatives containing oxidizing groups can be used instead of part of NH_4ClO_4 to prevent the degradation of propellant mechanical properties caused by excessive solid mass fraction of formulation components, such as perchlorate containing carborane groups.[68–70] The copolymer of carborane methyl acrylate can also replace part of the binder.[71–73] Carborane derivatives with epoxy groups can also be used as curing agents for carboxybutane propellants.

Li *et al.*[74] prepared a carborane methyl ammonium perchlorate, a good high-energy density burning rate catalyst, which has good compatibility with the common components of solid propellant. The propellant was prepared by HTPB (9.7% in mass fraction) addition. The density was 1.64 g cm^{-3} and the burning rate was 177.28 mm s^{-1} at 13.73 MPa. The carborane methyl ammonium perchlorate was replaced by NHC in equal mass fraction. The density of the sample was 1.58 g cm^{-3} and the burning rate was 131.56 mm s^{-1} at the same pressure (13.73 MPa).

1.3.3.2 High-performance Fuels

Carborane has the dual combustion characteristics of borane and hydrocarbon fuel, with a density of 1.0 g cm^{-3} and a combustion calorific value of 67.6 MJ kg^{-1}, which is more than 45% higher than that of cubane with the highest calorific value reported in the literature. The volumetric calorific value is 13% higher than that of cubane, which is significantly higher than that of common hydrocarbon fuels. Carborane does not have the glass-like boron oxide layer under combustion conditions, and boron atoms are dispersed into the physical space of the carborane cage to make boron more accessible during combustion, which is conducive to the rapid formation of hydrogen borate/boric acid (HOBO) and B_2O_3. This rapid oxidation method can release the potential energy of fuel and enhance the burning rate of high-performance propellant without delaying melting or limiting diffusion.

Wang *et al.*[75] used 10–45% of carborane ethers, aromatic hydrocarbons and polymer derivatives as the main fuels for fuel-rich solid propellant formulation alone or in combination. Such carborane propellants can be ignited normally under low pressure and high injection efficiency, showing high energy properties, excellent ignition and combustion performance. It was also found that carborane decomposes in the primary combustion to produce large numbers of highly active and electron deficient free radical fragments. The fragments are then disproportioned with the unsaturated small molecules with electron rich structure produced by the decomposition of hydrocarbon fuel to form carborane based highly active small molecules, thus reducing the carbon deposition phenomenon and improving the energy conversion efficiency greatly.

1.3.3.3 Others

The large molecular volume of carborane can prevent the close arrangement of polymers, play a plasticizing role, and improve the heat resistance and stability of polymers. Carborane modified polyurethane adhesive produced in Russia, heated to 600 °C, with mass loss of only 20%, can be used for aerospace engines and other equipment working in high temperature or high concentration acid–base environment. Because of its "super aromatic" cage-like structure, carborane is often introduced into the extended conjugated part of polymers to give excellent photoelectric properties and can be used in photoelectric materials. Moreover, the theoretical mass fraction of B in carborane is as high as 75%. It is considered as an excellent absorbing material, which can be used for neutron detection, neutron shielding and others. Carborane containing a copolymer has a high B content, and 10 B atoms can accept 15 O atoms during oxidation. Therefore, it is considered as a good atomic oxygen resistant material, which can prevent oxidation cracking of spacecraft and satellite surface coating and effectively extend its service life. Introducing a carborane structure into coordination polymers can form finer pores than those containing benzene rings. At present, carborane cages have been used as linkers to form highly porous metal organic frameworks (MOFs). These materials can also be used for gas storage and separation.

1.4 Application Progress of Hydroborate Salts in Propellants and Explosives

1.4.1 Synthesis, Structure and Properties of $[B_{10}H_{10}]^{2-}$

Hawthorne *et al.*[76] synthesized a colorless crystalline solid $[Et_3NH]_2B_{10}H_{10}$ (crude salt yield of 93%) for the first time through the reaction of *nido*-$B_{10}H_{14}$ (Figure 1.11) and Et_3N in boiling xylene. Zhang *et al.*[77] developed a simple

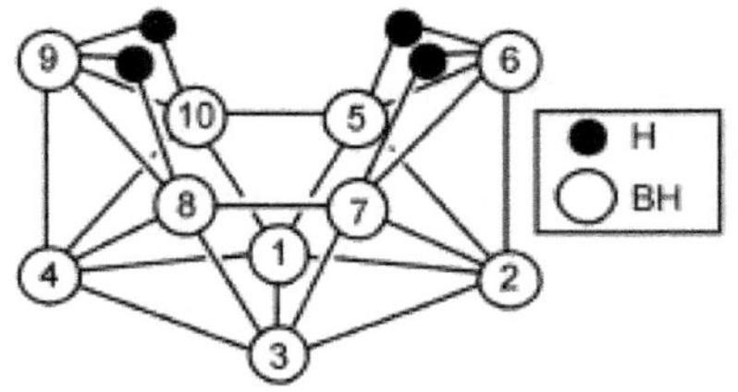

Figure 1.11 Schematic presentation of a *nido*-$B_{10}H_{14}$ skeleton. Reproduced from ref. 33.

one-step synthesis method of $[B_{10}H_{10}]^{2-}$ by using the pyrolysis reaction of Et_4NCl and KBH_4 in the presence of a small amount of water.

$B_{10}H_{10}{}^{2-}$ is aromatic due to the three-dimensional electron delocalization of a boron cage,[78] similar to aromatic organic systems and less similar to inorganic anions. Its salt is generally resistant to strong acid and alkali, and has unique thermal and hydrolytic stability compared with other neutral boranes. Moreover, $B_{10}H_{10}{}^{2-}$ has the advantages of low toxicity (equivalent to sodium chloride) and good water solubility. It has been reported that the calculated enthalpy of formation of $[Et_3NH]_2B_{10}H_{10}$ is -52.3 kJ mol^{-1}.[79]

1.4.2 Reaction of the *closo*-Decaborate Anion $B_{10}H_{10}{}^{2-}$ with Dichloroethane

The *closo*-decaborate anion is a pseudoaromatic system prone to numerous reactions involving replacement of exopolyhedral hydrogen atoms, including chlorination. It is known from published data that halogenation of the *closo*-decaborate anion with free halogen in an aqueous solution of the ammonium salt $(NH_4)_2B_{10}H_{10}$[80] results in mono- and dihalogenated anions $B_{10}H_{10-n}X_n{}^{2-}$, where X = Cl, Br, I and $n = 1, 2$, which were identified by NMR, IR, and Raman spectroscopy. The structure of monohalogenated derivatives of the *closo*-decaborate anion $[(C_5H_5N)_2CH_2][2\text{-}XB_{10}H_9]$, where X = Cl, Br, I, was determined more precisely by X-ray diffraction.[81,82] These products are formed upon the reaction of $B_{10}H_{10}{}^{2-}$ with chlorine, bromine, or iodine, or with *N*-halosuccinimide and can be separated from the unreacted starting anions or highly substituted halo derivatives by ion exchange chromatography. The reaction of the *closo*-dodecaborate anion $B_{12}H_{12}{}^{2-}$ with hydrogen halides in dichloroethane has been reported,[83] giving mono- and dichlorinated products.

1.4.2.1 Experimental

The $B_{10}H_{10}{}^{2-}$ anion was synthesized by a reported procedure[84] *via* 1,6-bis-(triethylaminedecaborane). The hydrogen halides used in the reaction were prepared by known procedures:[85] HCl was obtained by the reaction of sodium chloride with sulfuric acid; HBr and HI were obtained by the reaction of an aqueous solution of the respective hydrohalic acid with phosphorus oxide or by the reaction of bromine or iodine with tetralin. Prior to being supplied to the reactor, the hydrogen halide gas was passed through two absorption tubes with phosphorus oxide for complete drying.

1.4.2.1.1 Chlorination of the $B_{10}H_{10}^{2-}$ Anion. A solution of tetrabutylammonium decaborate (1.0 g) in dichloroethane (30 mL) was placed into a three-necked flask equipped with a bubbling device, a reflux condenser, a thermometer, and a stirrer, and the mixture was heated to 40–60 °C. Thoroughly dried hydrogen halide was bubbled continuously through the mixture for 1–5 h. The yellow reaction mixture was concentrated on a rotary evaporator until a precipitate formed. The reaction conditions, the reactants, and the products are summarized in Table 1.2.

1.4.2.1.2 Synthesis of $(Ph_3(NaphCH_2)P)_2B_{10}H_8Cl_2$. The reaction mixture (0.8 g) was dissolved in CH_3CN (25 mL). A solution of $(C_6H_5)_3(CH_2C_{10}H_7)PCl$ (2.0 g in 20 mL of H_2O) was added with stirring to the resulting solution. After 24 h, a white finely crystalline precipitate formed, which was filtered off and recrystallized from acetonitrile to give white crystals.

The elemental analysis for carbon, hydrogen, and nitrogen was carried out on a CHNS-3 FA 1108 analyzer (Carlo Erba). Boron was determined by electrothermal atomization atomic absorption spectroscopy on PerkinElmer spectrophotometers (model 2100 with HGA-700 and model 403 with HGA-72).[86] Elemental analysis data are presented in Table 1.3.

IR spectra were measured on an Infralum FT-02 spectrophotometer (the Lumeks Scientific and Production Company for Analytical Instrument) over the range of 400–4000 cm^{-1}. The samples were prepared as mineral oil mulls. The assignment for most informative absorption bands observed in the IR spectra of chlorinated *closo*-decaborates is presented in Table 1.4.

Table 1.2 Main products of the reactions of the $B_{10}H_{10}^{2-}$ anion salts with dichloroethane and hydrogen halides. Reproduced from ref. 15 with permission from Springer Nature, Copyright 2007.

	Temp.	
Time	40 °C	60 °C
Dichloroethane + hydrogen chloride ($C_2H_4Cl_2$ + HCl)		
1 h	$B_{10}H_9Cl^{2-}$	$B_{10}H_9Cl^{2-}$, $B_{10}H_8Cl_2^{2-}$
5 h	$B_{10}H_9Cl^{2-}$, $B_{10}H_8Cl_2^{2-}$, $B_{10}H_7Cl_3^{2-}$	$B_{10}H_8Cl_2^{2-}$, $B_{10}H_7Cl_3^{2-}$
Dichloroethane + hydrogen bromide ($C_2H_4Cl_2$ + HBr)		
1 h	$B_{10}H_9Cl^{2-}$	$B_{10}H_9Cl^{2-}$, $B_{10}H_8Cl_2^{2-}$, $B_{10}H_7Cl_3^{2-}$
5 h	$B_{10}H_9Cl^{2-}$, $B_{10}H_8Cl_2^{2-}$, $B_{10}H_7Cl_3^{2-}$	$B_{10}H_8Cl_2^{2-}$, $B_{10}H_7Cl_3^{2-}$, $B_{10}H_7Cl_2Br^{2-}$, $B_{10}H_6Cl_2Br_2^{2-}$, $B_{10}H_6Cl_3Br^{2-}$, $B_{10}H_5Br_5^{2-}$
Dichloroethane + hydrogen iodide ($C_2H_4Cl_2$ + HI)		
1 h	$B_{10}H_9Cl^{2-}$	$B_{10}H_8Cl_2^{2-}$, $B_{10}H_7Cl_3^{2-}$
5 h	$B_{10}H_8Cl_2^{2-}$, $B_{10}H_7Cl_3^{2-}$	$B_{10}H_7Cl_2I^{2-}$, $B_{10}H_6Cl_3I^{2-}$, $B_{10}H_6Cl_2I_2^{2-}$, $B_{10}H_5Cl_3I_2^{2-}$

Table 1.3 Elemental analysis data for chlorinated *closo*-decaborates. Reproduced from ref. 15 with permission from Springer Nature, Copyright 2007.

Compound	B, % Found/calcd	C, % Found/calcd	N, % Found/calcd	H, % Found/calcd
$[(C_4H_9)_4N]_2B_{10}H_9Cl$	16.92/16.96	60.21/60.28	4.34/4.39	12.78/12.81
$[(C_4H_9)_4N]_2B_{10}H_8Cl_2$	16.02/16.09	57.15/57.19	4.15/4.17	11.95/12.00

Table 1.4 Maxima of some absorption bands in the IR spectra of chlorinated *closo*-decaborates. Reproduced from ref. 15 with permission from Springer Nature, Copyright 2007.

Compound	ν(B–H), cm^{-1}	ν(BBCl), cm^{-1}
$[(C_4H_9)_4N]_2B_{10}H_9Cl$	2479	960
$[(C_4H_9)_4N]_2B_{10}H_8Cl_2$	2475	962
$(Ph_3(NaphCH_2)P)_2B_{10}H_8Cl_2$	2476	961

Table 1.5 ^{11}B NMR data for chlorinated *closo*-decaborates (in CH_3CN). Reproduced from ref. 15 with permission from Springer Nature, Copyright 2007.

Compound	Chemical shift, δ, ppm	Integrated intensity	Signal multiplicity	Assignment
$[(C_4H_9)_4N]_2B_{10}H_9Cl$	−2.3	2	d	B_a
	−7.2	1	s	B_{Cl}
	−20.7	4	d	B_e
	−24.5	2	d	
	−28.8	1	d	
$[(C_4H_9)_4N]_2B_{10}H_8Cl_2$	−4.3	1	d	B_a
$(Ph_3(NaphCH_2)P)_2B_{10}H_8Cl_2$	−7.8	1	s	B_{Cl}
	−21.4	1	d	B_e
	−25.1	1	d	
	−30.9	1	D	

The ^{11}B NMR spectra of acetonitrile solutions of the studied compounds were recorded on a Bruker AC 200 spectrometer operating at 64.297 MHz with internal deuterium lock. Boron trifluoride etherate was used as the external standard. The ^{11}B NMR data for chlorinated *closo*-decaborates are presented in Table 1.5.

Electrospray ionization mass spectra (microspray, 4500 V) of solutions of the compounds were recorded on a Bruker Esquire 3000 plus mass spectrometer in an acetonitrile (50%)–water mixture containing formic acid (0.2%). The average analytical concentration of the samples for ESI MS was a $(1.00 + 0.20)$ mg mL^{-1} solution. The molecular masses of the single-charged anions $(CatAn)^-$ determined from the spectra by analysis of reaction mixtures are given in Table 1.6.

Table 1.6 Data of the ESI mass spectra of some halogenated *closo*-decaborates for the $(CatAn)^{-}$ species. Reproduced from ref. 15 with permission from Springer Nature, Copyright 2007.

Compound	Exp.	Calcd
$[(C_4H_9)_4N]_2B_{10}H_9Cl$	394.93	395.10
$[(C_4H_9)_4N]_2B_{10}H_8Cl_2$	429.53	429.37
$[(C_4H_9)_4N]_2B_{10}H_7Cl_3$	464.45	463.33
$[(C_4H_9)_4N]_2B_{10}H_7Cl_2Br$	508.30	508.44
$[(C_4H_9)_4N]_2B_{10}H_6Cl_2Br_2$	588.13	587.34
$[(C_4H_9)_4N]_2B_{10}H_6Cl_3Br$	544.25	542.89
$[(C_4H_9)_4N]_2B_{10}H_5Br_5$	755.92	755.14
$[(C_4H_9)_4N]_2B_{10}H_7Cl_2I$	554.48	555.44
$[(C_4H_9)_4N]_2B_{10}H_6Cl_3I$	590.22	589.89
$[(C_4H_9)_4N]_2B_{10}H_6Cl_2I_2$	681.02	681.34
$[(C_4H_9)_4N]_2B_{10}H_5Cl_3Br_2$	715.89	715.79

X-ray diffraction studies were carried out on an Enraf-Nonius CAD-4 automated diffractometer (λMoK_a, graphite monochromator, ω scan mode). The absorption correction was applied by azimuthal scanning. The structure was solved by direct methods. The structure solution and refinement were carried out using SHELXS97 and SHELXL97 software.[87]

The set of diffraction data $I(hkl)$ at $T = 120(2)$ K was obtained by a standard procedure[88] on a Bruker AXS SMART 100 diffractometer (λMo, graphite monochromator, ω scan mode, $\theta_{max} = 30°$) from a 0.55 mm × 0.30 mm × 0.15 mm single crystal. A semiempirical correction for absorption was applied by the SADABS program.[89]

The crystal data, X-ray experiment details, and structure refinement data for $(Ph_3(NaphCH_2)P)_2B_{10}H_8Cl_2$ are presented in Table 1.7.

1.4.2.2 Results and Discussion

It was found that the reactions of the *closo*-decaborate anion with dichloroethane in the presence of hydrogen halides give chloro-substituted derivatives with different degrees of substitution. Analysis of the ^{11}B NMR spectra of the products (Table 1.8) and ESI mass spectra (Table 1.9) carried out at all stages of synthesis showed the presence of halogenated derivatives with degrees of substitution of mainly 1, 2 and 3.

The product ratio depends on the hydrogen halide used and on the synthesis temperature and time. The use of a heavier hydrogen halide and also an increase in the temperature and time of the synthesis promote accumulation of products with higher degrees of substitution.

The formation of the major chlorination products is always accompanied by the formation of derivatives with higher degrees of substitution, including those with different halides, which are formed when hydrogen bromide and hydrogen iodide are used.

Table 1.7 Selected crystal data, X-ray experiment details, and refinement data for $(Ph_3(NaphCH_2)P)_2B_{10}H_8Cl_2$. Reproduced from ref. 15 with permission from Springer Nature, Copyright 2007.

Compound	$(Ph_3(NaphCH_2)P)_2B_{10}H_8Cl_2$
Space group	*C*2/*c*
a, Å	24.229(4)
b, Å	11.81(2)
c, Å	22.223(2)
β, deg	122.191(11)
V, Å^3	5384.4(5)
ρ_{calcd}, g cm^{-3}	1.226
Z	4
μ(Mo), mm^{-1}	1.92
θ_{max}, deg	90
Number of independent reflection (*N*)	5107
Number of reflection with $I > 2\sigma(I)(N_{obs})$	1980
*R*1, w*R*2 for N_{obs}	0.081, 0.224
*R*1, w*R*2 for *N*	0.221, 0.305
GOOF	1.021

Table 1.8 HPLC data for the initial and some synthesized halogen derivatives of the *closo*-decaborate anion (elution with $CH_3CN : CH_2Cl_2 = 3 : 8$). Reproduced from ref. 15 with permission from Springer Nature, Copyright 2007.

Anion	Compound	Retention time, min
$B_{10}H_{10}^{2-}$	$[(C_4H_9)_4N]_2B_{10}H_{10}$	7.1
	$[(C_4H_9)_4N]_2B_{10}H_9Cl$	5.3
	$[(C_4H_9)_4N]_2B_{10}H_8Cl_2$	3.3
	$[(C_4H_9)_4N]_2B_{10}H_7Cl_3$	2.7

For none of the reactions were we able to select conditions under which one chlorinated product would be formed predominantly, because the yields of products present in the reaction mixtures in all reactions were commensurable. However, the products with lower substitution degrees (1 and 2) can be easily isolated from the reaction mixture by high performance liquid chromatography (HPLC) in a 3 : 8 acetonitrile–dichloroethane system, as their retention times are shorter than that of the initial unsubstituted anion (~7 min) or than those of more halogenated products, including the products with mixed halogens (>6 min) (Table 1.8).

The molecular masses of the single-charged anions obtained in all syntheses were estimated from ESI mass spectra (Table 1.9). As a rule, these spectra of *closo*-borate anions exhibit signals for single-charged $(CatAn)^-$ species. The results indicate, in all cases, the formation of $(C_4H_9)_4N[B_{10}H_9Cl]^-$, $(C_4H_9)_4N[B_{10}H_8Cl_2]^-$, and $(C_4H_9)_4N[B_{10}H_7Cl_3]^-$ anions.

The ^{11}B NMR spectra (Table 1.8) of mono- and dihalogenated products are the spectra of "classical" mono and disubstituted *closo*-decaborate anions.

Table 1.9 Effect of dodecahydrododecaborate salts on the properties of solid composite propellants. Reproduced from ref. 115.[a]

	w/%						
Samples	Binder	Nitrate plasticizers	Metal fuels (Al)	ρ/(g cm^{-3})	Volume impulse	Specific impulse (I_{sp})/s	Relative burning rate
0	9–10	16–22	0	0 1.55–1.59	11.2–11.5	195–205	2.10
1	9–10	10–18	1–17	1.7–1.95	14.3–15.5	205–235	0.20
2	9–10	5–10	1–17	1.7–1.98	14.5–15.5	200–240	—
3	9–10	5–10	2–15	1.7–1.9	14.1–15.6	210–240	4.00
4	9–10	5–10	8–20	1.7–1.8	14.4–15.8	220–235	2.60
5	9–10	5–10	2–15	1.7–2.1	13.0–15.7	202–235	2.10
6	9–10	10–18	8–18	1.6–1.7	11.2–11.5	235–255	—

[a]Binders are polyol + isocyanate + stabilizer + additive; the mass fraction of $B_{12}H_{12}^{2-}$ salts in the solid composite propellant is 2%.

The ^{11}B NMR spectrum of the monochlorinated derivative exhibits five signals, one from the apical boron atoms at −3.3 ppm, one from the boron atoms bound to substituents at −7.2 ppm, and three signals at −20.7 ppm, −24.5 ppm, and −28.8 ppm due to the equatorial boron atoms. The ^{11}B NMR spectrum of the compound $(Bu_4N)_2B_{10}H_8Cl_2$ shows five signals, one from the apical boron atoms (−4.3 ppm), one from the boron atoms bound to substituents (−7.8 ppm), and three signals at −21.4 ppm, −25.0 ppm, and −31.0 ppm due to the equatorial boron atoms. In both spectra without B–H decoupling, the signals for all boron atoms that are not bound to the substituents are split into symmetrical doublets.

The IR spectra of the halogenated *closo*-decaborate (Table 1.4), apart from the bands at 2570–2580 cm^{-1} typical of B–H stretches, contain the bands of the B–B–Cl bending modes at about 960 cm^{-1}.

For the dichloro-substituted *closo*-decaborate anion isolated by HPLC, white prismatic single crystals of the compound $(Ph_3(NaphCH_2)P)_2B_{10}H_8Cl_2$ were obtained. The X-ray reflections were collected from a 0.15 mm × 0.25 mm × 0.65 mm single crystal. The C, P, and Cl atoms were refined in the anisotropic approximation and B atoms, in the isotropic approximation. The occupancy of the chlorine positions was first refined independently. Since the sum of position occupancies (μ) of Cl(1) and Cl(2) atoms is equal to unity, the $\mu_{Cl(2)}$ value was included in the final refinement and the occupancy of the Cl(1) position was taken to be $1-\mu_{Cl(2)}$. All H atoms were refined by the riding model in idealized positions with isotropic thermal parameters $U_j(H)=1.2U_{eq}$ (or $1.2U_{iso}$) of the corresponding non-hydrogen atom.

The structure of the compound is built out of the $Ph_3(NaphCH_2)P^+$ cations and $B_{10}H_8Cl_2{}^{2-}$ anions. The anions are located on a twofold axis passing through the midpoints of the B(3)–B(7) and B(5)–B(9) edges. The Cl atom is disordered over two positions (Cl(2) and Cl(1)) bound to B(2) and B(1), respectively. The occupancy ratio for the positions is Cl(2) : Cl(1) = 0.8 : 0.2. The total occupancy of the Cl atom positions is equal to unity, which corresponds to the polyhedral $B_{10}H_8Cl_2{}^{2-}$ anion.

The great thermal parameters for boron atoms attest to probable overlap of the polyhedra with similar geometry. Presumably, the insignificant differences between the geometries of the decaborate polyhedral arise in different $B_{10}H_8Cl_2{}^{2-}$ anions under the influence of Cl atoms. Taking into account the Cl positions in the structure, four isomers are possible in which substituents can be located both at the vertices and in the belt of the *closo*-decaborate core. The mass fraction of the major 2,7-isomer is four times greater than the mass fraction of the other possible 1,10-, 1,2-, and 1,7-isomers. Determination of the exact structure of the isomers in the crystal from X-ray diffraction data is impossible.

The averaged decaborate polyhedron determined by the refinement has an irregular shape. The ranges of equivalent bonds are very broad: B_a–B_e 1.45–1.74(2) Å, B_e–B_e 1.75–2 and 1.60–1.82(2) Å in one belt and between the belts, respectively. The greatest anomalies are observed in the bonds involving the B(3) atom, which has the greatest thermal parameter: the B(2)–B(3) bond is virtually cleaved and amounts to 2.01(2) Å; the B(1)–B(3) and B(3)–B(7) bonds are shortened to 1.45(2) to 1.60(2) Å, respectively. In addition, the polyhedron is substantially flattened along the symmetrically bound edges, B(2)–B(3) and B(8)–B(7) (the B(1)B(2)B(3)/B(2)B(3)B(6) dihedral angle is 164.6°), and the average deviation of the equatorial B(2), B(3), B(4), and B(5) atoms from the root-mean-square plane reaches 0.041 Å. The B(2)–Cl(2) and B(1)–Cl(1) bond lengths are normal (1.846(9) and 1.86(2) Å, respectively).

The structures of the *closo*-decaborate anions in $(Ph_3(NaphCH_2)P)_2B_{10}H_8Cl_2$ determined by X-ray diffraction at 120(2) K and at room temperature (293 K) are identical to within experimental error. Figure 1.12 shows structures of the anions present in the $(Ph_3(NaphCH_2)P)_2B_{10}H_8Cl_2$ single crystal.

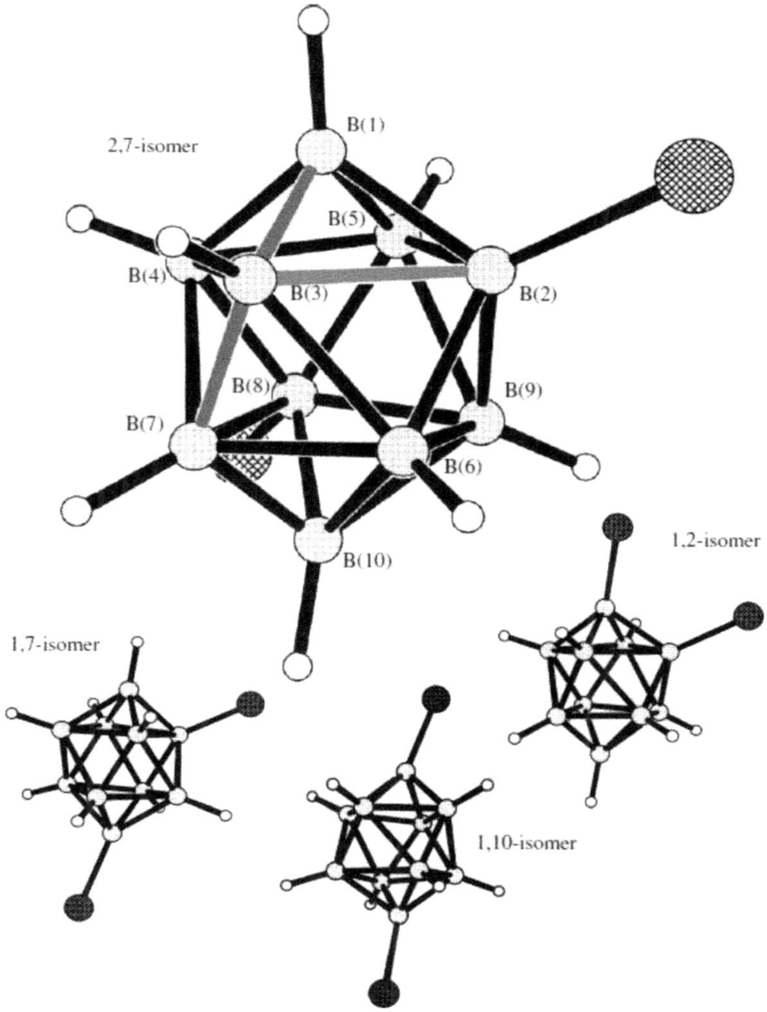

Figure 1.12 Structures of the anions present in the $(Ph_3(NaphCH_2)P)_2B_{10}H_8Cl_2$ single crystal. Reproduced from ref. 15 with permission from Springer Nature, Copyright 2007.

The geometry distortion of the *closo*-decaborate anion in an acid medium has been reported. In particular, the structure of $[(CH_3CN)_4]Cu_2B_{10}H_{10}$ obtained in the $Cu_2B_{10}H_{10}$–CH_3CN–CF_3COOH system was determined. According to X-ray diffraction, the *closo*-decaborate anion has a similar geometry distortion, one B–B bond being almost cleaved and two such bonds being shortened. Hence, a distortion of the geometry of the *closo*-decaborate anion in a highly acidic medium appears regular.

Thus, the reactions of the *closo*-decaborate anion $B_{10}H_{10}{}^{2-}$ with dichloroethane and hydrogen halides yield halogenated derivatives in which chlorine atoms from dichloroethane are mainly present as substituents. The mechanism of reactions resulting in these products is as yet unknown. However, the role of hydrogen halide molecules is evident. Performing analogous reactions at room temperature demonstrated that chlorination of the *closo*-decaborate anion does not take place but the anion is protonated to give $B_{10}H_{11}{}^{-}$, which is confirmed by ^{11}B NMR. This fact suggests that hydrogen halides are the protonating agents initiating halogenation, which proceeds *via* the protonated anion. Acid-catalyzed nucleophilic substitution of the *closo*-decaborate anion has been reported;[90] therefore, one may suggest that halogenation of the *closo*-decaborate anion by dichloroethane follows a similar mechanism (hydrogen halide initiation).

Apparently, the reaction also involves opening of the boron cluster anion. As noted above, the geometry of the *closo*-decaborate anion is distorted in highly acidic media, and the products can be isolated only in the presence of acids.[91] The *closo*-decaborate core in the compounds synthesized in this study has a similar unusual geometry, which is retained upon further chemical transformations.

1.4.3 Borane Anion *closo*-$[B_{10}H_{10}]^{2-}$ and Its Derivatives

1.4.3.1 Biomedical Science

The application of borane derivatives in biomedicine is mainly boron neutron capture therapy (BNCT) for tumors.[92] $B_{10}H_{10}{}^{2-}$ and its derivatives are promising candidate drugs; shortly after $B_{10}H_{10}{}^{2-}$ was synthesized, researchers began to study the application of its derivatives in BNCT. Heber and Diaz *et al.*[93a,b] studied the biodistribution of $Na_2{}^{10}B_{10}H_{10}$ (GB-10)[93c] during BNCT, and it was found that it could be used as a potential drug for BNCT treatment of head and neck tumors. Wu *et al.*[92g] reviewed the development and application of high molecular weight agents, including dendrimer derivatives of $B_{10}H_{10}{}^{2-}$. Shelly *et al.*[94a] studied the role of $Na_2{}^{10}B_{10}H_{10}$ model and $B_{20}H_{18}{}^{2-}$ in BNCT. Lü *et al.*[95] used $B_{10}H_{10}{}^{2-}$ derivatives with good water solubility and high boron content as boron containing reagents, and developed several synthesis methods of boronated nucleic acids. Tjarks[96] reviewed the progress of boron containing clusters in the design of boron nucleosides for BNCT. In addition, Bregadze *et al.*[97] proposed the use of radionuclide labeling with polyborate anions as carriers for targeted radionuclide therapy and diagnosis of cancer, and some metal derivatives have obvious antitumor activity.

1.4.3.2 Extraction of Radionuclides

$[1\text{-}LB_{10}H_9]^-$ type derivatives (L = amine or phosphine) can extract radioactive $^{137}Cs^+$ ions from acidic media, and their extracts can be dissolved in a variety of organic solvents of medium polarity, of which the most effective diluent is nitrobenzene.[98,99]

1.4.3.3 Development of New Functional Materials

Borane clusters have wide potential applications in the development of new materials with special properties.[100] Kaszynski *et al.*[101c] studied the application of closed boron clusters in liquid crystal materials and pointed out that the space and electronic effects of closed boron clusters are very effective structural factors and compatible with typical organic molecular structures. The NH_3 or CH_3CN solvate of $LnB_{10}H_{14}$ (Ln = Eu, Yb) is decomposed at 200 °C in vacuum to form $LnB_{10}H_{10}$, wherein the Yb compound is further decomposed at 600 °C to form the metal boride YbB_6.[102] Similarly, $Ln_2(B_{10}H_{10})_3$ (Ln = Ce, Gd) decomposes into LnB_4 and LnB_6 at 1000 or 1200 °C.[103] $Li_2B_{10}H_{10}$ is decomposed in vacuum (1073 K) to form Li_2B_9, while $Na_2B_{10}H_{10}$ is heated in vacuum (1273 K) to form Na_3B_{20}.[104] The photoluminescence and nonlinear optical properties of $B_{10}X_{10}{}^{2-}$ (X = H, Cl, Br) derivatives have also been studied.[105] $B_{10}H_{10}{}^{2-}$ derivatives can also catalyze the synthesis of polycarbonate.[106]

Hydrogen energy, as a kind of green energy and energy carrier with abundant reserves, wide sources and high energy density, is attracting extensive attention. Chemical hydrogen storage has great advantages in hydrogen storage capacity, and borohydride compounds have unique application prospects in chemical hydrogen storage.[107] Although $Na_2B_{10}H_{10}$ is relatively stable, it can also decompose and release hydrogen in the presence of appropriate catalyst. According to the information released on the official website of the US Department of Energy (DOE), the latest research shows that $B_{10}H_{10}{}^{2-}$ decomposes under the catalysis of some metal borides to generate $[B(OH)_4]^-$, $B(OH)_3$ and a large amount of H_2, which greatly expands the potential application range of $[B_{10}H_{10}]^{2-}$.

1.4.3.4 Object of Supramolecular Chemistry

$B_{10}H_{10}{}^{2-}$ can form a X–H···H–B type nonclassical hydrogen bond with its B–H bond, *i.e.* "dihydrogen bond",[108] and form a supramolecular system with alcohol in this way.[109] $[B_{10}H_{10}]^-$ (or $B_{10}I_{10}{}^{2-}$) and Lewis acidic carborane supported macrocyclic host compound $(1,2\text{-}C_2B_{10}H_8Et_2Hg)_4$[110] [12-mercury carborane-4, which is easy to combine with anions, so it is also known as "anti-crown ether"). Intermolecular interaction occurs to form a host guest complex. The two guests $B_{10}H_{10}{}^{2-}$ are bound to four mercury atoms through Hg–H–B3c–2e bonds among them; $B_{10}H_{10}{}^{2-}$ is easily substituted by halogen ions to form halogen ion binding species. $B_{10}H_{10}{}^{2-}$ and cyclic trimer $(o\text{-}C_6F_4Hg)_3$ were combined to obtain $[(o\text{-}C_6F_4Hg)_3[B_{10}H_{10}]^{2-}$ and $[(o\text{-}C_6F_4Hg)_3]_2[B_{10}H_{10}]^{2-}$.[111] A structural analysis showed that $B_{10}H_{10}{}^{2-}$ and $(o\text{-}C_6F_4Hg)_3$ were combined by a Hg–H–B bridge bond to form a (semi) sandwich structure.

Although borane chemistry has achieved very good results, this field is still full of opportunities. Borane and its derivatives continue to attract people's attention due to their unique structure, properties and application prospects. In addition to deeply revealing the basic principles that play a decisive role in the chemical behavior of electron deficient compounds, we also focus on exploring the potential applications of borane compounds. The reactivity of closed form $B_{10}H_{10}{}^{2-}$ has been widely studied, mainly including cage hydrogen substitution, cage opening rearrangement (degradation), cage opening metal insertion (degradation), *etc.* The unique three-dimensional aromaticity and thermal and hydrolytic stability of its skeleton are still worth further study. It is expected that more attention will be paid to the application of its unique properties in addition to continuing to explore novel derivatives.

1.5 Application Progress of Carborane Derivatives in Propellants and Explosives

$B_{12}H_{12}{}^{2-}$ salts are mainly used as high-energy and high-density combustion agent, fuel additive, gas generator and burning rate regulator in solid propellants and explosives.[112] For example, Shibata *et al.*[113] used $K_2B_{12}H_{12}$ in the solid propellant formulation containing triaminoguanidine. When the pressure is 49.02 MPa, the burning rate is 5400 mm s^{-1}, while when the pressure is 98.04 MPa, the burning rate is as high as 9200 mm s^{-1}, which is 1000 times that of M30 propellant. Johnson *et al.*[114] also used $K_2B_{12}H_{12}$ in a propellant formulation containing nitroguanidine. The results showed that the component reaction rate was accelerated and a large number of gases were produced in a short time. Table 1.9 lists the effect of dodecahydrododecaborate salts on the properties of solid composite propellants, $K_2B_{12}H_{12}$, $Cs_2B_{12}H_{12}$, $(Bu_4N)_2B_{12}H_{12}$, and KIO_4 were solid oxidizers. It can be found that the density increased by 0.15–0.51 g cm^{-3}, the volume specific impulse increased by 3511–10 646 N s kg^{-1}, and the specific impulse increased by 10–50 s when the mass fraction of the $B_{12}H_{12}{}^{2-}$ salts is 0.5–5.0%. It can also be found from the data in Table 1.10 that when the mass fraction of borohydride is 2% and the mass ratio of borohydride to KIO_4 is 10 : 90, the burning rates of the three borohydride propellants are 4089.4, 787.4, and 736.6 mm s^{-1}, respectively. When the mass fraction of $Cs_2B_{12}H_{12}$ increased in the formulation, the burning rate increased rapidly. When the mass ratio of $Cs_2B_{12}H_{12}$ to KIO_4 is 30 : 70, the burning rate of propellant is as high as

Table 1.10 Burning rate of the mixture of dodecahydrododecaborate salts with different oxidizers. Reproduced from ref. 115.

	$r/(\mathrm{mm\,s^{-1}})$		
Catalysts + oxidizers	10 : 90	20 : 80	30 : 70
$Cs_2B_{12}H_{12} + KIO_4$	787.4	1803.4	25 400.0
$Cs_2B_{12}H_{12} + KNO_3$	—	431.8	17 881.6
$Cs_2B_{12}H_{12} + Bi_2O_3$	20.3	279.4	—
$Cs_2B_{12}H_{12} + Sr(NO_3)_2$	10.7	279.4	—
$Cs_2B_{12}H_{12}$ + PSAN	—	20.3	—
$K_2B_{12}H_{12} + KIO_4$	4089.4	—	—
$((C_4H_9)_4N)_2B_{12}H_{12} + KIO_4$	736.6	—	—

25 400 $\mathrm{mm\,s^{-1}}$. In the meantime, it should be noted that it may be better to replace $Cs_2B_{12}H_{12}$ with cheap $K_2B_{12}H_{12}$.

Saldin *et al.*[116–120] invented the preparation method of $B_{12}H_{12}{}^{2-}$ chitin salt ($[[(C_6O_4H_9NH_3)_2B_{12}H_{12}]$). It is observed that this salt can ignite quickly and burn completely. It is an explosion-proof, non-toxic and non-toxic gas emission, heat-resistant, moisture-resistant, high-energy material with excellent mechanical properties and bond cooperation, which is suitable for gun powder pyrotechnics, explosives, igniter of propellants and components of primary explosives.[121–123] For example, the composition of $B_{12}H_{12}{}^{2-}$ chitin salt and the complex composed of transition metals, especially Cu^{2+}, CO^{2+}, Ni^{2+}, Zn^{2+}, or Mn^{2+} and nitrate (or perchlorate) of magnesium or aluminum can be used as energy enhancing ignition additives for pyrotechnics, *etc.* Moreover, $B_{12}H_{12}{}^{2-}$ chitin salt containing an effective amount of chitin chromate (oxidant) is used as an ignition additive for pyrotechnics, so that no harmful gas is produced during pyrotechnic combustion.[124] Adducts of $B_{12}H_{12}{}^{2-}$ chitin salt with perchloric acid or ammonium perchlorate or chitin perchlorate are used as non-toxic storage stabilizers in pyrotechnic formulations with heat resistance and good mechanical properties.[125]

Furthermore, Hanumantha Rao[23,24] found that $B_{12}H_{12}{}^{2-}$ nitrogen heterocyclic ammonium salts, including imidazole salt, 1,2,4-triazole salt, 1,2,3-triazole salt, tetrazole salt, pyrazole salt, oxazole salt, isoxazole salt, furazan salt, and furazan oxide salt, *etc.* with high bulk density and low molecular weight of combustion products are suitable for use as insensitive energetic materials and combustion accelerators in solid rocket propellants. These compounds help to enhance the high-density specific impulse of propellant, increase the missile payload, and improve the combustion performance and the specific impulse of rocket propellants.

1.6 Conclusions

(1) The boron and hydrogen compounds mainly include open cage borane, closed cage type ionic hydroborate, carborane and metal carborane. Borohydride has been paid attention to by aviation and

rocket workers because of its high combustion calorific value. It is an effective and practical way to obtain a high burning rate of solid propellant by using borohydrides as burning rate regulators.
(2) The combustion agents, including ionic hydroborates, carborane and its derivatives, metal carboranes, borane energetic ionic liquids (salts) for their stable properties and ease of chemically modification, are promising candidates for the burning rate increasing of solid propellants.
(3) Although borane chemistry has developed rapidly and made some progress in biomedicine, new materials, molecular devices and other emerging fields, there are relatively few reports on the research of borohydride as a burning rate regulator of solid propellants. How to apply the current research results to this field to achieve a new breakthrough in solid propellant burning rate can be an important research direction.

References

1. S. Wengang, L. Xuli and L. Shangwen, *Cruise Missile*, 1996, **11**, 40–43.
2. R. Xiulun and Y. Rongjie, *Cruise Missile*, 2006, 44–50.
3. O. Haifeng and W. Zhixian, *Sci. Technol. Inf. Dev. Econ.*, 2006, **16**(9), 168–170.
4. Z. Yuanyang and W. Minwen, *Chin. J. Explos. Propellants*, 19893), 32–37.
5. T. Songqing and D. Hongxun, *J. Propul. Technol.*, 19832), 35–51.
6. J. M. Makhlouf and G. T. Hefferdan, *Method of preparing tetraethylammonium decahydrodecaborates*, US3373202, 1968.
7. D. C. Sayles, *Method for preparation of a carboranyl burning rate accelerator precursor*, US4150057, 1979.
8. B. Spielvogel and K. Cook, *Method of production of* $B_{10}H_{10}{}^{2-}$ *ammonium salts and methods of production* $B_{18}H_{22}$, US200510169828A, 2005.
9. G. B. Dunks and K. P. Ordonez, *Process for the synthesis of decaborane (14)*, US4115521, 1978.
10. W. Weiqiang and X. Yunna, *Chin. J. Energ. Mater.*, 2012, **20**(1), 132–136.
11. S. V. Ivanov and B. Casas, *Process for producing boranes*, US771815482, 2010.
12. C. Fu-Tai, T. Hui-Min and L. Yun-Jun, *Chin. J. Explos. Propellants*, 2000, **27**(3), 19–21.
13. W. J. Casteel and S. V. Ivanov, *Process for the fluorination of boron hydrides*, US678100581, 2004.
14. D. V. Peryshkov, A. A. Popov and S. H. Strauss, *J. Am. Chem. Soc.*, 2009, **131**, 18393–18403.
15. V. V. Drozdova, K. Y. Zhizhin and E. A. Malinina, *et a1.*, *Russ. J. Inorg. Chem.*, 2007, **52**(7), 996–1001.
16. S. A. Shackelford, *Heterocyclic salts of HBS & HCS*, US7521S6481, 2009.
17. J. F. Valliant, K. J. Guenther and A. S. King, *et al.*, *Coord. Chem. Rev.*, 2002, **232**, 173–230.

18. M. C. Parrott, E. B. Marchington and J. F. Valliant, *et al.*, *J. Am. Chem. Soc.*, 2005, **127**, 12081–12089.
19. R. N. Grimes, *J. Chem. Educ.*, 2004, **81**(5), 658–672.
20. T. Songqing and D. Hongxun, *Chem. Propellants Polym. Mater.*, 2004, **2**(1), 8–11.
21. W. Weiqiang, X. Yunna and Y. Jianming, *et al.*, *Chin. J. Energ. Mater.*, 2012, **20**(1), 132–136.
22. V. I. Saldin and V. V. Sukhovei, *Chitosan chromate synthesis for use as oxidant in energy intensibe orpyrotechnic compositions burning without emissions*, RU2439081, 2012.
23. V. I. Saldin, V. M. Buznik and Y. M. Nikolenko, *et al.*, *Adducts of chitosan dodecahydro-closo-dodecaborate with perchloric acid or ammonium perchlorate*, RU2394840, 2010.
24. M. Hanumantha Rao and K. Muralidharan, *Polyhedron*, 2016, **115**, 105–110.
25. M. Hanumantha Rao and K. Muralidharan, *Dalton Trans.*, 2013, **42**(24), 8854–8860.
26. S. A. Shackelford, *Borane and carborane salts with heterocyclic amine cations for use in propellants, fuel additives, and hydrogen storage material*, US7521564, 2009.
27. W. E. Hill, *Fast burning double-base propellant*, US3791893, 1974.
28. L. Ruibin and L. Zhanxiong, *Chin. J. Synth. Chem.*, 2010, **18**(2), 229–231.
29. R. L. Barnes and D. Grafstein, *High temperature bis(epoxyalkyl)carborane adhesives*, US3669993, 1972.
30. H. G. Adolph and L. A. Nock, *Copolyformals of 1,2-bis(2-hydroxyethyl)-1,2-dicarbadodecaborane and polynitroalkyl diols*, USH000786, 1990.
31. S. Wengang, S. Tiegang and Z. Gudong, *J. Solid Rocket Technol.*, 1995, **18**(3), 24–26.
32. G. Guoyiqibayi, Z. Rui and Y. Hong, *Chin. J. Inorg. Chem.*, 2010, **26**(5), 733–743.
33. N. Yong, C. Haiyan and M. Linling, *Chin. J. Org. Chem.*, 2009, **29**(6), 822–834.
34. S. M. Vuga, *Propellants, Explos., Pyrotech.*, 1991, **16**, 293–298.
35. Z. Ren, L. Jiahua and W. Wujun, 1994, 62–65.
36. W. Pang, L. T. DeLuca, A. A. Gromov and A. S. Cumming, *Innovative Energetic Materials: Properties, Combustion, Application*, Springer Nature, 2020.
37. G. V. Nechai, F. S. Sokolovskii and S. V. Chuiko, *Russ. J. Phys. Chem. B*, 2009, **3**(3), 458–468.
38. S. V. Chuiko and F. S. Sokolovskii, *Russ. J. Phys. Chem. B*, 2009, **3**(6), 926–935.
39. I. B. Sivaev and V. I. Bregadze, *Collect. Czech. Chem. Commun.*, 1999, **64**, 783–801.
40. I. B. Sivaev and V. I. Bregadze, *J. Organomet. Chem.*, 2000, **614–615**, 27–36.
41. L. Na, C. Chunpeng and G. Zhiyong, *et al.*, *Chin. J. Energ. Mater.*, 2010, **18**(4), 467–475.

42. M. Nieuwenhuyzen, K. R. Seddon and F. Teixidor, *Inorg. Chem.*, 2009, **48**(3), 889–901.
43. P. F. Keitz, A. Jahangir and F. J. Lopez-Tapia, *et al.*, *Substituted aryl amides as IP antagonists*, US727949882, 2007.
44. M. Weigele, G. P. Luke and T. K. Sawyer, *et al.*, *Novel heterocycles*, US200210103161A1, 2002.
45. S. A. Shackelford, J. L. Belletire and J. A. Boatz, *Org. Lett.*, 2010, **12**(12), 2714–2717.
46. S. A. Shackelford, J. L. Belletire and J. A. Boat, *Org. Lett.*, 2009, **11**(12), 2623–2626.
47. C. Haonan, W. Rui and X. Jinwu, *et al.*, *J. Solid Rocket Technol.*, 2019, **42**(5), 585–191.
48. L. Juyou, D. Yongmei and W. Hong, *et al.*, *Chin. J. Synth. Chem.*, 2015, **23**(9), 883–892.
49. L. Ruibin, *Preparation, properties and applications of novel earborane derivatives*, PhD thesis, Suzhou University, Suzhou, 2010.
50. Y. Zhu and N. S. Hosmane, *J. Organomet. Chem.*, 2017, **849–850**, 286–292.
51. C. Shan and C. Gong, *Chin. J. Synth. Chem.*, 2011, **19**(4), 513–515.
52. X. Yunna, L. Jian and Y. Jianming, *et al.*, *Synthesis of nhexyl carborane*, CNl02558206A, 2012.
53. E. Cooke and P. Anderson, *Burn rate modification with caborane polymers*, AD-E403956, 2017.
54. Z. Lin, J. Qibai and Y. Hong, *J. Inorg. Chem.*, 2014, **30**(10), 2246–2251.
55. S. Y. Kim, S. Y. Ma and S. O. Kang, *et al.*, *J. Organomet. Chem.*, 2018, **865**, 100–108.
56. M. S. Cheung, A. Hoishan Chan and Z. Xie, *Organometallics*, 2004, **23**(3), 517–526.
57. P. Dozzo, R. A. Kasar and S. B. Kahl, *Inorg. Chem.*, 2005, **44**(22), 8053–8057.
58. M. Weihua, C. Ruijiao and S. Yingwei, *et al.*, *Chin. J. Org. Chem.*, 2018, **38**(6), 1327–1340.
59. X. Xinbin, *J. Org. Chem.*, 2018, **38**(11), 3078–3085.
60. L. Huanhuan, Y. A. N. Hong and L. U. Changsheng, *J. Inorg. Chem.*, 2017, **33**(8), 1313–1329.
61. Z. Qiu, Y. Quan and Z. Xie, *J. Am. Chem. Soc.*, 2013, **135**(33), 12192–12195.
62. A. M. Spokoyny, *Pure Appl. Chem.*, 2013, **85**(5), 903–919.
63. D. Zhao and Z. Xie, *Coord. Chem. Rev.*, 2016, **314**, 14–33.
64. C. Sayles David, *Non-nitroglycerin containing composite modified double base propellant*, US4707199, 1987.
65. Z. Qinghua, L. Xuejun and W. Lili, *et al.*, *Chem. Propellants Polym. Mater.*, 2011, **9**(6), 56–59.
66. S. Wengang and S. Tiegang, *J. Solid Rocket Technol.*, 1995, **18**(3), 24–27.
67. L. Ruibin and L. Zhanxiong, *Synth. Chem.*, 2010, **18**(2), 229–231.
68. C. Sayles David, *Methyltricarboranylmethyl perchlorate*, US3980710, 1976.

69. C. Sayles David, *Ultrahigh burning rate propellants containing an organic perchlorate oxidizer*, US3986906, 1976.
70. C. Sayles David, *Carboranyldiferrocenylmethyl perchlorate*, US4026912, 1977.
71. M. Baldwin, *Propellant composition with a hydroxyl-terminated copolymer-binder-catalyst of butadiene and acrylate*, US3845139, 1973.
72. M. Baldwin, *Method for preparing hydroxyl-terminated copolymers of butadiene and acrylate*, US3845139, 1974.
73. C. W. Huskins, *Co-polymers of butadiene and carboranyl methacrylate*, US4097662, 1978.
74. L. Zhanxiong, L. Anqi and Z. Ying, *et al.*, *Preparation of ammonium perchlorate containing carborane*, CN106928265B, 2019.
75. W. Rui, X. Jinwu and L. Weidong, *et al. A carborane propellant and its preparation method*, CN109305869A, 2019.
76. M. F. Hawthorne and R. L. Pilling, *Inorg. Synth.*, 1967, **9**, 16.
77. (a) G.-M. Zhang and H. Zhu, *Acta Chim. Sin.*, 1978, **36**, 315 (in Chinese); (b) G.-M. Zhang, H. Zhu and Q. J. Ding, *Wuhan Univ. J. Nat. Sci.*, 1981, **2**, 101 (in Chinese).
78. (a) R. B. King, *Chem. Rev.*, 2001, **101**, 1119; (b) Z. Chen and R. B. King, *Chem. Rev.*, 2005, **105**, 3613.
79. M. T. Nguyen, M. H. Matus and D. A. Dixon, *Inorg. Chem.*, 2007, **46**, 7561.
80. W. Preetz and H.-G. Srebny, *Z. Naturforsch., B: J. Chem. Sci.*, 1984, **39**, 6.
81. W. Preetz and C. Nachtigal, *Z. Anorg. Allg. Chem.*, 1995, **621**, 1632.
82. W. H. Knoth, H. C. Miller and J. C. Sauer, *et al.*, *Inorg. Chem.*, 1964, **3**(2), 159.
83. V. V. Drozdova, E. A. Malinina, I. N. Polyakova, G. A. Razgonyaeva, K. Yu Zhizhin and N. T. Kuznetsov, *Zh. Neorg. Khim.*, 2007, **52**(1), 57 [*Russ. J. Inorg. Chem*, 2007, **52**(1), 52].
84. H. C. Miller, N. E. Miller and E. L. Muetterties, *J. Am. Chem. Soc.*, 1963, **85**(23), 3885.
85. *Handbuch der präparativen anorganischen Chemie*, ed. G. Brauer, Ferdinand Enke, Stuttgart, 1975–1981; Mir, Moscow, 1985.
86. L. I. Ochertyanova, K. Yu Zhizhin and V. N. Mustyatsa, *et al.*, *Neorg. Mater.*, 2004, **40**(1), 188.
87. G. M. Sheldrick, SHELXS97 and SHELXL97, Program for the Solution ind Refinement of Crystal Structures, Univ. of Göttingen, 1997.
88. SMART (control) and SAINT (integration) Software, Version 5.0. Bruker AXS Inc., Madison, 1997.
89. G. M. Sheldrick, SADABS. Program for Scaling and Correction of Area Detector Data, Univ. of Göttingen, Germany, 1997.
90. K. Y. Zhizhin, E. A. Malinina and M. V. Lisovskii, *et al.*, *Zh. Neorg. Khim.*, 2002, **47**(8), 1279 [*Russ. J. Inorg. Chem*, 2002, **47**(8), 1168].
91. E. A. Malinina, K. Yu Zhizhin and V. N. Mustyatsa, *et al.*, *Zh. Neorg. Khim.*, 2003, **48**(7), 1102 [*Russ. J. Inorg. Chem.*, 2003, **48**(7), 993].
92. (a) M. F. Hawthorne, *Angew. Chem., Int. Ed.*, 1993, **32**, 950; (b) A. H. Soloway, W. Tjarks, B. A. Barnum, F. Rong, R. F. Barth, I. M. Codogni and J. G. Wilson, *Chem. Rev.*, 1998, **98**, 1515; (c) D. Gabel,

S. Mai and O. Perleberg, *J. Organomet. Chem.*, 1999, **581**, 45; (d) M. F. Hawthorne and A. Maderna, *Chem. Rev.*, 1999, **99**, 3421; (e) R. F. Barth, J. A. Coderre, M. G. H. Vicente and T. E. Blue, *Clin. Cancer Res.*, 2005, **11**, 3987; (f) M. V. Backer, T. I. Gaynutdinov, V. Patel, A. K. Bandyopadhyaya, B. T. S. Thirumamagal, W. Tjarks, R. F. Barth, K. Claffey and J. M. Backer, *Mol. Cancer Ther.*, 2005, **4**, 1423; (g) G. Wu, R. F. Barth, W.-L. Yang, R. Lee, W. Tjarks, M. V. Backer and J. M. Backer, *Anti-Cancer Agents Med. Chem.*, 2006, **6**, 1; (h) D. S. Wilbur, M. K. Chyan, D. K. Hamlin, R. L. Vessella, T. J. Wedge and M. F. Hawthorne, *Bioconjugate Chem.*, 2007, **18**, 1226.

93. (a) E. Heber, V. A. Trivillin, D. Niqq, E. L. Kreimann, M. E. Itoiz, R. J. Rebaqliati, D. Batistoni and A. E. Schwint, *Arch. Oral Biol.*, 2004, **49**, 313; (b) A. Diaz, K. Stelzer, G. Laramore and R. Wiersema, *In Research and Development in Neutron Capture Therapy*, ed. W. Sauerwein, and R. Moss and A. Wittig, Monduzzi Editore, Bologna, 2002, p. 993; (c) B. Spielvogel and K. Cook, US2005169828, 2005, [*Chem. Abstr.*, 2005, 143, 196293].
94. (a) K. Shelly, D. A. Feakes, M. F. Hawthorne, P. G. Schmidt, T. A. Krisch and W. F. Bauer, *Proc. Natl. Acad. Sci. U. S. A.*, 1992, **89**, 9039; (b) S. Masunaga, S. Kasaoka, K. Maruyama, D. Nigg, Y. Sakurai, K. Nagata, M. Suzuki, Y. Kinashi, A. Maruhashi and K. Ono, *Int. J. Radiat. Oncol., Biol., Phys.*, 2006, **66**, 1515.
95. Y.-X. Lü, H.-X. Ding and H.-M. Lü, *Acta Chim. Sin.*, 1990, **48**, 820 (in Chinese).
96. W. Tjarks, *J. Organomet. Chem.*, 2000, **37**, 614.
97. V. I. Bregadze, I. B. Sivaev and S. A. Glazun, *Anticancer Agents Med. Chem.*, 2006, **6**, 75.
98. (a) D. Naoufal, B. Grüner, B. Stibr, P. Selucký, B. Bonnetot and H. Mongeot, *J. Radioanal. Nucl. Chem.*, 2005, 145. (b) D. Naoufal, B. Bonnetot, H. Mongeot and B. Grüner, *Collect. Czech. Chem. Commun.*, 1999, **64**, 856.
99. 99. R. Bernard, D. Cornu, B. Grüner, J. F. Dozol, P. Miele and B. Bonnetot, *J. Organomet. Chem.*, 2002, **657**, 83.
100. M. F. Hawthorne, *Mol. Struct. Energ.*, 1988, **5**, 225.
101. (a) S. Pakhomov, P. Kaszynski and V. G. Young, Jr., *Inorg. Chem.*, 2000, **39**, 2243; (b) A. Miniewicz, A. Samoc, M. Samoc and P. Kaszynski, *J. Appl. Phys.*, 2007, **102**, 33108; (c) P. Kaszynski and A. G. Douglass, *J. Organomet. Chem.*, 1999, **581**, 28.
102. J. P. White and S. G. Shore, *Inorg. Chem.*, 1992, **31**, 2756.
103. H. Itoh, Y. Tsuzuki and S. Naka, *Mater. Res. Bull.*, 1987, **22**, 1259.
104. M. Panda PhD, Dissertation, University Hamburg, Hamburg, 2006.
105. (a) P. Schouwink, H. Berlepsch, L. Dahne and R. F. Mahrt, *Chem. Phys.*, 2002, **285**, 113; (b) V. V. Volkov, E. A. Il'inchik and O. P. Yuryeva, *Chem. Sustainable Dev.*, 2000, **8**, 185; (c) T. N. Gerasimova, N. A. Orlova, V. V. Shelkovnikov, Z. M. Ivanova, R. V. Markov, A. Plekhonov, T. M. Polyanskaya and V. V. Volkov, *Chem. Sustainable Dev.*, 2000, **8**, 109.

106. J. P. Lemmon, US2004002577, 2004, [*Chem. Abstr.*, 2004, **140**, 60161].
107. G. J. Kubas, *Chem. Rev.*, 2007, **107**, 4152.
108. R. Custelcean and J. E. Jackson, *Chem. Rev.*, 2001, **101**, 1963.
109. E. S. Shubina, E. V. Bakhmutova, A. M. Filin, I. B. Sivaev, L. N. Teplitskaya, A. L. Chistyakov, I. V. Stankevich, V. I. Bakhmutov, V. I. Bregadze and L. M. Epstein, *J. Organomet. Chem.*, 2002, **657**, 155.
110. (a) X.-G. Yang, C. B. Knobler and M. F. Hawthorne, *J. Am. Chem. Soc.*, 1993, **115**, 4904; (b) X.-G. Yang, C. B. Knobler, Z. Zheng and M. F. Hawthorne, *J. Am. Chem. Soc.*, 1994, **116**, 7142; (c) Z. Zheng, C. B. Knobler, M. D. Mortimer, G. Kong and M. F. Hawthorne, *Inorg. Chem.*, 1996, **35**, 1235.
111. E. S. Shubina, I. A. Tikhcojoa, E. V. Bakhmutova, F. M. Dolgushin, M. Y. Anlipin, V. I. Bakhmutov, I. B. Stvaev, L. N. Tepltskaya, I. T. Chtzhevsky, I. V. Pisareva, V. I. Bregadze, L. M. Epstem and V. B. Shur, *Chem. – Eur. J.*, 2001, **7**, 3783.
112. T. Song-qing and D. Hong-xun, *J. Propul. Technol.*, 1983, 35–51.
113. A. Shibata, A. Miyake and T. Ogawa, *Kayaku Gakkaishi*, 2000, 85–93.
114. S. Johnson, L. H. Barr and B. E. Smith, *Nitroguanidine-based propellants and gas generators with potassium-borohydride catalysts for inflation of safety devices*, WO9916731, 1999.
115. S. Zixing, S. Lili and Y. Rongjie, *Chin. J. Explos. Propellants*, 2017, **40**(3), 1–16.
116. V. I. Saldin, V. V. Sukhovey and L. N. Ignativeva, *et al.*, *Theor. Found. Chem. Eng.*, 2010, **44**(4), 467–470.
117. V. I. Saldin, V. V. Sukhovei and L. N. Ignativeva, *et al.*, *Khim. Technol.*, 2009, 193–196.
118. V. I. Saldin, V. M. Buznik and and V. V. Sukhovei, *Method to manufacture of salts of dodecahydro-closo-dodecaboric acid*, RU2323879, 2008.
119. V. I. Saldin, L. N. Ignativeva and Y. M. Nikolenko, *J. Struct. Chem.*, 2006, **47**(1), 35–40.
120. V. I. Saldin and and A. P. Babii, *Dodecahydroborate adducts of chitosan as high caloric, rapid burning material*, RU2172745, 2001.
121. V. I. Saldin and and A. P. Babii, *Synthesis of polyhedral closo-complexes of hydroborates and chitosan as potential pyrotechnic components*, RU2158221, 2000.
122. V. I. Saldin, L. N. Ignativeva and V. M. Buznik, *Preparation of chitosanium perchlorate for energy absorbing compositions with chitosanium dodecahydro-closo-dedecaborate*, RU231577, 2008.
123. V. I. Saldin and and V. V. Sukhovei, *Adducts of chitosan dodecahydro-closo-dodecaborate with nitrates or perchlorates of magnesium or aluminum and method for production thereof*, RU2596741, 2016.
124. W. Q. Pang, X. Z. Fan and F. Q. Zhao, *et al.*, *J. Solid Rocket Technol.*, 2013, **36**(5), 647–653 (in Chinese).
125. W. P. C. de Klerk, A. E. D. M. vander Heijden and W. H. M. Veltmans, *J. Therm. Anal. Calorim.*, 2001, **64**, 973–985.

CHAPTER 2

Synthesis and Characterization of Organic Combustion Agents

2.1 Introduction

Polyhedron boron hydrogen compounds, a class of compounds with closed cage structure, because they have the advantages of low toxicity, high calorific value and high burning rate, have been used in solid rocket propellant.[1–4] The present study they mainly contain $B_{10}H_{10}^{2-}$ and $B_{12}H_{12}^{2-}$ ionic salts,[5–7] to study the compound synthesis and thermal stability.[8–10] A variety of $B_{10}H_{10}^{2-}$ and $B_{12}H_{12}^{2-}$ ion salts were synthesized and the thermal stability was investigated by Mutterties.[11] Kuznetsov[12] studied the thermal decomposition of the compound in the presence of oxygen and observed B–H by infrared spectra molecules. Isaenko[13] found that the two thermal decomposition processes of an anion and a cation may form a B–N bond in $B_{12}H_{12}[(CH_3)_4N]_2$. To improve the propellant burning rate for this class of compound, Goddard[14] considered that the instability of the hydrogen atom connected to the top of the boron cage by boron oxide cage can increase the decomposition reaction, release a large amount of heat at the same time, hasten the propellant combustion, and increase the burning rate. Kaste[15] speculated that the mechanism of $K_2B_{12}H_{12}$ and $B_{12}H_{12}[(CH_3)_4N]_2$ on the thermal decomposition of cyclotrimethyltrinitroamine (RDX) was the interaction between the B–H bond of polyhedral borohydride and the nitro group, which promoted the fracture of the nitrogen–nitrogen bond and the nitrogen–oxygen bond in the ammonium nitrate molecule. Up to now, researchers have done a lot of research on the thermal behavior of polyhedral borohydride and its interaction with ammonium nitrate compounds, but less on the thermal decomposition mechanism and kinetics of these compounds, and the results play an important role in the research on the burning rate promotion mechanism of propellants.

High-energy Combustion Agents of Organic Borohydrides: Synthesis, Characterization and Applications
By XiaoHong Zhang, WeiQiang Pang, Luigi T. DeLuca and Yu Zhao

Published by the Royal Society of Chemistry, www.rsc.org

Tetraethylammonium decahydrododecaborate ($B_{10}H_{10}[(C_2H_5)_4N]_2$, BHN) is a kind of energetic organic boronhydride with simple and good thermal stability polyhedron properties. Its thermal stability in nitrogen has been studied in ref. 16. The thermal decomposition of the compound occurs only in argon, and the heat release is small. In oxygen, it decomposes first, then oxidizes, and releases a lot of heat at the same time. In this chapter, several $B_{10}H_{10}$ salts are synthesized and characterized, the heat of combustion of the compound determined by the oxygen bomb method, and its thermal stability in argon and oxygen atmosphere compared by DSC–TG analysis. Furthermore, the mechanism of thermal decomposition is discussed by *in situ* Fourier transform infrared spectroscopy (FTIR) and the thermogravimetry–differential scanning calorimetry–Fourier transform infrared–mass spectrometry (DSC–TG–FTIR–MS) technique.

2.2 Synthesis and Characterization of $B_{10}H_{10}[(C_2H_5)_4N]_2$

2.2.1 Experimental

2.2.1.1 Reagents and Instruments

Sodium borohydride and tetraethyl ammonium chloride (purity 99%, industrial grade product of Shandong Sanheyuan Chemical Co. Ltd); methanol, dichloromethane and tetrahydrofuran (analytically pure, Chengdu Kelong Chemical Reagent Factory); paraffin oil (chemical purity, Aladdin Reagent Co.); other reagents are commercially available.

Zf-II three purpose ultraviolet spectrometer (Shanghai Anting electronic instrument factory); Nexus 870 Fourier transform infrared spectrometer (thermoelectric Ni Gaoli company, USA); AV 500 superconducting nuclear magnetic resonance instrument (500 MHz, Bruker Company, Switzerland); Vario-el-3 element analyzer (Elementar, Germany); Sta449f3 and STA449C DTA-TG analyzers.

TG–DSC analyzer (Netzsch, Germany); 5700 Fourier transform infrared spectrometer (Nicolet, USA).

2.2.1.2 The Experimental Process

Bis(tetraethylammonium) decahydroborate is an ionic hydroborate with high calorific value. It can be used as a high burning rate regulator and an important precursor for the synthesis of carborane based on burning rate regulators. $B_{10}H_{10}[(C_2H_5)_4N]_2$ was synthesized from sodium borohydride and tetraethylammonium chloride by hydroborate thermolysis. The structure and thermal properties of $B_{10}H_{10}[(C_2H_5)_4N]_2$ were characterized.

$$NaBH_4 + Et_4NCl \rightarrow Et_4NBH_4 \tag{2.1}$$

$$10(C_2H_5)_4NBH_4 \rightarrow B_{10}H_{10}\left[(C_2H_5)_4N\right]_2 + 8(C_2H_5)_3N + 8C_2H_6 + 10H_2 \tag{2.2}$$

2.2.1.3 *Synthesis of Bis(tetraethylammonium) Decahydrodecaborate*

Under the protection of nitrogen, 65 g of tetraethylammonium tetrahydroborate and 60 mL of paraffin oil were added into the pyrolysis reactor using a tail gas absorption device. The oil bath was heated to 185 °C and the reaction time was about 16 h. After cooling to room temperature, the reaction mixture was filtered. The filter cake was washed and dried to obtain white solid product. After recrystallization with water and methanol, 6.9 g of bis-tetraethylammonium decahydroborate was obtained in 39% yield. IR (KBr), $\bar{\nu}$/cm^{-1}: 2991 (ν_{C-H}), 2450 (ν_{B-H}), 1460 (ν_{C-H}), 1401 (ν_{C-H}), 1309 (ν_{C-N}), 1184 (ν_{C-N}), 1034 (ν_{B-B}), 1008 (ν_{B-B}); ^{1}H NMR ($CDCl_3$, 500 MHz), δ: 3.3713–3.7674 (m, 24H), 5.7198–5.7635 (m, 16H); 5.5403–6.3836 (q, 2H), 2.3279–3.0925 (m, 8H); ^{11}B NMR (DMSO, 500 MHz), δ: −28.669 (d, 8B, $J = 130$ Hz); −0.834 (d, 2B, $J = 130$ Hz); elements analysis ($C_{16}H_{50}B_{10}N_2$ calculated value, %): C 50.58 (50.50), H 10.54 (10.50), N 7.32 (7.36).

2.2.1.4 *Structural Characterization*

The technology of infrared *in situ* pyrolysis pool Fourier transform infrared spectroscopy was used in the process of thermal decomposition of the condensed phase. The sample weight is 0.7 mg, KBr tablet. The resolution is 4 cm^{-1}, the heating rate is 10 °C min^{-1}, and the detection temperature is 20–455 °C; DTGs detector.

2.2.1.5 *Thermal Behavior Analysis*

A Q-200 differential scanning calorimeter was used to test the thermal stability of tetraethylammonium decahydroborate in dynamic argon or oxygen atmosphere at a flow rate of 100 mL min^{-1}. The measurement temperature was 50–970 °C. The heating rate was 10 °C min^{-1}, sample weight: 1.0 mg; aluminum sample with dish.

The heating rates of the thermal decomposition kinetics were 5, 10, 15 and 20 °C min^{-1}, respectively, the other conditions are the same as the thermal stability study.

TG–DSC–FTIR–MS was used to analyze the decomposition gas products. Under the dynamic argon atmosphere, the flow rate of argon was 75 mL min^{-1}, the heating rate was 10 °C min^{-1}, and the sample volume was about 1.5 mg.

2.2.2 Characterization of $B_{10}H_{10}[(C_2H_5)_4N]_2$

2.2.2.1 *Structure and Nuclear Magnetic Resonance Spectra Analysis*

As shown in Figure 2.1, a molecule of tetraethylammonium decahydrodecaborate is composed of one decahydrodecaborate anion and two quaternary ammonium cations; the anion of decarboxy acid is a closed three-dimensional cage structure with a ten top, 16 faces and double cap square anti prism. It can be seen from Figure 2.1 that the boron atoms in the anionic structure of decahydrodecaboric acid are mainly divided into equatorial boron and top boron due to their different positions. Equatorial boron is eight boron atoms of a tetragonal anti prism in the middle of a boron cage, and the top boron is two boron atoms of a double capped top.

In the hydrogen and boron NMR spectra of the compound, the hydrogen and boron atoms in the cage structure are divided into two groups. The four peaks of the chemical shift δ 5.5403–6.836 in the hydrogen spectrum and the two peaks of the chemical shift δ −28.669 in the boron spectrum correspond to eight hydrogen and boron atoms on the equator. The four peaks of chemical shift δ 5.5403–6.3836 in the hydrogen spectrum and the two peaks of chemical shift δ −0.834 in the boron spectrum correspond to two hydrogen atoms and two boron atoms at the top of the polyhedron.

2.2.2.2 *SEM of Prepared $B_{10}H_{10}[(C_2H_5)_4N]_2$ Particles*

A scanning electron microscope (SEM) was used to observe the micromorphology of prepared $B_{10}H_{10}[(C_2H_5)_4N]_2$ powder, as shown in Figure 2.2. It can be seen that the $B_{10}H_{10}[(C_2H_5)_4N]_2$ prepared particle is a polyhedral compound with particle size of 50–200 μm.

2.2.2.3 *Nuclear Magnetic Resonance Spectral Analysis*

The nuclear magnetic resonance spectrum of a $B_{10}H_{10}[(C_2H_5)_4N]_2$ sample is shown in Figure 2.3. It can be seen that ^{1}H NMR ($CDCl_3$, 500 MHz), δ: 3.37–3.77 (m, 24H), 5.72–5.76 (m, 16H), 5.54–6.38 (q, 2H), and 2.33–3.09 (m, 8H); ^{11}B NMR (DMSO, 500 MHz), δ: −28.67 (d, 8B, J = 130 Hz), −0.83 (d, 2B, J = 130 Hz).

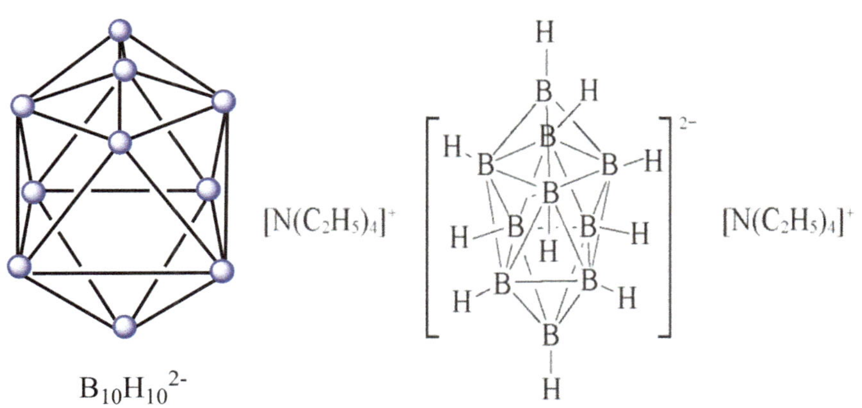

Figure 2.1 Chemical structure of $B_{10}H_{10}[(C_2H_5)_4N]_2$. Reproduced from ref. 17.

Figure 2.2 SEM micrograph of $B_{10}H_{10}[(C_2H_5)_4N]_2$.

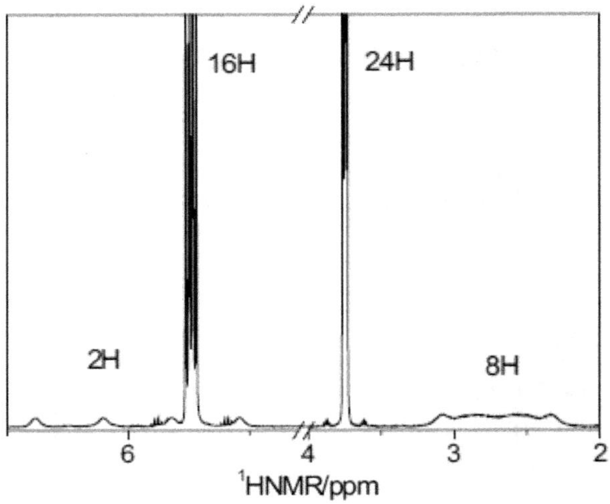

Figure 2.3 ^{1}H NMR of $B_{10}H_{10}[(C_2H_5)_4N]_2$ sample.

2.2.2.4 *Infrared Spectroscopy (IR) and Elemental Analysis*

Figure 2.4 shows the IR spectra of $B_{10}H_{10}[(C_2H_5)_4N]_2$ samples at different temperatures. It can be seen from Figure 2.4 that the IR spectral analysis of the sample shows: IR (KBr), ν/cm^{-1}: 2991 ($\nu_{C\text{-}H}$), 2450 ($\nu_{B\text{-}H}$), 1460 ($\nu_{C\text{-}H}$),

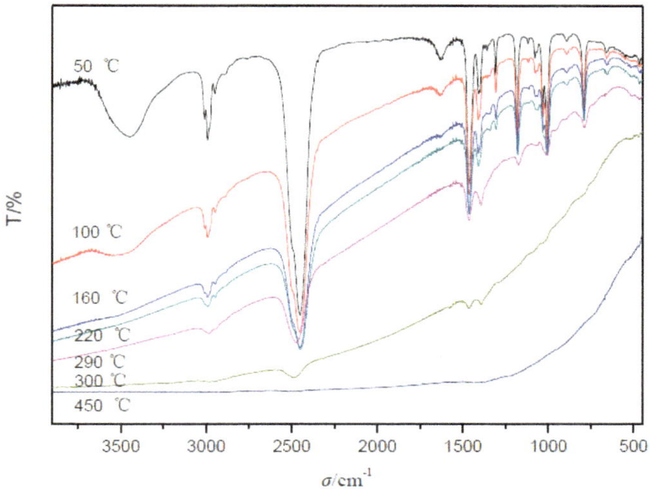

Figure 2.4 Infrared spectra (IR) of $B_{10}H_{10}[(C_2H_5)_4N]_2$ at different temperatures. Reproduced from ref. 17.

1401 (ν_{C-H}), 1309 (ν_{C-N}), 1184 (ν_{C-N}), 1034 (ν_{B-B}), 1008 (ν_{B-B}). Comparing the elemental analysis of ($C_{16}H_{50}B_{10}N_2$): calculated B 28.5%, C 50.5%, H 10.5%, and N 7.4%; measured values B 29.5%, C 50.6%, H 10.5%, and N 7.3%.

2.2.2.5 Heat of Combustion Analysis

The heat of combustion of the compound $B_{10}H_{10}[(C_2H_5)_4N]_2$ is 53.076 MJ kg^{-1} measured by the oxygen bomb method, showing that the heat of combustion value of the compound is higher than that of Al (30.22 MJ kg^{-1}) and Mg (25.0 MJ kg^{-1}). The reason maybe that the polyhedral boron hydrogen structure in the molecule decomposes and releases heat during the combustion process, and the boron, hydrogen, carbon, and nitrogen atoms in the molecule are fully oxidized to release a large amount of heat.[18]

2.2.2.6 Thermal Decomposition Analysis

DSC–TG can be used to characterize the thermal behavior of the energetic materials in different atmospheres and temperatures.[19] In order to study the thermal decomposition characteristics of the compound and the thermal behavior in oxygen atmosphere, the DSC and TG curves were measured with argon and oxygen as carrier gases, respectively. As shown in Figures 2.5 and 2.6, in argon atmosphere there are three exothermic peaks on the DSC curve (Figure 2.5), the initial temperature is 294.8 °C, the peak values are 298.2, 303.2, and 311.9 °C, respectively, and the total exothermic heat is 112.8 J g^{-1}. On the corresponding TG curve, at 295.6 °C, the compound began to lose weight, and the thermal decomposition of the compound decreased to the

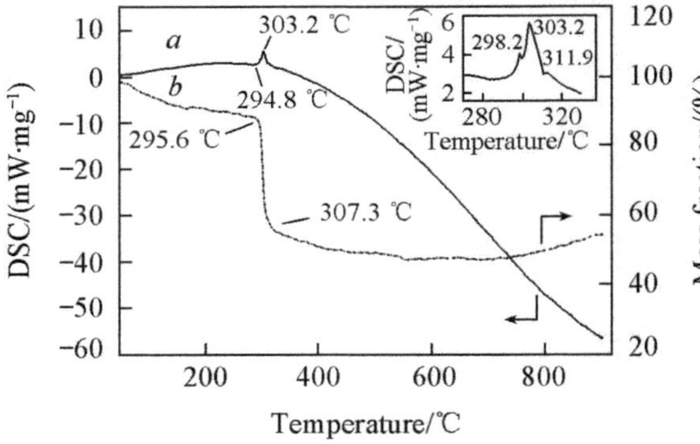

Figure 2.5 DSC (a) and TG (b) curves of $B_{10}H_{10}[(C_2H_5)_4N]_2$ in Ar. Reproduced from ref. 17.

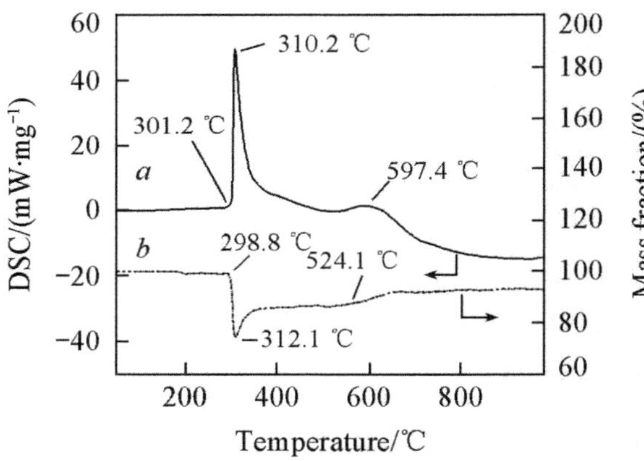

Figure 2.6 DSC (a) and TG (b) curves of $B_{10}H_{10}[(C_2H_5)_4N]_2$ in O_2. Reproduced from ref. 17.

lowest, 307.3 °C, while the mass did not change significantly. Combined with the analysis of the two curves, it can be inferred that the thermal decomposition of the compound can be divided into three consecutive steps; the exothermic of the intermediate step is the largest, and the total weight loss in the decomposition process is 29.8%, which may be due to the tetraethylammonium group in the molecule and the protohydrogen on the boron cage. This is caused by the dissociation of small gas molecules from the sample cell.

It can be seen from Figure 2.6 that there are two exothermic peaks on the DSC curve in oxygen atmosphere. The first exothermic peak has an initial temperature of 301.2 °C and a peak temperature of 310.2 °C with a peak area of 4967 $J g^{-1}$, followed by a broader exothermic peak with a peak temperature of 597.4 °C, and the peak area is 2031 $J g^{-1}$. On the corresponding TG curve, the thermal decomposition of the sample began to lose weight at

298.8 °C, and reached the minimum at 312.1 °C. The mass of the residue was 75.4%, and then increased rapidly to 86.5%. At 524.1 °C, the weight of the residue increased again. At 970 °C, the mass of the residue was 92.5%, and only lost 7.5% in mass fraction. This was consistent with the theoretical weight loss (7.4%) when all boron in the sample was oxidized to B_2O_3; the rest of them escape as the gaseous products, and the boron produced by decomposition reacts with oxygen to release a lot of heat and increase weight at the same time.

Comparing the DSC and TG curves of the compound in two different atmospheres, it can be seen that the initial exothermic temperature and the initial weight loss temperature of the compound in argon atmosphere are lower than those in oxygen atmosphere. Therefore, it is speculated that the compound first decomposes in oxygen, then oxidizes rapidly, and releases a lot of heat at the same time. The exothermic heat of the latter is much higher than that of the former in the DSC curve, which is due to the fact that, in argon, the exothermic temperature of the compound is higher than that of the former compounds that only decompose and release decomposition heat. In oxygen, the decomposition products, such as boron, react with oxygen to form B_2O_3 and release high reaction heat.

The DSC curves of the pure $B_{10}H_{10}[(C_2H_5)_4N]_2$ sample at the pressures of 0.1, 1, 2, and 4 MPa, and the DSC curves of the $B_{10}H_{10}[(C_2H_5)_4N]_2$ sample at a heating rate of 5, 10, 15, and 20 K min^{-1}, are shown in Figures 2.7 and 2.8, respectively. The DSC curves of the pure $B_{10}H_{10}[(C_2H_5)_4N]_2$ sample obtained at various pressures and heating rates showed an exothermic change.

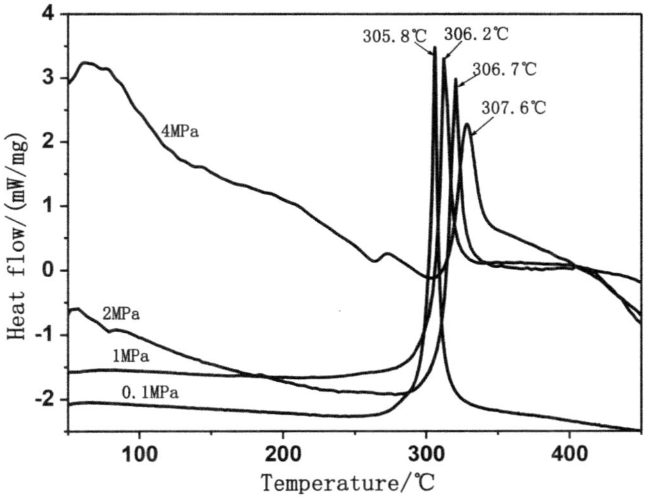

Figure 2.7 The DSC curves of $B_{10}H_{10}[(C_2H_5)_4N]_2$ at pressures of 0.1, 1, 2, and 4 MPa (heating rate: 10 K min^{-1}, sample pass: 0.35–0.46 mg). Reproduced from ref. 20 with permission from John Wiley & Sons, Copyright © 2013 Wiley-VCH Verlag GmbH & Co. KGaA, Weinheim.

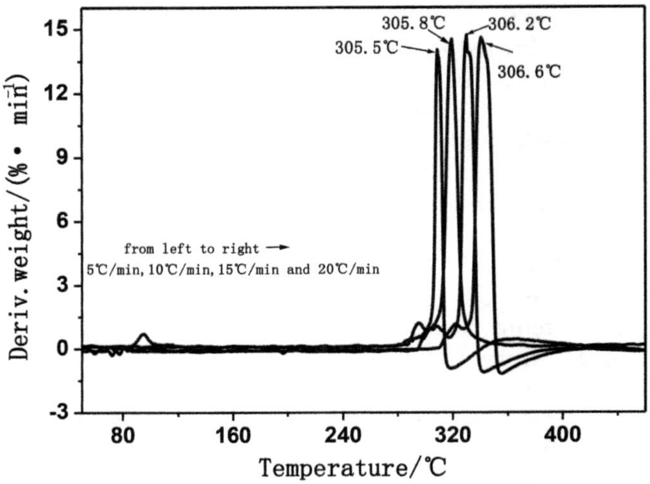

Figure 2.8 The DSC curves of $B_{10}H_{10}[(C_2H_5)_4N]_2$ at the heating rates of 5, 10, 15, and 20 K min^{-1} (pressure: 0.1 MPa; sample mass: 0.35–0.46 mg). Reproduced from ref. 20 with permission from John Wiley & Sons, Copyright © 2013 Wiley-VCH Verlag GmbH & Co. KGaA, Weinheim.

The exothermic peak became very sharp when the pressure was changed from 4 MPa to 0.1 MPa. The peak temperature of $B_{10}H_{10}[(C_2H_5)_4N]_2$ sample is 305.8 °C at 0.1 MPa or 306.2 °C at 1 MPa and the peak temperature of the $B_{10}H_{10}[(C_2H_5)_4N]_2$ sample is 305.8 °C at the heating rate of 10 K min^{-1} or 306.6 °C at the heating rate of 20 K min^{-1}. The DSC curves showed no endothermic change before the onset of exothermic change, and it was indicated that this compound was decomposed in solid state.

2.2.2.7 *Thermal Decomposition Kinetics Analysis*

The thermal decomposition peak temperatures (T_p) were obtained by means of derivative thermogravimetry method (DTG), and they were fitted by using Kissinger (2.3) and Ozawa (2.4) equations; the apparent activation energy, factor, and linear coefficient are shown in Table 2.1.

$$\ln\left(\frac{\beta_i}{T_i^2}\right) = -\frac{E_k}{R}\cdot\frac{1}{T_i} + \ln\left(\frac{A_k R}{E_k}\right), \quad (I = 1-4) \tag{2.3}$$

$$\lg\beta_i = \ln\left(\frac{AE}{RG(a)}\right) - 2.315 - 0.4567\frac{E}{RT_i}. \tag{2.4}$$

In the equations, β_i is heating rate, K min^{-1}; R is the gas constant, 8.314 J (mol K^{-1})$^{-1}$; T_i is the peak temperature, K; E_k is apparent activation energy, J mol^{-1}; $G(a)$ is the integral form of the mechanism function.

Table 2.1 Kinetic parameters of the two thermal decomposition stages of $B_{10}H_{10}[(C_2H_5)_4N]_2$. Reproduced from ref. 21.

			E/(kJ mol^{-1})		lg(A_k/s^{-1})	r	
Stage	β/(K min^{-1})	T_p/K	Kissinger	Ozawa	Kissinger	Kissinger	Ozawa
1	2.5	532.4	150.9	152.2	12.11	0.9970	0.9974
	5.0	541.5					
	10.0	522.8					
	20.0	564.5					
2	2.5	553.8	161.7	162.8	13.54	0.9976	0.9979
	5.0	563.7					
	10.0	574.2					
	20.0	586.7					

Table 2.2 Apparent activation energies of $B_{10}H_{10}[(C_2H_5)_4N]_2$ at different reaction depths. Reproduced from ref. 21.

	E/(kJ mol^{-1})	
a	Stage 1	Stage 2
0.1	160.6	162.3
0.2	159.2	162.1
0.3	159.9	161.7
0.4	158.9	160.6
0.5	161.4	18.8
0.6	162.5	156.4
0.7	163.4	154.6
0.8	165.9	152.5
0.9	168.2	152.5
1.0	170.8	157.3

From the data in Table 2.1 it can be seen that the activation energies of the first stage of thermal decomposition of $B_{10}H_{10}[(C_2H_5)_4N]_2$ calculated by the Kissinger and Ozawa methods from DTG peak temperatures are 150.9 kJ mol^{-1} and 152.2 kJ mol^{-1}, respectively. The activation energies of the second stage are 161.7 kJ mol^{-1} and 162.8 kJ mol^{-1}, respectively. The difference between the activation energies calculated by the two methods mentioned above is small, and the correlation coefficients are higher than 0.9970, indicating that the calculation results are more accurate. The average activation energies calculated by the two methods are 151.6 kJ mol^{-1} and 162.3 kJ mol^{-1}, respectively.

The mass loss of the TG curve is regarded as the reaction depth, a. Because the reaction process is divided into two stages, and the mass loss of each stage is corrected, so that the mass loss of each stage is 0–100%. The temperature value T_i of the reaction depth with equal conversion under different heating rates β_i is regressed according to the Ozawa equation, and the reaction depth change of activation energy under different reaction depths were obtained, and the results are shown in Table 2.2. It can be seen that the apparent activation energy is in the range (163.1 ± 7.7) kJ mol^{-1} and (157.9 ± 5.4) kJ mol^{-1} for the two stages of thermal decomposition, respectively, which means that the apparent activation energy obtained by the equal conversion method is slightly larger than that data calculated by the DTG peak temperature in the first thermal decomposition stage, while it is in the same range as that obtained by the DTG peak temperature in the second thermal decomposition stage.

In order to determine the thermal decomposition dynamic mechanism function of $B_{10}H_{10}[(C_2H_5)_4N]_2$, the heating rate of 2.5 °C min^{-1} was selected. According to the Coats–Redfern eqn (2.5), the thermal decomposition data were replaced by the thermal decomposition mechanism function of 26 common solids;[22,23] for the same mechanism function, there was a linear correlation between $\ln\frac{G(a)}{T^2}$ and $-\frac{1}{T}$ at different reaction depths, a.

$$\ln\frac{G(a)}{T^2} = -\frac{E}{RT} + \ln\frac{AR}{\beta E}. \tag{2.5}$$

By fitting the mechanism function and comparing the activation energy calculated by the Kissinger and Ozawa methods, it can be seen that the most probable mechanism function of the first thermal decomposition stage of $B_{10}H_{10}[(C_2H_5)_4N]_2$ is $G(a) = [-\ln(1-a)]^{1/3}$, that is the reaction follows the Avrami–Erofeyev equation of nucleation and growth ($n = 3$), the most probable mechanism function of the second thermal decomposition stage of $B_{10}H_{10}[(C_2H_5)_4N]_2$ is $G(a) = a^{1/4}$, and the decomposition process follows the law of power series (Mampel power).

According to the characteristic quantity of the DSC curve of the compound at different heating rates, the kinetic parameters of the thermal decomposition reaction $E_a = 112.06$ kJ mol^{-1}, $\ln A = 30.79$ s^{-1}, $r = 0.9998$ were obtained by the Kissinger method.[24] According to the above activation energy and temperature range, $E_a/RT > 20$ can be calculated. Therefore, the heat decomposition reaction mechanism equation $[f(a)]$ can also be inferred by using the Coat–Redfern equation.[25] The corresponding heat decomposition reaction mechanism is a spherical symmetric three-dimensional diffusion process, and the non-isothermal kinetic equation is $da/dt = 10^{13.37}\exp(-1.347847 \times 10^4/T)[1-(1-a)^{1/3}]$.

2.2.2.8 *Thermal Decomposition Products Analysis*

Because the decomposition heat and mass loss of thermal analysis are usually the total process of decomposition, the reaction mechanism cannot be obtained by thermal analysis data alone. In order to study the thermal decomposition process, the thermal decomposition products were analysed. The change in characteristic functional groups during the pyrolysis of condensed phase can be tracked in real time by pyrolysis *in situ* cell FTIR; TG–DSC–FTIR–MS can be used to determine the change of gaseous products in the decomposition process. These two techniques have been applied to the study of the thermal decomposition mechanism.[26]

(1) Thermal decomposition process of the condensed phase. The molecular structure of bis(tetraethylammonium decahydroborate) is composed of a double capped anti tetraprism closed caged $B_{10}H_{10}^{2-}$ ion and tetraethylammonium cations. Therefore, the infrared characteristic peaks are mainly B–H bonds on a caged structure and the characteristic vibration of alkyl groups in cations. The changes of groups in condensed phase during heating process were tracked by pyrolysis *in situ* cell-FTIR, the infrared spectra at different temperatures and the curve of the characteristic absorption peak intensity of each functional group with temperature are shown in Figure 2.9. In Figure 2.9, curve b (2992 cm^{-1}) and curve c (1459 cm^{-1}) are the characteristic absorption peaks of ν_{C-H} and δ_{C-H}, respectively. The curves d (2452 cm^{-1}) and a (1182 cm^{-1}) are the characteristic absorption peaks of ν_{B-H} and δ_{B-H}, respectively. From the variation trend of the characteristic absorption peak intensity of functional groups with temperature, it can be seen that the peak heights of the B–H bond and the C–H bond decrease rapidly and almost disappear at 290–300 °C, indicating that the cage structure of boron hydrogen polyhedron in the compound decomposes rapidly. The B–H bond and C–H bond in the condensed phase almost disappear at the same time, resulting in almost zero light transmittance. The decomposition process may be that hydrogen atoms and ethyl groups produced by B–H bond breaking on the boron cage form hydrogen or ethane escaping from the sample pool. At the same time, the cage structure of the boron cage is unstable and disintegrated due to the loss of hydrogen protons, and boron atoms agglomerate to form elemental boron; as a result, the transmittance of the sample decreases gradually.

Figure 2.10 shows the SEM-EDS spectra of the intermediate and final product of the thermal decomposition of $B_{10}H_{10}[(C_2H_5)_4N]_2$. It can be seen that the

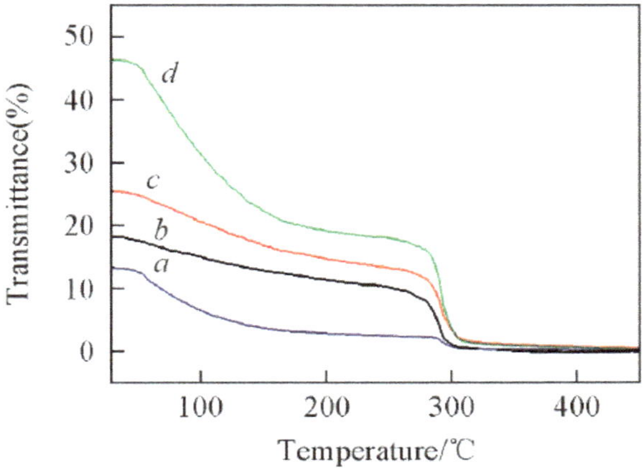

Figure 2.9 IR characteristic transmittance of condensed phase of $B_{10}H_{10}[(C_2H_5)_4N]_2$ thermal decompositions *vs.* temperature. (a) 1182 cm^{-1}; (b) 2992 cm^{-1}; (c) 1459 cm^{-1}; (d) 2452 cm^{-1}. Reproduced from ref. 17.

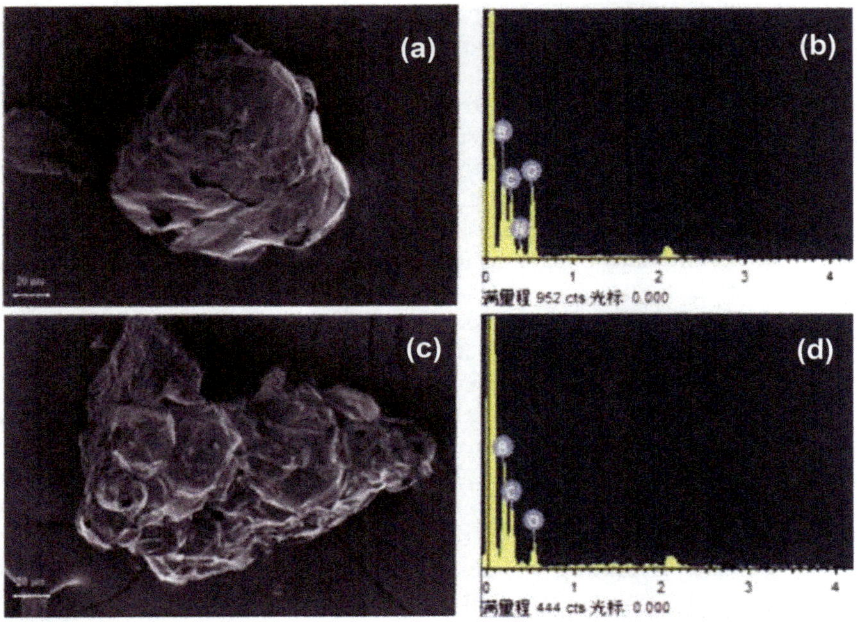

Figure 2.10 SEM-EDS spectra of the intermediate and final products of thermal decomposition of $B_{10}H_{10}[(C_2H_5)_4N]_2$. (a) SEM image of the intermediate products; (b) EDS of the intermediate products; (c) SEM image of the final products; (d) EDS of the final products. Reproduced from ref. 21.

intermediate products and the final products of $B_{10}H_{10}[(C_2H_5)_4N]_2$ thermal decompositions have melted and agglomerated. After thermal decomposition, the polyhedral crystal structure of $B_{10}H_{10}[(C_2H_5)_4N]_2$ is destroyed. The intermediate products are mainly composed of C (36.1%), B (57.6%), and N (6.3%), while the final products are composed of C (52.6%) and B (42.3%). A small amount of oxygen element (4.7%) appears as well in the sample.

(2) Composition of the gaseous products of thermal decomposition. The composition of the thermal decomposition gas products of the $B_{10}H_{10}[(C_2H_5)_4N]_2$ compound was analysed by DSC–TG–FTIR–MS. The thermal decomposition mechanism of the title compound was studied by FTIR and MS information of the gas products at the synchronous temperature of DSC–TG. In the process of heating up, the infrared spectrum of gas product is detected in real time by using the fast speed scanning function of an infrared spectrometer. The infrared spectrum of the gas phase decomposition product of bis(tetraethylammonium decahydroborate) in the process of heating up is shown in Figure 2.11. It can be seen that the spectrum at 192 °C is the same as the initial atmosphere, and there are absorption peaks at 2361 cm^{-1} and 1505 cm^{-1}, which are characteristic absorption peaks of CO_2. In the spectrum at 215 °C, there are absorption peaks at 2421 cm^{-1} and 1416 cm^{-1}, which are characteristic absorption peaks of C_2H_6. After that, the peak value gradually increases and then decreases to 300 cm^{-1}.

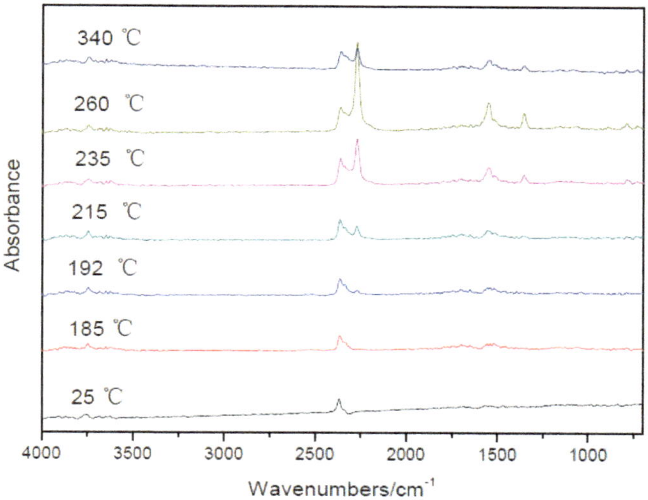

Figure 2.11 IR spectra of gas products of $B_{10}H_{10}[(C_2H_5)_4N]_2$ thermal decomposition at various temperatures. Reproduced from ref. 17.

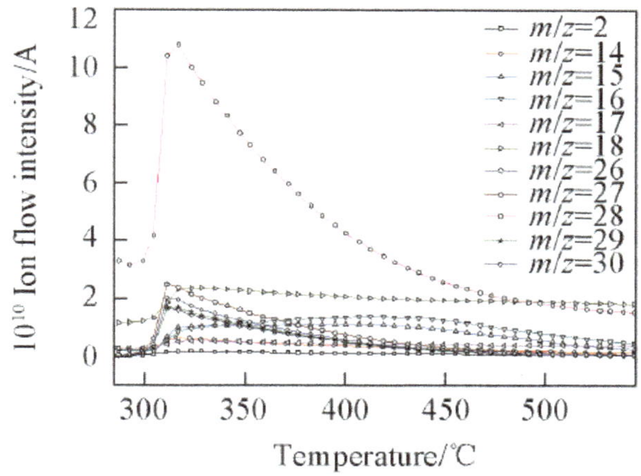

Figure 2.12 MS ion flow intensity of gas products of $B_{10}H_{10}[(C_2H_5)_4N]_2$ thermal decompositions *vs.* temperature. Reproduced from ref. 17.

Corresponding to the decomposition temperature of DSC and TG, it is indicated that ethane is the most abundant product in gas phase decomposition.

Due to the high sensitivity of the mass spectrometer, more decomposition products can be captured. It can be seen from Figure 2.12 that corresponding to the decomposition temperature of DSC and TG curves, the MS ions appearing in the gas phase products mainly include *m*/*z* 2, 14, 15, 16, 17, 18, 26, 27, 28, 29, 30, and 41 fragments at 300 °C. The corresponding gaseous products are mainly C_2H_6 (with characteristic mass numbers of 30, 29, 28,

27, and 26), NH_3 (with characteristic mass numbers of 17, 16, 15, and 14), and H_2 (with characteristic mass number of 2). The peak values of ion current intensity are mostly at 310 °C, and the peak value of C_2H_6 is higher than that of ammonia and hydrogen.

2.2.2.9 *Thermal Decomposition Mechanism of $B_{10}H_{10}[(C_2H_5)_4N]_2$*

Combined with the above analysis results, the reaction mechanism of the decomposition process of ditetraethylammonium decahydrodecaborate is speculated, as shown in Scheme 2.1. In the process of thermal decomposition, tetraethylammonium decahydrodecaborate decomposes rapidly at 303.2 °C (Scheme 2.2). The gas phase decomposition products are mainly ethane, a small amount of ammonia and hydrogen, and the solid phase product is boron. When it is used in solid propellant, the compound decomposes at a certain temperature, and the high active boron reacts with the oxidant in the propellant to release a lot of heat. In addition, there are more gases in the thermal decomposition products or combustion products, which can increase the gas work ability of the propellant.

The calculated mass loss of hydrogen on the boron cage is 2.6%, which is close to the actual mass loss of 2.9% in the first stage of thermal decomposition, indicating that the first stage of thermal decomposition is mainly the dehydrogenation process of the boron cage. At the end of the first step of the decomposition reaction, a tetraethylammonium cation, an intermediate product of melt decomposition, will be formed. In the second stage of thermal decomposition, a tetraethylammonium cation undergoes proton transfer through the Hoffman elimination reaction[27] to produce ethylene and triethylamine (Et_3N), and then Et_3N is further decomposed into H_2, C_2H_6, NH_3 and carbon©). At the same time, the TG curve shows that the mass loss in the second stage is 39.4%, which is close to the calculated value of 43.4% of the gas mass loss in the second stage. The final solid product is composed of amorphous B and C, except that a small amount of oxygen in the final product is ignored.

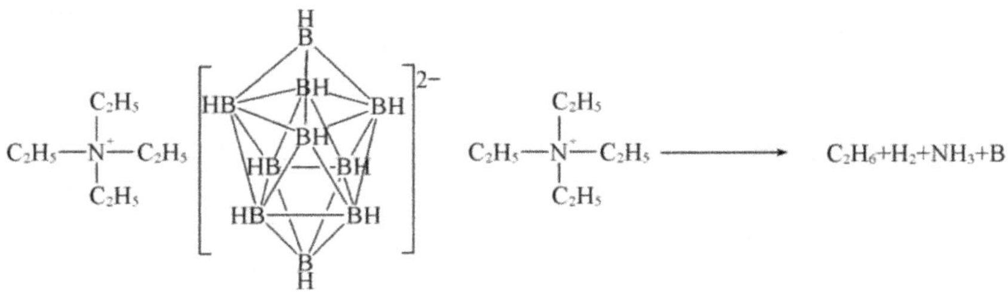

Scheme 2.1 Thermal decomposition mechanism of $B_{10}H_{10}[(C_2H_5)_4N]_2$. Reproduced from ref. 17.

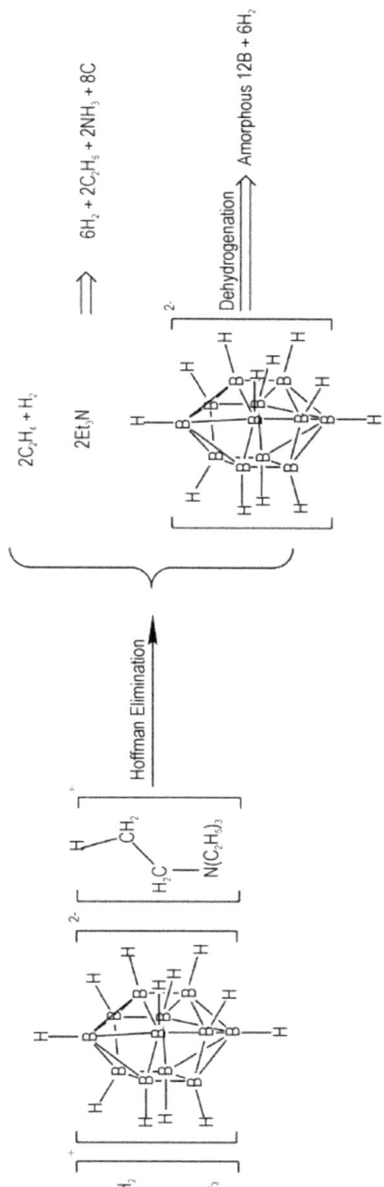

Scheme 2.2 Thermal decomposition mechanism of $B_{12}H_{12}[(C_2H_5)_4N]_2$. Reproduced from ref. 28.

2.3 Synthesis and Characterization of Bis(dialkyl-5-aminotetrazolium) Dodecaborates

The preparation methods of icosahedral closed cage borohydride anion $B_{12}H_{12}{}^{2-}$ and the thermal behavior, thermochemistry and energy properties of related compounds in the study of explosives and explosives are summarized. It is pointed out that the dodecahydrododecaboric acid anion compounds have high decomposition temperature, low toxicity, low sensitivity, and good compatibility with common energetic components.[29] In particular, the preparation process of $B_{12}H_{12}{}^{2-}$ ion is simple and can be synthesized in one pot. These characteristics have a great attraction for researchers of solid propellants and explosives.

Nitrogen rich heterocyclic compounds play an important role in the research of high-energy density materials.[30,31] The sensitivity of tetrazolium compounds with high nitrogen content, high density, high enthalpy of formation and large amount of gas generation is lower than that of RDX and HMX. Most of the gas products are nitrogen, which can achieve the effect of less smoke or smoke-free. The fuel gas does not contain HCl, which also has the characteristics of low characteristic signal, insensitivity and environmental protection. Therefore, it has attracted researchers' attention in the green research field on high-energy density materials such as explosives and solid propellants.[32–34] A new kind of high-energy materials with high enthalpy of formation and good thermal stability can be obtained by combining high-energy tetrazole cations with high-energy borohydride anions.

2.3.1 Synthesis of Bis(dialkyl-5-aminotetrazolium) Dodecaborates

Nitrogen-rich high-boron compounds can be prepared by the double decomposition reaction of dodecahydrododecaboric acid metal salt and halogenated dialkylated 5-aminotetrazole salt which are soluble in water. After these two reactants are mixed in water and react for 1–2 h, they were cooled to room temperature, and the colorless crystals of the expected products can be obtained. The yield of the expected product is generally not less than 90%.

Figure 2.13 shows the synthesis of bis(dialkyl-5-aminotetrazolium) dodecahydrododecaborates.

*2.3.1.1 Synthesis of Bis(1,4-dimethyl-5-aminotetrazole) Dodecahydrododecaborate Salt (**1**)*

0.66 g $K_2B_{12}H_{12}$ and 1.45 g iodized 1,4-dimethyl-5-aminotetrazolium salt were dissolved in an appropriate amount of deionized water respectively, and these two were mixed for 1.5 h, then heated at 65 °C for 0.5 h, cooled and

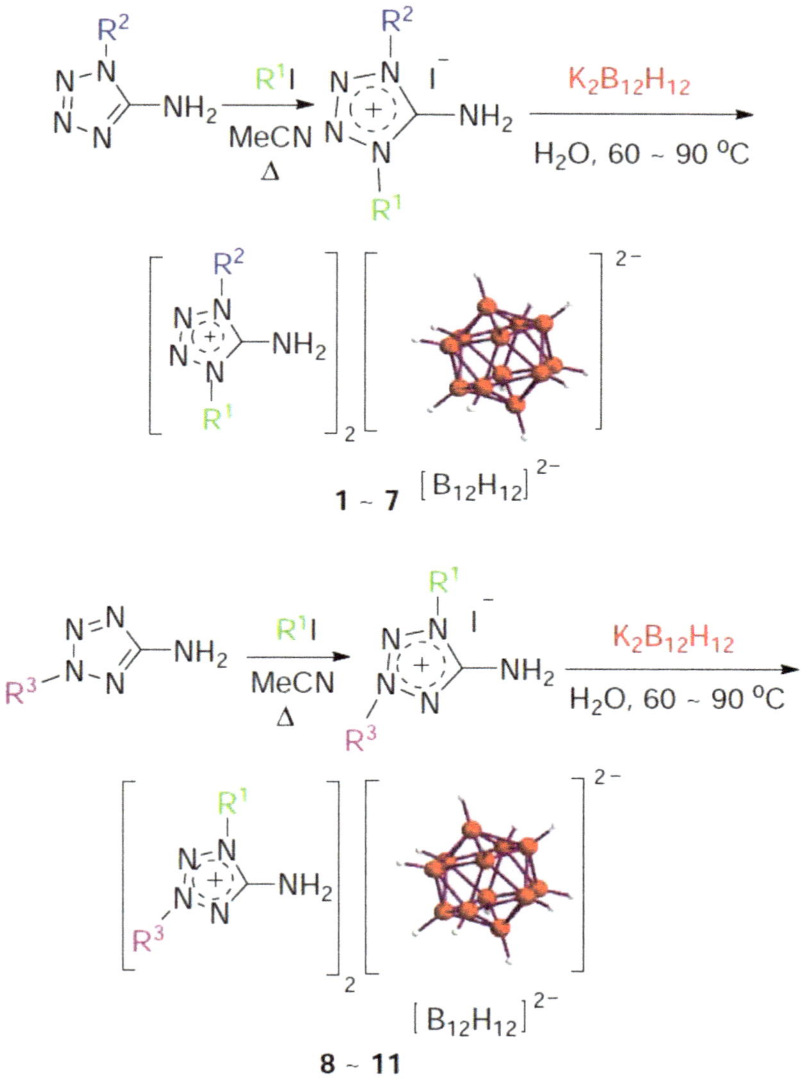

(1) R[1]=Me, R[2]=Me; (2) R[1]=Me, R[2]=Et; (3) R[1]=Et, R[2]=Et; (4) R[1]=Me, R[2]=*n*-Bu; (5) R[1]=Et, R[2]= *n*-Bu; (6) R[1]=Me, R[2]= *n*-Hex; (7) R[1]=Et, R[2]= *n*-Hex; (8) R[1]=Me, R[3]=Me; (9) R[1]=Me, R[3]=Et; (10) R[1]=Et, R[3]=Et; (11) R[1]=Et, R[3]=Me.

Figure 2.13 Synthesis of bis(dialkyl-5-aminotetrazolium) dodecahydrododecaborates. Reproduced from ref. 35.

crystallized to obtain white crystal (**1**), drying at 80 °C under reduced pressure, the yield was 72%. ^{1}H NMR (DMSO-d_6, 600 MHz) δ: 9.09 (s, 4H, $2NH_2$), 3.87 (s, $4NCH_3$), 1.3–0.4 (platform, 12H, $B_{12}H_{12}$); ^{13}C NMR (DMSO-d_6, 150 MHz) δ: 149.0 (ring C), 34.5(N–C); ^{11}B NMR (DMSO-d_6, 200 MHz) δ: −15.5; IR ν: 3341 (m), 3294 (m), 3248 (m), 3116 (m), 2487 (vs), 1681 (vs), 1606 (m), 1532 (m), 1437 (m), 1185 (w), 1046 (s), 771 (s), 716 (s) cm^{-1}; ESI-MS *m*/*z*: 114 $[C_3H_8N_5]^+$, 71.35 ($1/2[B_{12}H_{12}]^{2-}$). Anal. calcd for $C_6N_{10}H_{28}B_{12}$: C 19.47, H 7.63, N 37.85; found C 19.08 H 7.58, N 37.99.

2.3.1.2 *Synthesis of Bis(1-methyl-4-ethyl-5-aminotetrazole) Dodecahydrododecaborate Salt (2)*

0.86 g of $K_2B_{12}H_{12}$ and 2.00 g of iodized 1-methyl-4-ethyl-5-aminotetrazole salt were dissolved in an appropriate amount of deionized water respectively. These two were mixed and reacted for 2 h, then heated at 80 °C for 0.5 h, cooled and crystallized to obtain flake white crystal (**2**), drying at 80 °C under reduced pressure, the yield was 77%. ^{1}H NMR (DMSO-d_6, 600 MHz) δ: 9.08 (s, 4H, $2NH_2$), 4.26 (q, $J = 6$ Hz, 4H, $2N^4–CH_2$), 3.86 (s, 6H, $2N^1–CH_3$), 1.40 (t, $J = 6$ Hz, 6H, $2N^4–CCH_3$), 1.4–0.4 (sym. near platform, 12H, $B_{12}H_{12}$). ^{13}C NMR (DMSO-d_6, 150 MHz) δ: 148.3 (ring C), 43.1 (NCH_3), 34.5 (NCH_2), 13.3 ($NCCH_3$); ^{11}B NMR (DMSO-d_6, 200 MHz) δ: −15.5; IR ν: 3366 (m), 3291 (m), 3241 (m), 3162 (m), 2980 (w), 248l (w), 2455 (vs), 1677 (vs), 1524 (m), 1443 (m), 1054 (m), 746 (m), 710 (m) cm^{-1}; ESI-MS *m*/*z*: 128 $[C_4H_{10}N_5]^+$, 70.55 ($1/2[B_{12}H_{12}]^{2-}$). Anal. calcd for $C_8N_{10}H_{32}B_{12}$: C 24.13, H 8.10, N 35.18; found C 23.42, H 7.93, N 36.41.

2.3.1.3 *Synthesis of Bis(1,4-diethyl-5-aminotetrazole) Dodecahydrododecaborate Salt (3)*

0.61 g $K_2B_{12}H_{12}$ and 1.50 g iodized 1,4-diethyl-5-aminotetrazolium salt were dissolved in an appropriate amount of deionized water respectively, mixed for 2 h, heated at 80 °C for 0.5 h, cooled and crystallized to obtain white crystal (**3**) with a yield of 78%. M.p. 184.0–203.9 °C; ^{1}H NMR (DMSO-d_6, 150 MHz) δ: 9.07 (s, 4H, $2NH_2$), 4.26 (q, $J = 6$ Hz, 8H, $4N–CH_2$), 1.42 (q, $J = 6$ Hz, 12H, $4N–CCH_3$), 1.4–0.4 (platform, 12H, $B_{12}H_{12}$); ^{13}C NMR (DMSO-d_6, 150 MHz) δ: 147.7 (ring C), 43.1 ($N–CH_2$), 13.3 ($N–CCH_3$); ^{11}B NMR (DMSO-d_6, 200 MHz) δ: −15.5; IR ν: 3353 (m), 3286 (m), 3231 (m), 3165 (m), 2982 (w), 2449 (vs), 1670 (vs), 1514 (m), 1452 (m), 1043 (s), 724 (m), 710 (s) cm^{-1}; ESI-MS *m*/*z*: 142 $[C_5H_{12}N_5]^+$, 71.45 ($1/2[B_{12}H_{12}]^{2-}$). Anal. calcd for $C_{10}N_{10}H_{36}B_{12}$: C 28.18, H 8.5l, N 32.86; found C 25.85, H 8.22, N 30.35.

2.3.1.4 *Synthesis of Bis(1-methyl-4-*n*-butyl-5-aminotetrazole) Dodecahydrododecaborate Salt (4)*

0.78 g $K_2B_{12}H_{12}$ and 2.00 g iodized 1-methyl-4-*n*-butyl-5-aminotetrazole salt were dissolved in an appropriate amount of deionized water respectively, mixed for 1.5 h, and cooled to obtain white crystal (**4**) with a yield of 91%. M.p. 230.9 °C; ^{1}H NMR (DMSO-d_6, 600 MHz) δ: 9.08 (s, 4H, $2NH_2$), 4.23 (t, $J=6$ Hz, 4H, $2N^4$–CH_2), 3.86 (s, 6H, $2N^1$–CH_3), 1.79–1.74 (m, 4H, $2N^4$–CCH_2); 1.38–1.33 (m, 4H, $2CH_2$), 0.92 (t, $J=6$ Hz, 6H, $2CH_3$), 1.4–0.4 (platform, 12H, $B_{12}H_{12}$); ^{13}C NMR (DMSO-d_6, 150 MHz) δ: 148.5 (ring C), 47.3 (N^4–CH_2), 34.6 (N^1–CH_3), 29.7 (N^4–CCH_2), 19.3 (CH_2), 13.8 (CH_3); ^{11}B NMR (DMSO-d_6, 200 MHz) δ: −15.5; IR ν: 3367 (m), 3394 (m), 3241 (m), 3168 (m), 2960 (w), 2875 (w), 2471 (vs), 1677 (vs), 1060 (s), 1034 (s), 761 (s), 708 (s) cm^{-1}; ESI-MS *m*/*z*: 156 $[C_6H_{14}N_5]^+$, 70.95 (1/2$[B_{12}H_{12}]^{2-}$). Anal. calcd for $C_{12}N_{10}H_{40}B_{12}$: C 31.73, H 8.88, N 30.84; found C 29.96, H 8.37, N 29.20.

2.3.1.5 *Synthesis of Bis(1-ethyl-4-*n*-butyl-5-aminotetrazole) Dodecahydrododecaborate Salt (5)*

0.56 g $K_2B_{12}H_{12}$ and 1.50 g iodized 1-ethyl-4-*n*-butyl-5-aminotetrazole salt were dissolved in an appropriate amount of deionized water respectively, mixed for 1.5 h, and cooled to obtain white crystal (**5**) with a yield of 93%. M.p. 204.9 °C; ^{1}H NMR (DMSO-d_6, 600 MHz) δ: 9.08 (s, 4H, $2NH_2$), 4.27–4.22 (m, 8H, $2N^4$–CH_2 + $2N^1$–CH_2), 1.80–1.75 (m, 4H, $2N^4$–CCH_2), 1.42 (t, $J=6$ Hz, 6H, $2N^1$–CCH_3), 1.37–1.33 (m, 4H, $2CH_2$), 0.93 (t, $J=6$ Hz, 6H, $2CH_3$), 1.4–0.4 (sym. near platform, 12H, $B_{12}H_{12}$); ^{13}C NMR (DMSO-d_6, 150 MHz) δ: 147.9 (ring C), 47.3 (N^4–CH_2), 43.2 (N^1–CH_2), 29.6 (N^4–CCH_2), 19.3 (CH_2), 13.8 (CH_3), 13.2 (N^1–CCH_3); ^{11}B NMR (DMSO-d_6, 200 MHz) δ: −15.5; IR ν: 3361 (m), 3288 (m), 3234 (m), 3169 (m), 2959 (w), 2876 (w), 2463 (vs), 1673 (vs), 1047 (s), 746 (m), 707 (s) cm^{-1}; ESI-MS *m*/*z*: 170 $[C_7H_{16}N_5]^+$, 71.40 (1/2$[B_{12}H_{12}]^{2-}$). Anal. calcd for $C_{14}N_{10}H_{44}B_{12}$: C 34.86, H 9.20, N 29.04; found C 34.42, H 9.13, N 28.66.

2.3.1.6 *Synthesis of Bis(1-methyl-4-hexyl-5-aminotetrazole) Dodecahydrododecaborate Salt (6)*

1.24 g $K_2B_{12}H_{12}$ and 3.50 g iodized 1-methyl-4-hexyl-5-aminotetrazole salt were dissolved in an appropriate amount of deionized water at 90 °C respectively, mixed for 1.5 h, cooled, and collected solids then dried under reduced pressure to obtain white crystal (**6**) with a yield of 93%. M.p. 245.6 °C; ^{1}H NMR (DMSO-d_6, 600 MHz) δ: 9.08 (s, 4H, $2NH_2$), 4.22 (t, $J=6$ Hz, 4H, $2N^4$–CH_2), 3.87 (s, 6H, $2N^1$–CH_3), 1.77 (d, $J=6$ Hz, 4H, $2N^4$–CCH_2); 1.33–1.28 (m, 12H, 2 $CH_2CH_2CH_2$), 0.88 (q, $J=6$ Hz, 6H, $2CH_3$), 1.4–0.4 (sym. near platform, 12H, $B_{12}H_{12}$); ^{13}C NMR (DMSO-d_6, 150 MHz) δ: 148.5 (ring C), 47.6 (N^4–CH_2), 34.6 (CH_3), 31.0 (CH_2), 27.6 (N^4–CCH_2), 25.6 (N^1–CH_3), 22.4 (CH_2), 14.3 (CH_3); ^{11}B NMR (DMSO-d_6, 200 MHz) δ: −15.5; IR ν: 3364 (m), 3292 (m), 3240 (m), 3163 (m), 2954 (w), 2928 (w), 2860 (w), 2484 (vs), 1677 (vs), 1061 (s), 1034 (s), 705 (m) cm^{-1}; ESI-MS *m*/*z*: 184 $[C_8H_{18}N_5]^+$, 71.35 (1/2$[B_{12}H_{12}]^{2-}$). Anal. calcd for $C_{16}N_{10}H_{48}B_{12}$: C 37.65, H 9.48, N 27.45; found C 37.57, H 9.50, N 27.53.

2.3.1.7 Synthesis of Bis(1-ethyl-4-hexyl-5-aminotetrazole) Dodecahydrododecaborate Salt (7)

1.05 g $K_2B_{12}H_{12}$ and 3.10 g iodized 1-ethyl-4-hexyl-5-aminotetrazole salt were dissolved in an appropriate amount of deionized water at 80 °C respectively, mixed for 2 h, and cooled to precipitate crystal (7) with a yield of 91%. M.p. 215.8 °C; ^{1}H NMR (DMSO-d_6, 600 MHz) δ: 9.07 (s, 4H, $2NH_2$), 4.27–4.21 (m, 8H, $2N^4$–$CH_2 + 2N^1$–CH_2), 1.80–1.77 (m, 4H, $2N^4$–CCH_2); 1.42 (t, $J=6$ Hz, 6H, $2N^1$–CCH_3), 1.31–1.28 (m, 12H, $2CH_2CH_2CH_2$), 0.89–0.86 (m, 6H, $2CH_3$), 1.4–0.4 (sym. near platform, 12H, $B_{12}H_{12}$); ^{13}C NMR (DMSO-d_6, 150 MHz) δ: 147.8 (ring C), 47.6 (N^4–CH_2), 43.2 (N^1–CH_2), 31.0 (CH_2), 27.6 (CH_2), 25.7 (N^4–CCH_2), 22.4 (CH_2), 14.3 (CH_3), 13.2 (N^1–CH_3); ^{11}B NMR (DMSO-d_6, 200 MHz) δ: −15.5; IR ν: 3358 (m), 3289 (m), 3235 (m), 3166 (m), 2931 (w), 2959 (w), 2469 (vs), 1667 (vs), 1061 (m), 1039 (m), 746 (w), 709 (m) cm^{-1}; ESI-MS *m/z*: 198 $[C_9H_{20}N_5]^+$, 70.85 (1/2$[B_{12}H_{12}]^{2-}$). Anal. calcd for $C_{18}N_{10}H_{52}B_{12}$: C 40.15, H 9.74, N 26.02; found C 39.89, H 9.77, N 25.83.

2.3.1.8 Synthesis of Bis(1,3-dimethyl-5-aminotetrazole) Dodecahydrododecaborate Salt (8)

2.05 g $K_2B_{12}H_{12}$ and 4.50 g iodized 1,3-dimethyl-5-aminotetrazole salt were dissolved in an appropriate amount of deionized water respectively, heated and mixed for 1.5 h, cooled and crystallized to obtain product (**8**) with a yield of 79%. ^{1}H NMR (DMSO-d_6, 600 MHz) δ: 8.27 (s, 4H, $2NH_2$), 4.36 (s, 6H, $2N^1$–CH_3), 3.93 (s, 6H, $2N^3$–CH_3), 1.4–0.3 (platform, 12H, $B_{12}H_{12}$); ^{13}C NMR (DMSO-d_6, 150 MHz) δ: 158.5 (ring C), 43.1 (N^3–C), 34.7 (N^1–C); ^{11}B NMR (DMSO-d_6, 200 MHz) δ: −15.5; IR ν: 3420 (m), 3334 (m), 3286 (w), 3212 (w), 2466 (vs), 1658 (s), 1059 (m), 719 (m), 671 (m) cm^{-1}; ESI-MS *m/z*: 114 $[C_3H_8N_5]^+$, 71.80 (1/2$[B_{12}H_{12}]^{2-}$). Anal. calcd for $C_6N_{10}H_{28}B_{12}$: C 19.47, H 7.63, N 37.85; found C 19.16, H 7.51, N 37.73.

2.3.1.9 Synthesis of Bis(1-methyl-3-ethyl-5-aminotetrazole) Dodecahydrododecaborate Salt (9)

1.30 g $K_2B_{12}H_{12}$ and 3.00 g iodized 1-methyl-3-ethyl-5-aminotetrazole salt were dissolved in an appropriate amount of deionized water respectively, heated and mixed for 1.5 h, cooled and crystallized to obtain product (**9**) with a yield of 76%. ^{1}H NMR (DMSO-d_6, 600 MHz) δ: 8.29 (s, 4H, $2NH_2$), 4.36 (s, 6H, $2N^1$–CH_3), 4.32–4.31 (m, 4H, $2N^3$–CH_2), 1.42 (t, $J=6$ Hz, 6H, $2N^3$–CCH_3), 1.4–0.4 (sym. near platform, 12H, $B_{12}H_{12}$); ^{13}C NMR (DMSO-d_6, 150 MHz) δ: 157.8 (ring C), 43.6 (N^3–CH_2), 33.2 (N^1–CH_3),13.2 (N^3–CCH_3); ^{11}B NMR (DMSO-d_6, 200 MHz) δ: −15.5; IR ν: 3396 (m), 3309 (m), 3269 (w), 3195 (w), 2989 (w), 2474 (s), 1645 (vs), 1058 (s), 710 (m), 656 (m) cm^{-1}; ESI-MS *m/z*: 128 $[C_4H_{10}N_5]^+$, 70.55 (1/2$[B_{12}H_{12}]^{2-}$). Anal. calcd for $C_8N_{10}H_{32}B_{12}$: C 24.13, H 8.10, N 35.18; found C 23.88, H 8.16, N 33.99.

*2.3.1.10 Synthesis of Bis(1,3-diethyl-5-aminotetrazole) Dodecahydrododecaborate Salt (**10**)*

0.82 g $K_2B_{12}H_{12}$ and 2.00 g iodized 1,3-diethyl-5-aminotetrazole salt were dissolved in deionized water as 1 : 2 in molar ratio, heated and mixed for 2 h, cooled and crystallized to obtain transparent sheet crystal (**10**) with a yield of 76%. M.p. 199.1 °C; ^1H NMR (DMSO-d_6, 600 MHz) δ: 8.29 (s, 4H, $2NH_2$), 4.65–4.69 (m, 4H, $2N^1$–CH_2), 4.33–4.30 (m, 4H, $2N^3$–CH_2), 1.54 (t, $J = 6$ Hz, 6H, $2N^1$–CH_3), 1.43 (t, $J = 6$ Hz, 6H, $2N^3$–CH_3), 1.4–0.3 (platform, 12H, $B_{12}H_{12}$); ^{13}C NMR (DMSO-d_6, 150 MHz) δ: 157.8 (ring C), 52.1 (N^3–CH_2), 43.6 (N^1–CH_2), 13.3 (N^1–CCH_3), 13.1 (N^3–CCH_3); ^{11}B NMR (DMSO-d_6, 200 MHz) δ: −15.5; IR ν: 3403 (m), 3379 (m), 3308 (m), 3196 (w), 2454 (vs), 1635 (vs), 1055 (s), 725 (w) cm^{-1}; ESI-MS *m*/*z*: 142 $[C_5H_{12}N_5]^+$, 70.55 (1/2$[B_{12}H_{12}]^{2-}$). Anal. calcd for $C_{10}N_{10}H_{36}B_{12}$: C 28.18, H 8.51, N 32.86; found C 27.32, H 8.56, N 31.33.

*2.3.1.11 Synthesis of Bis(1-ethyl-3-methyl-5-aminotetrazole) Dodecahydrododecaborate Salt (**11**)*

1.51 g $K_2B_{12}H_{12}$ and 3.50 g iodized 1-ethyl-3-methyl-5-aminotetrazole salt were dissolved in an appropriate amount of deionized water respectively, heated at 80 °C and mixed for 2 h, cooled and crystallized to obtain product (**11**) with a yield of 75%. ^1H NMR (DMSO-d_6, 600 MHz) δ: 8.27 (s, 4H, $2NH_2$), 4.67 (d, $J = 6$ Hz, 4H, $2N^1$–CH_2), 3.93 (s, 6H, $2N^3$–CH_3), 1.53 (t, $J = 6$ Hz, 6H, $2N^1$–CH_3), 1.4–0.4 (sym. near platform, 12H, $B_{12}H_{12}$); ^{13}C NMR (DMSO-d_6, 150 MHz) δ: 158.5 (ring C), 52.0 (N^3–CH_3), 34.8 (N^1–CH_2), 13.4 (N^1–CCH_3); ^{11}B NMR (DMSO-d_6, 200 MHz) δ: −15.5; IR ν: 3445 (m), 3331 (s), 3295 (w), 3215 (w), 2985 (w), 2467 (vs), 1652 (vs), 1056 (s), 710 (m), 672 (m) cm^{-1}; ESI-MS *m*/*z*: 128 $[C_4H_{10}N_5]^+$, 70.95 (1/2$[B_{12}H_{12}]^{2-}$). Anal. calcd for $C_8N_{10}H_{32}B_{12}$: C 24.13, H 8.10, N 35.18; found C 24.02, H 8.20, N 34.69.

2.3.2 Characterization of Bis(dialkyl-5-aminotetrazolium) Dodecaborates

The synthesized bis(dialkyl-5-aminotetrazole) dodecahydrododecaborates were characterized by FT-IR, ^1H NMR, ^{13}C NMR, ^{11}B NMR, ESI-MS and elemental analysis. The ESI-MS of all new compounds showed ion peaks corresponding to cations and $B_{12}H_{12}{}^{2-}$. Their FT-IR showed strong bands corresponding to a B–H stretching vibration and B_{12} boron cage absorption at 2450–2480 and 1060 cm^{-1}, respectively. Their ^{11}B NMR spectra were a sharp resonant single peak corresponding to $B_{12}H_{12}{}^{2-}$ presented at δ −15.5. The composition of dialkylated 5-aminotetrazole cations has little effect on the chemical shift of $B_{12}H_{12}{}^{2-}$. The ^{13}C NMR shows that the chemical shift of the ring carbon atom in the cation is mainly determined by the position of the substituent, and is almost independent of the composition of the substituent. For dodecylboronic acid bis(1,3-dialkyl-5-aminotetrazole) salt, the chemical shift of the ring carbon atom is δ (158.0 $\pm$ 0.5).

For dodecahydrododecaborate bis(1,4-dialkyl-5-aminotetrazole) salt, the chemical shift of the ring carbon atom is δ (148.3 ± 0.7). The ^{1}H NMR spectra of bis(dialkyl-5-aminotetrazole) dodecyl borate were δ 0.4–1.4 and show the characteristic band of $B_{12}H_{12}{}^{2-}$ (an approximate low resonance platform). The amino resonance at the 5-position of the tetrazole ring occurs at low field (greater than δ 8). It is almost independent of the composition of the substituents on the ring, indicating that the amino nitrogen atom participates in the generation of tetrazole ring cations, and the lone pair electrons on the nitrogen atom transfer to the tetrazole ring, so that the amino protons are in a strong unshielded environment. However, the position of the substituent in the tetrazole ring cation has a significant effect on the chemical shift of the amino group. The amino resonance of 1,3-dialkyl-5-aminotetrazolium salt occurred at δ 8.29, while the amino resonance of 1,4-dialkyl-5-aminotetrazolium salt occurred at δ 9.07, indicating that the 1,4-dialkyl substituted tetrazolium ring has a stronger shielding effect on amino protons than the 1,3-dialkyl substituted tetrazolium ring. On the other hand, comparing the NMR spectra of dodecylhydrododecaborate with similar iodized salts, the chemical shifts of carbon and proton resonance in the cations of the two compounds are almost the same, revealing that $B_{12}H_{12}{}^{2-}$ and I^- have almost the same influence on the cationic groups.

2.3.3 Thermal Behavior of Bis(dialkyl-5-aminotetrazolium) Dodecaborates

The thermal behavior of bis(diaIkyl-5-aminotetrazolium) dodecaborates was investigated using TG-TDG and DSC techniques; the TG–DTG results are shown in Figures 2.14, 2.15 and Table 2.3. T_5% represents the initial thermal decomposition temperature of each substance, and T_{max} represents

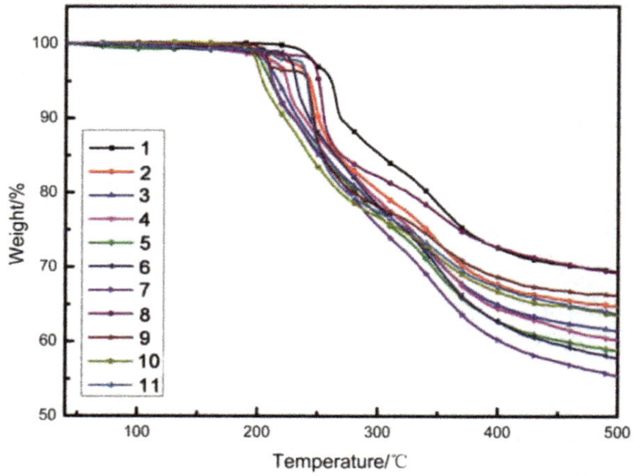

Figure 2.14 TG curves of bis(dialkyl-5-aminotetrazolium)dodecahydrododecaborates. Reproduced from ref. 35.

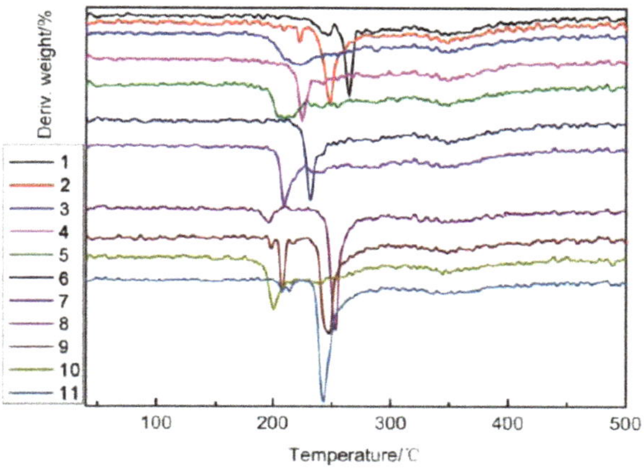

Figure 2.15 DTG curves of bis(dialkyl-5-aminotetrazolium)dodecahydrododecaborates. Reproduced from ref. 35.

Table 2.3 TG–DTG data of bis(dialkyl-5-aminotetrazolium)-dodecahydrododecaborates. Reproduced from ref. 35.

Compound	T_5%/°C	T_{max1}/°C	T_{max2}/°C	T_{max3}/°C	Residue weight/%
1	260.8	246.7	264.5	349.0	69.39
2	243.3	222.4	248.3	348.2	64.75
3	216.7	223.1	287.6	348.9	61.47
4	224.1	224.8	284.7	349.0	60.23
5	210.6	209.7	241.0	349.3	58.82
6	231.9	231.8	284.7	348.7	57.80
7	212.4	209.5	240.0	343.3	55.44
8	250.4	196.4	253.0	343.8	69.23
9	241.2	207.8	247.8	348.3	66.15
10	204.0	200.6	241.0	344.3	63.54
11	240.9	207.7	242.9	344.2	63.74

the temperature corresponding to the maximum thermal decomposition rate. It can be seen that their thermal decomposition process shows three stages: the residual mass at 500 °C is between 55% and 70%, and the residual material may be boron containing compounds and other thermally stable solid substances. Their thermal stability is good, T_5% is higher than 200 °C and the T_{max} is between 200–270 °C, which is related to the position and composition of the alkyl group.

Figure 2.16 and Table 2.4 show the DSC data of bis(diaIkyl-5-amino-tetrazolium) dodecaborates. It can be found that the five salts of 1,4-dimethyl, 1-methyl-4-ethyl, 1,3-dimethyl, 1-methyl-3-ethyl and 1-ethyl-3-methyl did not show a melting endothermic peak but directly had a strong decomposition effect, indicating that they were rapidly decomposed before crystallization and melting, and the other six salts had an obvious

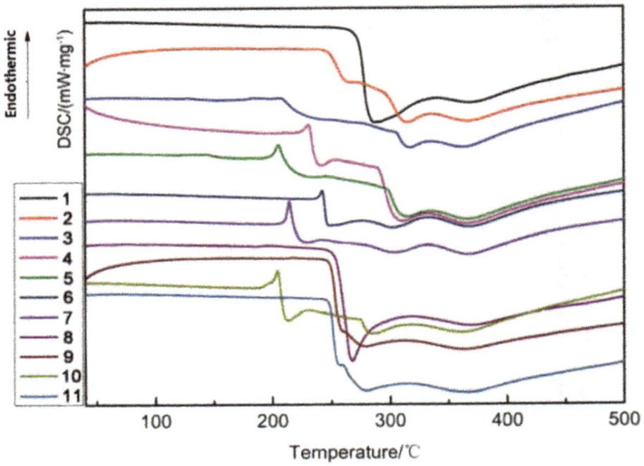

Figure 2.16 DSC curves of bis(dialkyl-5-aminotetrazolium) dodecaborates. Reproduced from ref. 35.

Table 2.4 DSC data of bis(dialkyl-5-aminotetrazolium) dodecaborates. Reproduced from ref. 35.

Compound	M.p./°C
1	—
2	—
3	203.9
4	230.9
5	204.9
6	245.6
7	215.8
8	—
9	—
10	204.2
11	—

endothermic peak corresponding to their melting points. The results showed that the rapid decomposition occurred after crystallization melting, but the thermal decomposition occurred soon after melting, indicating that the melting points of the six compounds were close to their respective decomposition temperatures.

The best thermal stability was in dodecahydrododecaborate bis(1,4-dimethyl-5-aminotetrazole) salt, followed by dodecahydrododecaborate bis(1,3-dimethyl-5-aminotetrazole) salt. The thermal stabilities of the dodecahydrododecaborate bis(1-ethyl-3-methyl-5-aminotetrazole) salt, the dodecahydrododecaborate bis(1-methyl-3-ethyl-5-aminotetrazole) salt and the dodecahydrododecaborate bis(1-methyl-4-ethyl-5-aminotetrazole) salt were similar, but the thermal stability of the compound dodecahydrodidecaborate bis(1,3-diethyl-5-aminotetrazole) salt was the worst.

In addition, among the six compounds showing melting points, the melting points of **3** and **5** compounds are less than $T_5\%$, respectively, but the temperature difference is not significant (about 10 °C), indicating that the initial decomposition occurs after crystallization melting. The melting points of **4**, **6**, **7**, and **10** compounds are slightly higher than their $T_5\%$, indicating that the initial decomposition and crystallization melting of these substances intersect, but the thermal effect of crystallization melting can be controlled.

There are many kinds of borohydride compounds, especially the closed cage borohydride anion compound $B_nH_n^{2-}$ ($n = 6, 7, 8, 9, 10, 11, 12$), has high thermal and chemical stability. $B_{12}H_{12}^{2-}$, with a regular icosahedral cage structure, is a kind of borohydride compound with high thermal stability and chemical stability.[36–39] It has a wide application prospects in the field of composite solid propellants for its advantages, *i.e.* high heat of combustion value, complete combustion, low relative molecular weight of combustion products, not easily to deflagrated, and it can produce high jet power.[40,41] The physico-chemical performance of $B_{12}H_{12}K_2$ (BH-12-K), $B_{12}H_{12}[C_7H_{15}N_4]_2$ (BH-12-U) and $B_{12}H_{12}[CH_9N_6]_2$ (BH-12-G) were characterized by means of thermal decomposition technique and mechanical sensitivity apparatus. The molecular structures of these three types of dodecahydrododecaborates are shown in Figure 2.17.

The physico-chemical properties of three kinds of dodecahydrododecaborate (BH-12-K, BH-12-U and BH-12-G) were investigated, and the results are shown in Table 2.5 and Figure 2.18. It can be seen that the densities of these three kinds of dodecahydrododecaborate are higher than that of $B_{10}H_{10}[(C_2H_5)_4N]_2$, and the density of BH-12-K is the highest. The impact sensitivity and friction sensitivity of BH-12-K are higher than 40 J and 360 J, respectively, indicating that it is insensitive to impact and friction stimuli (Table 2.5). There is no mass loss due to thermal decomposition of

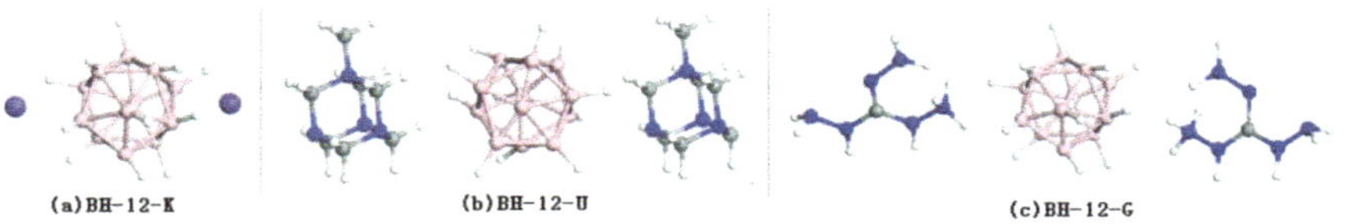

Figure 2.17 The molecular structures of three types of dodecahydrododecaborates (BH-12-K, BH-12-U and BH-12-G). Reproduced from ref. 42.

Table 2.5 The physico-chemical properties of three kinds of dodecahydrododecaborate (BH-12-K, BH-12-U and BH-12-G). Reproduced from ref. 42.

Sample	ρ/(g cm^{-3})	Impact sensitivity/J	Friction sensitivity/J
BH-12-K	1.223	>40	>360
BH-12-U	1.199	>40	144
BH-12-G	1.221	>40	216

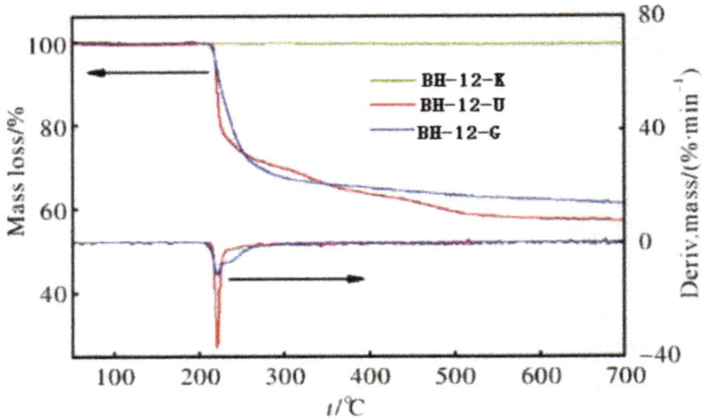

Figure 2.18 TG–DTG curves of three kinds of dodecahydrododecaborate (BH-12-K, BH-12-U and BH-12-G). Reproduced from ref. 42.

Table 2.6 The parameters of thermal mass loss of three kinds of dodecahydrododecaborate (BH-12-K, BH-12-U and BH-12-G). Reproduced from ref. 42.

Samples	T_1/°C	T_2/°C	Δm/%
BH-12-K	>700.0	>700.0	100.0
BH-12-U	219.9	221.6	57.6
BH-12-G	220.5	220.0	61.0

BH-12-K, indicating that it has good thermal stability at 700 °C. While the thermal decomposition of BH-12-U and BH-12-G is similar, the mass loss of both samples began at 220 °C, the decomposition rate reached the maximum, and the final mass loss for both samples was 57–60% (Table 2.6).

2.4 Synthesis and Characterization of Cationic Polynitrogen Heterocyclium Decahydrodecaborates (CPHD)

2.4.1 Synthesis of Cationic Polynitrogen Heterocyclium Decahydrodecaborates

The synthesis route of cationic polynitrogen heterocyclium decahydrodecaborates is shown in Figure 2.19.

The structure characterization of the target compounds is shown in Table 2.7. It can be seen that the structure data of the products are consistent with the theoretical analysis method, and the products are the target compounds.

1-*H*-1,2,3-Triazole, 1,2,4-triazole and imidazole are aromatic π electron rich five membered nitrogen heterocycles with weak basicity. They react with

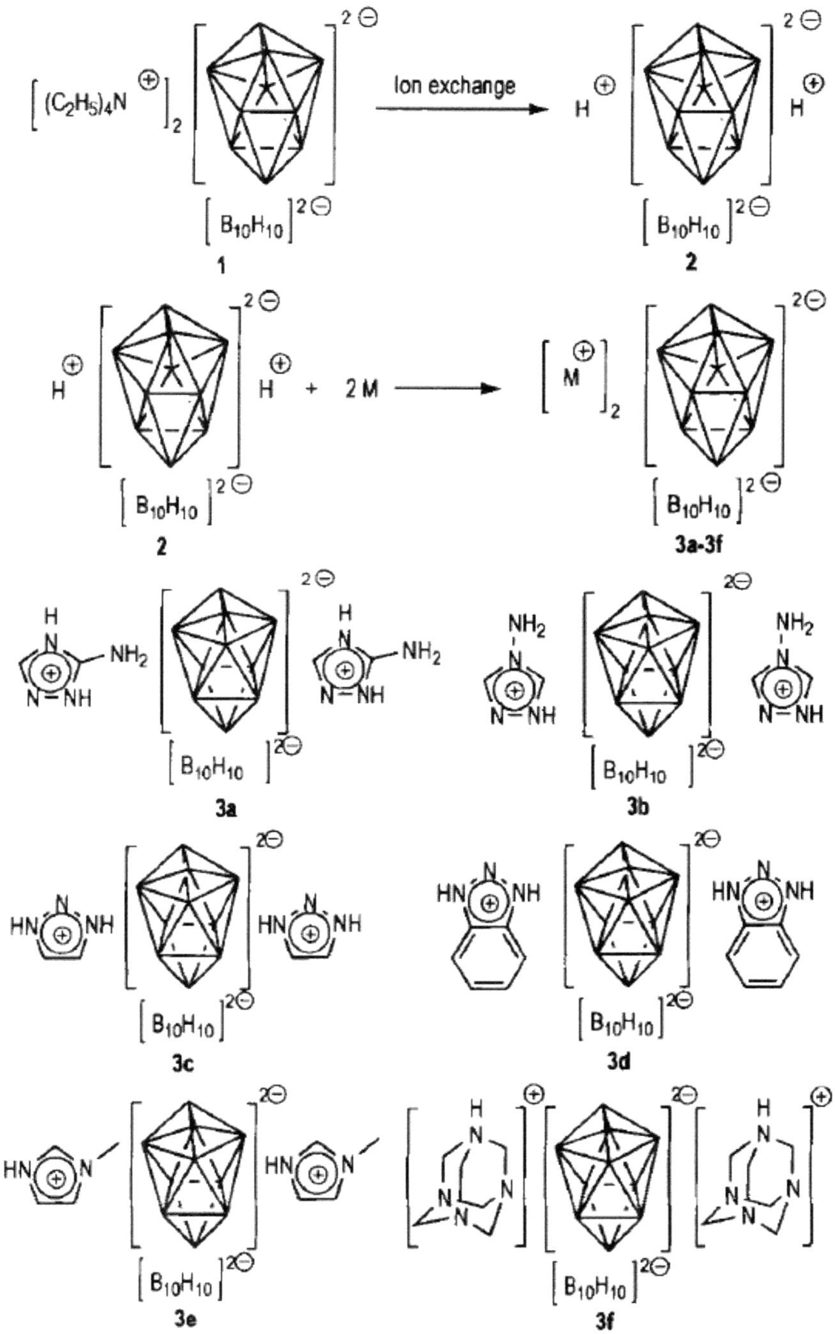

Figure 2.19 Synthesis of the heterocyclium decahydrodecaborates. Reproduced from ref. 43.

Table 2.7 The structure characterization of the target compounds. Reproduced from ref. 43.[a]

Compound	IR (KBr)/cm^{-1}	^{1}H NMR	^{13}C NMR	Elemental analysis/%
[3-Amino-1,2,4-triazolium]$_2$ [*closo*-$B_{10}H_{10}$] (**3a**)	3099, 2989, 2481, 1694, 1641, 1569, 1050, 1022, 950, 865, 642	7.842 (br, s, 4H), 8.197 (s, 2H), 2.506–3.337 (m, 2H), −0.551 to 0.187 (m, 8H)	140.99, 152.26	C 16.55 (16.67 H 6.23 (6.25) N 38.76 (38.89)
[4-Amino-1,2,4-triazolium]$_2$ [*closo*-$B_{10}H_{10}$] (**3b**)	3127, 2461, 1633, 1526, 1455, 1382, 1194, 1074, 975, 863, 619	8.429 (s, 4H), 6.005 (br, s, 6H), 2.498–3.362 (m, 2H), −0.533 to 0.199 (m, 8H)	144.56	C 16.53 (16.67) H 6.23 (6.25) N 38.72 (38.89)
[1-*H*-1,2,3-Triazolium]$_2$ [*closo*-$B_{10}H_{10}$] (**3c**)	3141, 2905, 2850, 1499, 2444, 1561, 1456, 1130, 1084, 1028, 969, 778	10.424 (br, s 8H), 7.970 (s, 4H), 2.790–3.348 (m, 2H), −0.544 to 0.172 (m, 8H)	130.52	C 18.43 (18.60) H 6.96 (6.98) N 32.36 (32.48)
[Benzotriazolium]$_2$ [*closo*-$B_{10}H_{10}$] (**3d**)	3149, 2466, 1614, 1473, 1377, 1288, 1244, 1117, 1021, 999, 970, 895, 750, 624	9.584 (br, s, 8H), 7.913–7.931 (m, 4H), 7.444–7.462 (m, 4H), 3.101–3.217 (m, 2H), −0.528 to 0.214 (m, 8H)	139.22, 125.87, 115.36	C 40.12 (40.20) H 6.12 (6.14) N 23.29 (23.46)
[1-Methylimidazolium]$_2$ [*closo*-$B_{10}H_{10}$] (**3e**)	3133, 2493, 1579, 1541, 1381, 1276, 1077, 1009, 754, 624	14.116 (br, s, 2H), 7.649 (s, 2H), 7.689 (s, 2H), 3.870 (s, 6H), 2.804–3.361 (m, 2H), −0.476 to 0.177 (m, 8H)	136.33, 123.60, 120.31, 35.90	C 33.64 (33.80) H 8.42 (8.45) N 19.54 (19.72)
[Urotropinium]$_2$ [*closo*-$B_{10}H_{10}$] (**3f**)	2953, 2873, 2464, 1461, 1369, 1265, 1238, 1008, 814, 673	8.075 (br, s, 2H), 4.817 (s, 24H), 2.778–3.326 (m, 2H), −0.519 to 0.174 (m, 8H)	71.79	C 35.86 (36.00) H 8.97 (9.00) N 27.89 (28.00)

[a]The data in parentheses of the elemental analysis is the theoretical value.

strong acids to form monovalent cations with positive charges delocalized on the five membered rings.[44] Urotropine is a cage structure composed of six methylene groups and four secondary amine groups, which forms monovalent cations with strong acids.[45]

Among the infrared characteristic absorption peaks of decahydrodecaborate, the peak about 3100 cm^{-1} is the C–H stretching vibration of aromatic nitrogen heterocyclic cations. The C–H stretching vibration of methylene in a urotropine cation is 2953 cm^{-1} and 2873 cm^{-1}. The stretching vibration of B–H on a boron cage in decahydrodecaborate is between 2450 and 2500 cm^{-1}. The force of cations with different structures and boron cage anions is different, resulting in different peaks.

The decahydrodecaborate anion is a cage structure of double capped tetraprism type. According to the different positions of the boron atoms on the cage, it is divided into equatorial boron and top boron. Therefore, the hydrogen atoms on the boron cage are divided into two groups, with the numbers of 8 and 2, respectively. The chemical shifts of hydrogen atoms on boron cages are different due to different cation structures. The chemical shifts of eight hydrogen atoms on equatorial boron are between −0.5 and 0.3, which are multiple broad peaks. The chemical shifts of the two hydrogen atoms on the top boron are between 2.4 and 4.0, and the splitting is divided into four peaks. In addition, the active hydrogen in the molecule shows a short and wide peak in the NMR spectrum, with a chemical shift between 8 and 15, and its position changes due to the different forces of cations on it.

There is a single peak of 24 hydrogen atoms at 4.817 and a group of carbon with a carbon spectrum of 71.79 in the nuclear magnetic hydrogen spectrum of diurotropine decahydrodecaborate, indicating that the positive charge in the cation is highly delocalized in the cage structure, which is consistent with the results of other urotropine salts in the literature.[46]

2.4.2 Characterization of Cationic Polynitrogen Heterocyclium Decahydrodecaborates (CPHD)

The thermal properties of the target compound (**3a–3f**) were analyzed by DSC and TG. The results are shown in Figure 2.20. It can be seen that compounds **3a**, **3b** and **3e** all have the thermal characteristic curve of melting first and then decomposition. The melting peak temperatures were 116.5, 72.8, and 201.5 °C, respectively. The peak decomposition temperatures were 232.6, 274.6, and 219.4 °C, respectively. Among them, compound **3b** has the lowest melting point, can be melted into liquid at 72.8 °C, and has a high thermal decomposition temperature. It is an energetic ionic liquid with good thermal stability. In the TG curves of compounds **3a** and **3e**, the weight gain phenomenon occurs after the decomposition loss. According to the research results of literature,[17] it may be caused by the reaction of the decomposition product boron with trace oxygen in the carrier gas. Compound **3b** has no weight gain, which may be related to its thermal properties of melting before decomposition. The thermal decomposition mechanism will be further studied later.

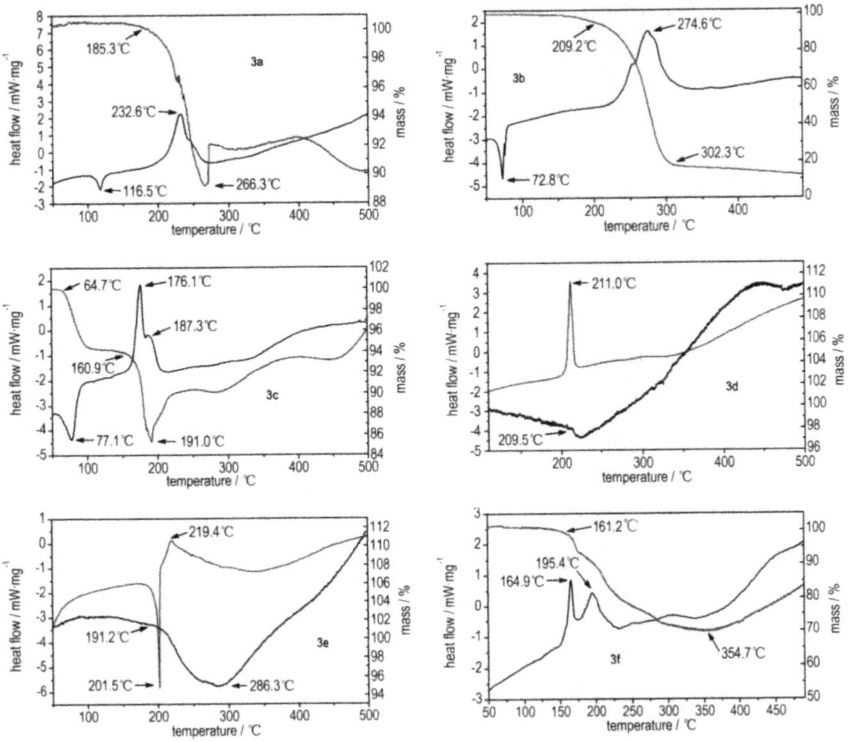

Figure 2.20 DSC–TG curves of compounds **3a–3f**. Reproduced from ref. 43.

In the DSC–TG curve of compound **3c**, the endothermic peak at 77.1 °C corresponds to 6% weight loss, and the exothermic peak at 176.1 °C corresponds to 10% weight loss. The former is speculated to be the evaporation and heat absorption of the adsorbed water of the compound, while the latter corresponds to the decomposition and heat release process of the compound, and the shoulder peak at 187 °C corresponds to a 4% weight gain, which may be caused by the reaction of the decomposition product of boron with trace oxygen in the carrier gas.

The DSC–TG curves of compounds **3d** and **3e** are similar. There is no melting peak in both compounds, and they decompose directly after heating. The peak decomposition temperature of compound **3d** is at 211.0 °C. In the DSC curve of **3e**, there are two decomposition exothermic peaks, which are 164.9 and 195.4 °C, respectively, corresponding to two rapid weight loss stages in the TG curve, indicating that the thermal decomposition of the compound is divided into two main steps, and then gradually increases in weight, corresponding to the oxidation of the decomposition product boron.

The thermal decomposition temperatures of the above six compounds from high to low is: [bis(4-amino-1,2,4-triazolium) decahydrodecaborate

(**3b**)] > [bis(3-amino-1,2,4-triazolium) decahydrodecaborate (**3a**)] > [bis(1-methyl-imidazolium) decahydrodecaborate (**3e**)] > [bis(benzotriazolium) decahydrodecaborate (**3d**)] > [bis(1-*h*-1,2,3-triazolium) decahydrodecaborate (**3c**)] > [bis-urotropine decahydrodecaborate (**3f**)]. It can be seen that the cationic structure has a significant effect on the stability of the compounds. The stability of cations is related to the delocalization degree of positive charge and the density of the electron cloud. For aromatic five membered heterocyclic cations, the positive charge is highly delocalized and the density of the electron cloud is high. Although the positive charge in the cation structure of urotropine is delocalized in the cage structure, the density of the carbon nitrogen single bond electron cloud is low, so the thermal decomposition temperature is low.

The cationic main structure of compounds **3a** and **3b** is 1,2,4-triazole. The thermal stability of the amino group at position 4 is higher than that of the amino group at position 3. This is because the existence of N–H at the position of 4 in 3-amino-1,2,4-triazole makes it less alkaline than 4-amino-1,2,4-triazole,[36] so the stability of the ionic salt formed with strong acid decahydrodecaboric acid is low.

The thermal stability of compound **3d** is higher than that of **3c**, because the benzene ring with high electron cloud density and good thermal stability is introduced into the former molecule, and the benzene ring is super conjugated with the cation structure of 1,2,3-triazole, so it has higher stability.

2.5 Synthesis and Characterization of Dihydrazine-1,2,4,5-tetrazine Decahydrodecaborate

Polyhedral borohydride and carborane high-energy compounds have been used in the field of high-energy solid propellants and explosives with excellent properties. Polyhedral borohydride has been widely researched by researchers worldwide because of its easy synthesis, low toxicity, and wide application fields. The reported compounds are metal salts of decahydrodecaboric acid, such as $K_2B_{10}H_{10}$, $Cs_2B_{10}H_{10}$. It has low enthalpy of formation, small gas production and an obvious combustion characteristic signal.

2.5.1 Synthesis of Dihydrazine-1,2,4,5-tetrazine Decahydrodecaborate

The composition route is shown in Figure 2.21.

2.5.1.1 Synthesis of 3,6-dihydrazine-1,2,4,5-tetrazine

Add 3,6-bis(3,5-dimethylpyrazole)-1,2,4,5-tetrazine (6.0 g, 22.2 mmol) and 60 mL of acetonitrile in a three port flask in turn, heat and stir until

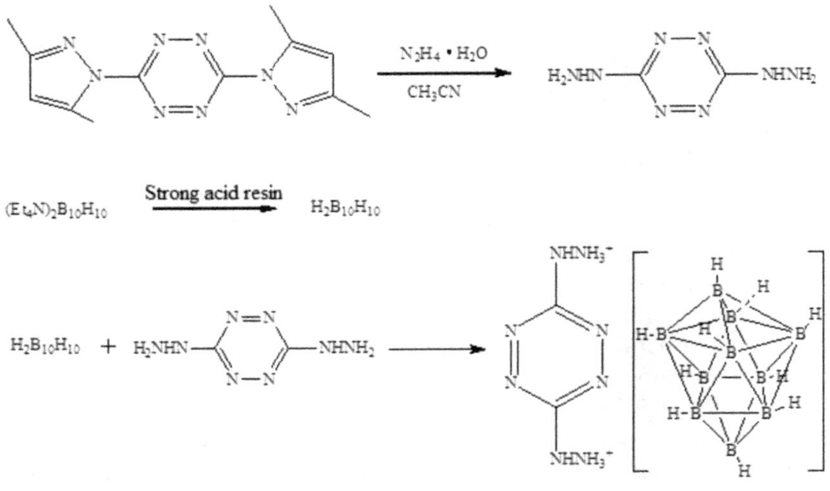

Figure 2.21 Synthesis of dihydrazine-1,2,4,5-tetrazine decahydrodecaborate. Reproduced from ref. 47.

completely dissolved, add 2.5 mL of hydrazine monohydrate drop by drop, wait until hydrazine monohydrate is added, raise the temperature to 80 °C and reflux for 1 h, cool the solution to room temperature, filter, wash the filter cake with acetonitrile three times, and dry to obtain 2.81 g of dark red powder, with a yield of 88.8%, M.P. 157–158 °C.

IR (KBr, ν/cm^{-1}): 3292, 3224, 3027, 1639, 1541, 1053, 941 (N=N). Anal. calcd for $C_2H_6N_8$: C, 16.90; H, 4.23; N, 78.87; found C, 16.90; H, 4.30; N, 77.80.

2.5.1.2 *Synthesis of Dihydrazine Tetrazine Decahydrodecaborate (I)*

Add $[N(C_2H_5)_4]_2B_{10}H_{10}$ (1.0 g, 2.6 mmol) into 20 mL of water, stir until it is completely dissolved, pour it into a strong acid cation exchange column, add water for rinsing until the effluent is neutral, and concentrate the collected rinsing solution under reduced pressure at 35 °C to 40–50 mL for standby.

Add 3,6-dihydrazine-1,2,4,5-tetrazine (0.37g, 2.6 mmol) to the above aqueous solution of decahydrodecaboric acid in batches under stirring conditions. After 3,6-dihydrazine-1,2,4,5-tetrazine is completely dissolved, continue stirring for 1 h, evaporate the reaction solution at 40 °C under reduced pressure, and dry it in a vacuum oven at 60 °C for 6 h to obtain 0.54 g of earthy yellow solid, with a yield of 80%.

IR (KBr, ν/cm^{-1}): 3210, 2471 (B–H), 1679, 1625, 1442 (tetrazine ring skeleton), 1193, 1023, 820, 547. ^{1}H NMR (D_2O, 500 MHz) δ: 2.8629–3.6537 (m, 2H); −0.4981 to 0.7185 (m, 8H). Anal. calcd for $C_2H_{18}N_8B_{10}$: C, 9.16; H, 6.87; N, 42.74; found C, 9.12; H, 6.89; N, 42.94.

2.5.2 Characterization of Dihydrazine-1,2,4,5-tetrazine Decahydrodecaborate

The heat of combustion value of the synthesized dihydrazine tetrazine decahydrodecaborate was tested by the oxygen bomb combustion method; the result is 37.5 kJ g^{-1}. At the same time, the heat of combustion value of cesium decahydrate decaborate $Cs_2B_{10}H_{10}$ was determined to be 36.2 kJ g^{-1}. The heat of combustion value of the target compound is slightly higher than that of $Cs_2B_{10}H_{10}$. At the same time, because it contains more N–N bonds, N=N bonds, C–N bonds and C=N bonds, it can be inferred that its formation enthalpy is high and has high energy.

2.6 Synthesis and Characterization of Energetic Caged-carborane Derivatives

Carborane is a kind of tetrahedral closed cage structure compound formed by the reaction of borohydride and alkynes. Unlike borohydride compounds, which are extremely sensitive to heat, moisture and various chemical reactions, carborane and its derivatives have high thermal, hydrolytic and oxidative stability, and have been attracting much attention from aviation and rocket workers. As a high burning rate regulator or binder for solid propellants, a series of stable carborane derivatives have been developed. They can be compatible with solid propellant components, in particular to enable composite propellants and double base propellants to obtain good ballistic properties such as high or ultra-high burning rate.

2.6.1 Synthesis of Energetic Caged-carborane Derivatives

1-Nitromethyl carborane (**2**) and 1-azido methyl carborane (**3**) were synthesized by nucleophilic substitution reaction of 1-bromomethyl carborane (**1**) with sodium nitrite and sodium azide, respectively, in DMSO or DMF (Figure 2.22).

*2.6.1.1 Synthesis of 1-nitromethyl Carborane (**2**)*

Add 10.36 g (1.5 mmol), 0.21 g (3 mmol) of sodium nitrite and 10 mL of DMSO to the reaction flask in turn, and react at room temperature under stirring for 10 h. Evaporate the solvent under reduced pressure (70 °C/0.92 kPa), and use *n*-hexane (2 × 5 mL) washed with chloroform (2 × 5 mL). The extracted liquid was combined, dried with anhydrous sodium sulfate, and the residue is 20.11 g light green transparent liquid, with a yield of 35.5%. 1H NMR δ: 1.23–2.96 (m, 10H, BH), 3.97 (s, 2H, CH_2NO_2), 7.26 (s, ^{1}H, CH); FT-IR ν: 2530 (B–H), 1022 (B–B), 1654 (C=C), 1549 (NO_2), 3006, 2919 (C–H) cm^{-1}.

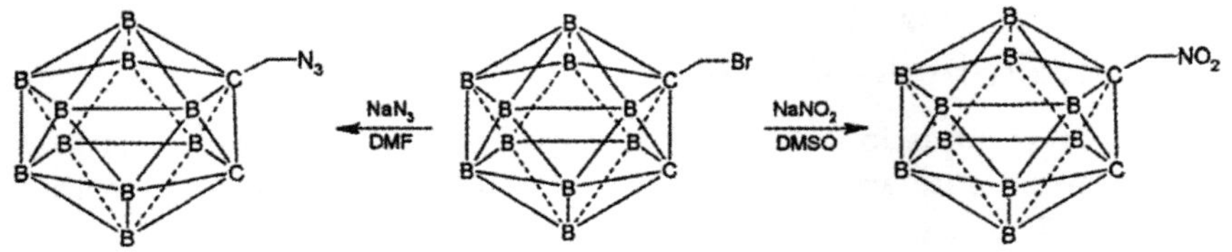

(3) (1) (2)

Figure 2.22 Synthesis of 1-nitromethyl carborane (**2**) and 1-azido methyl carborane (**3**). Reproduced from ref. 48.

2.6.1.2 *Synthesis of 1-azido Methyl Carborane (3)*

Add 10.6 g (1.5 mmol), 0.12 g (1.8 mmol) of sodium azide, 10 mL of DMF and 2 mL of water into the reaction flask in turn, and then react at room temperature for 30 h under stirring. The reaction solution was separated by silica gel chromatography column (eluent: *n*-hexane), and the eluent was evaporated and dissolved. The residue was 380 mg of colorless transparent liquid, with a yield of 25.8%. ^1H NMR δ: 1.24–2.97 (m, 10H, BH), 3.32 (s, 2H, CH_2N_3), 7.27 (s, 1H, CH); FT-IR ν: 2526 (B–H), 1063 (B–B), 1657 (C=C), 2115 (N_3), 2931 (C–H) cm^{-1}.

The bromine methyl group in the structure is strongly adsorbed by the cage carborane parent structure, resulting in the greatly reduced C–Br bond reaction activity. It is not only unable to quaternize with tertiary amines, but it is also difficult to replace with nucleophiles. In some weak polar solvents, such as methanol or acetone, the nucleophilic substitution reaction of bromomethyl carborane with azido was attempted, and the reaction did not occur even if it was stirred at room temperature for one week. Only in some aprotic strong polar solvents (such as DMF and DMSO) can nitro and azido groups be introduced into cage carborane derivatives through nucleophilic substitution reaction, but the yield is low.

2.6.2 Characterization of Energetic Caged-carborane Derivatives

The DSC curves of synthesized compounds are shown in Figure 2.23. It can be seen that the decomposition peaks of 1–3 are at 162.2, 168.7, and 187.7 °C, respectively, and the decomposition heats are 130.4 J g^{-1}, 245.6 J g^{-1} and 91.1 J g^{-1}, respectively. Both 1 and 2 are cage derivatives with energetic groups, which give off more heat during thermal decomposition.

From the H NMR results of 1-nitromethyl carborane (**2**) and 1-azido methyl carborane (**3**), it can be seen that due to the extreme lack of electricity in the cage carborane parent structure, it has a deshielding effect on the methylene bonded to the cage structure, which shifts the chemical shift of hydrogen protons in $-CH_2-$ towards the low field direction, making the degree shift of H in $-CH_2Br$, $-CH_2NO_2$ and $-CH_2N_3$ groups 4.01, 3.97 and 3.32, respectively.

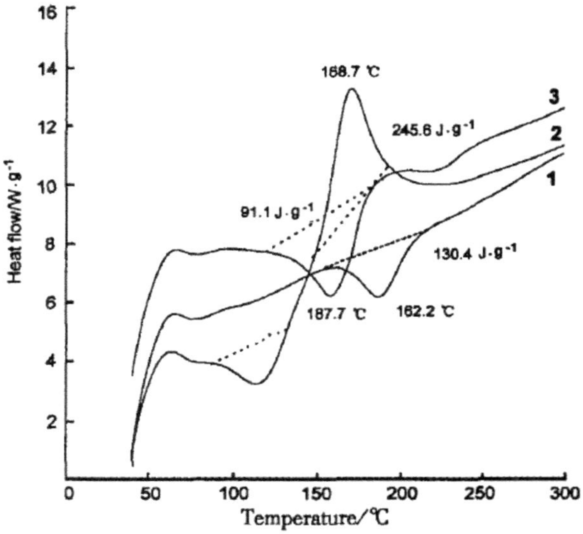

Figure 2.23 Thermal decomposition curves of synthesized compounds. Reproduced from 48.

2.7 Synthesis and Characterization of Copper and Thallium(I) Coordination Compounds $[Cu_2[B_{10}H_{10}]]_n$ and $Tl_2[B_{10}H_{10}]$

Boron cluster anions $[B_nH_n]^{2-}$ ($n = 10, 12$)[49,50] form an independent area of boron chemistry, which combines coordination, inorganic, organic, and organoelement chemistry. Coordination compounds of this class can participate in reactions of the substitution of terminal hydrogen atoms with various functional groups forming a large number of substituted derivatives.[51–53] In addition, boron clusters and their substituted derivatives act as ligands forming metal complexes.[54–59] Among the great number of complexes with boron cluster anions, solvent-free complexes $[M_x(B_nH_n)]$ (M is a metal atom, $x = 1$ (for metals in the oxidation state +2) or 2 (for metals in the oxidation state + 1), $[B_nH_n]^{2-}$ is a boron cluster anion or its substituted derivative) are very few in number. Metal complexes with boron cluster anions which were isolated later confirmed that this method of binding between a metal atom and a boron cluster is the predominant one in complexes with coordinated boron cluster anions. For example, the solvent-free cobalt(II) and nickel(II) complexes $[M(B_{12}H_{12})]_n$ (M = Co, Ni, Mn, Fe) were obtained.[60] The synthesis and properties of thallium(I) salts with the $[B_{10}H_{10}]^{2-}$ and $[B_{12}H_{12}]^{2-}$ anions were also reported.[61,62] Later, nickel compound $[Ni(B_{12}H_{12})]$ was obtained,[63] and silver(I) complexes $[Ag_2(B_{10}H_{10})]$[64] and $[Ag_2(B_{12}H_{12})]$[65] were synthesized.[66] In this section, two polymer complexes $[Cu_2(B_{10}H_{10})]_n$ and $Tl_2[B_{10}H_{10}]$ were synthesized, the structures of the compounds are compared, the coordination features of the boron cluster anion are studied, and the nature of the metal bonds with the boron cluster formed in both structures is discussed.

2.7.1 Synthesis and Structure of Complex $[Cu_2(B_{10}H_{10})]_n$

Polymer complex $[Cu_2(B_{10}H_{10})]_n$ was synthesized by the reaction between copper(II) sulfate and the *closo*-decaborate anion according to the reaction:[67]

$$Cu^{2+} + [B_{10}H_{10}]^{2-} \xrightarrow{H_2O} [Cu_2(B_{10}H_{10})]_n. \quad (2.6)$$

In this case, the yield of the compound is low, since part of the anion is spent on the redox reaction with copper(II) atoms, which leads to partial degradation of the boron cluster anion.

Note that salt $Na_2[B_{10}H_{10}]$ was used to obtain the target compound, due to the use of *closo*-decaborate with a large organic cation. For example, $(Et_3NH)_2[B_{10}H_{10}]$, can result in anionic complex $[(Et_3NH)Cu(B_{10}H_{10})]_n$. Complexes $\{Cat[Cu(B_{10}H_{10})]\}_n$ have been studied in detail.[64,68] In the IR spectrum of the compound **I**, a strongly split band of stretching vibrations of BH bonds ν(BH) is observed with maxima at 2565, 2539, 2512, and 2144 cm^{-1}, indicating the involvement of the boron cluster anion in the metal coordination.

In contrast to data reported,[69] the hydrogen atoms of the boron cluster in compound **I** are localized in the difference Fourier synthesis and refined independently in the isotropic approximation. The crystallographically independent part of the orthorhombic cell (space group *Pbca*) of complex $[Cu_2(B_{10}H_{10})]_n$ contains two copper atoms and one anion. A 2D polymer is formed in the crystal structure of the complex due to the formation of four pairs of three-center two-electron bonds CuHB of the boron cage with four copper atoms. Coordination is carried out along two opposite axial edges of one of the anion vertices (B1B2 and B1B4) and the two adjacent edges of the other (B8B10 and B9B10). The Cu–B and Cu–H bond lengths are given in Table 2.8 and are consistent with the data reported for the complex.

The Cu–H contacts are shown in Figure 2.24 (highlighted in color). Analysis of the Hirschfeld surface of the $[B_{10}H_{10}]^{2-}$ anion shows that H···H contacts with neighboring clusters account for 70.8% of the anion surface, H–Cu contacts account for 23.8% of the surface, and B–Cu contacts correspond to 5.4% (Figure 2.25).

2.7.2 Synthesis and Structure of Compound $Tl_2[B_{10}H_{10}]$

Thallium(I) salt was obtained by reacting thallium nitrate with salt $(Et_3NH)_2[B_{10}H_{10}]$ according to the reaction:

$$2Tl^{+} + [B_{10}H_{10}]^{2-} \xrightarrow{H_2O} [Tl_2(B_{10}H_{10})]_n. \quad (2.7)$$

Table 2.8 Bond lengths Cu–B and Cu–H in complex **I**. Reproduced from ref. 70 with permission from Springer Nature, Copyright [2022].

Bond	Length, Å	Bond	Length, Å
Cu(1)–B(1)	2.226(2)	Cu(1)–H(1)	2.033
Cu(1)–B(2)	2.161(3)	Cu(1)–H(2)	1.804
Cu(1)–B(9)[a]	2.160(2)	Cu(1)–H(9)[a]	1.823
Cu(1)–B(10)[a]	2.232(2)	Cu(1)–H(10)[a]	2.062
Cu(2)–B(1)	2.228(2)	Cu(2)–H(1)	2.097
Cu(2)–B(4)	2.148(2)	Cu(2)–H(4)	1.768
Cu(2)–B(8)[b]	2.143(2)	Cu(2)–H(8)[b]	1.730
Cu(2)–B(10)[b]	2.320(2)	Cu(2)–H(10)[b]	2.248

[a] $-1/2 + x, 1/2 - y, 1 - z$.
[b] $-1/2 + x, +y, 3/2 - z$.

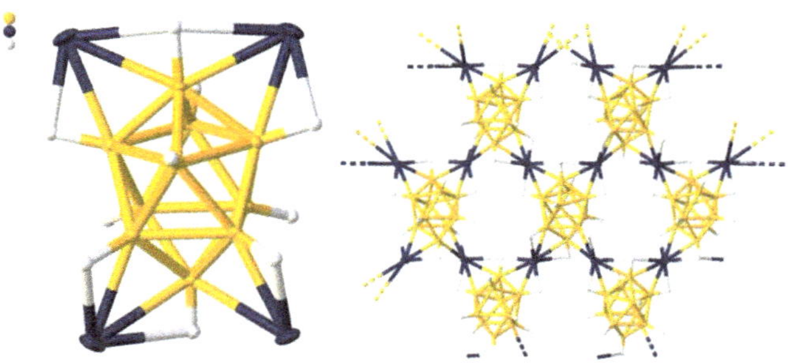

Figure 2.24 Monomer unit of complex **I** (left) and part of the 2D polymer network (right). Reproduced from ref. 70 with permission from Springer Nature, Copyright [2022].

The compound is formed in high yield (67%).

The IR spectrum of the compound shows ν(BH) bands with maxima at 2523 cm^{-1} and 2458 cm^{-1}, suggesting that there is no coordination between the metal cation and the boron cluster anion, since the low-frequency band $\nu(BH)_{MHB}$ of coordinated BH groups of the boron cluster is not found, the maximum of which is usually observed in the region 2300–2200 cm^{-1}.

Salt II is iso-structural to the previously described salts of the *closo*-decaborate anion with alkali metals Na, K, Rb,[71] similarly to salt $Tl_2[B_{12}H_{12}]$,[61,62] which is iso-structural to K, Rb, and Cs salts of the *closo*-dodecaborate anion.[63] The crystallographically independent part of the monoclinic unit cell (space group $P2_1/n$) includes the $[B_{10}H_{10}]^{2-}$ anion and two cations that are in a common position (Figure 2.26). The *closo*-decaborate anion is surrounded by eight Tl^+ cations (Figure 2.25) with distances Tl···B 3.17–3.65 Å (Figure 2.25) and Tl···H 2.67–3.44 Å. As can be seen on the Hirschfeld surface of the boron cluster anion (Figure 2.25a), the thallium cations are located above the centers of the faces of the *closo*-decaborate anion, and the lengths of the contacts Tl···B are less than the sum of their van der Waals radii (red color in Figure 2.25b). In this case, the Tl···B contacts account for 1.9% of the anion

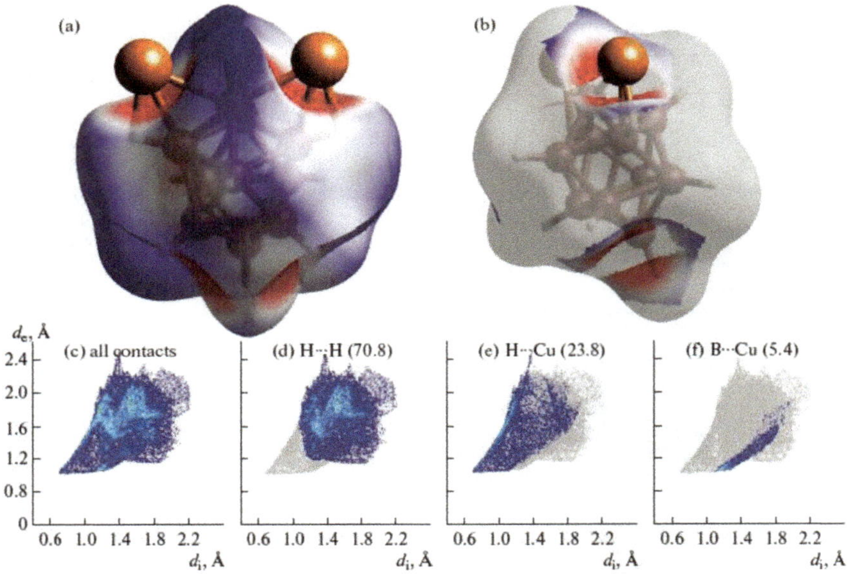

Figure 2.25 (a) d_{norm} map plotted on the Hirschfeld molecular surface of the $[B_{10}H_{10}]^{2-}$ anion for compound **I** and (b) Cu–H interaction boundaries on this surface; (c) 2D unfolding of the Hirschfeld surface of the anion and the boundaries of the (d) H···H, (e) H···Cu, and (f) B···Cu contacts. Reproduced from ref. 70 with permission from Springer Nature, Copyright [2022].

surface, with the Tl–B distance less than the sum of their van der Waals radii. The Tl···H contacts account for 33.9%, and the remaining 64.2% is accounted for by the H···H contacts of neighboring anions with the shortest distance of ~2.15 Å. In each case, the thallium cation is surrounded by four faces of the boron cage (Figure 2.26) of four different $[B_{10}H_{10}]^{2-}$ anions.

An analysis of the Hirschfeld surface of two compounds shows that the proportion of the M–B contact in the copper complex significantly increases, in contrast to salt II (5.4% compared to 1.9%), while the proportion of M–H contacts in complex **I** is 23.8%, and in salt II it is equal to 33.9%, which is related to the large ionic radius of Tl^+.

Note that complex compounds with boron cluster anions are used as extractants of heavy metal salts, components for obtaining thermal protective coatings, in medicine, catalysis, *etc.*[72–74]

2.8 Conclusions

(1) The thermal stability of ditetraethylammonium decahydrodecaborate in argon and oxygen atmosphere was studied. The results showed that the initial decomposition temperature of ditetraethylammonium decahydrodecaborate was 294.8 °C and 301.2 °C, respectively.

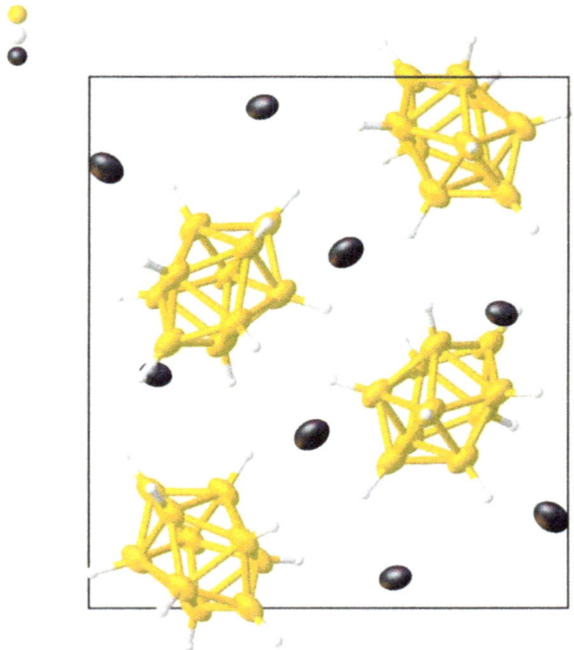

Figure 2.26 Crystallographic structure of compound **II**, view along axis *a*. Reproduced from ref. 70 with permission from Springer Nature, Copyright [2022].

(2) The decomposition heat release in argon is only 112.8 J g^{-1}. In oxygen, 301.2% oxygen first decomposes and oxidizes rapidly, giving off a large amount of heat (4967 J g^{-1}), and then oxidizes further at 597.4% oxygen, giving off 2031 J g^{-1}.
(3) The thermal decomposition kinetics of the compound in argon atmosphere was studied, and the non-isothermal kinetic equation was obtained as $da/dt = 10^{13.37} \exp(-1.347847 \times 10^4/T)[1-(1-a)^{1/3}]$. The main reaction mechanism of the thermal decomposition of the compound was a spherical symmetric three-dimensional diffusion process.
(4) The change trend of the condensed phase and gas phase products in the process of thermal decomposition was studied by *in situ* cell FTIR and TG–DSC–FTIR–MS, respectively. It was speculated that the thermal decomposition mechanism was the rapid decomposition of compounds at 303.2 °C, the main products of the gas phase decomposition were ethane, ammonia and hydrogen, and the solid phase product was boron.

References

1. S. Q. Tang and H. X. Ding, *J. Propul. Technol.*, 1983, **4**(2), 35–51.
2. A. A. Juhasz, *J. Energ. Mater.*, 1988, **6**(1), 81–106.

3. Y. Y. Zheng and M. W. Wang, *Chin. J. Explos. Propellants*, 1989, **12**(3), 32–37.
4. W. Q. Wang, Y. N. Xue, J. M. Yang, Y. N. Li, Q. W. Yu, S. N. Mei and J. Lv, *Chin. J. Energ. Mater.*, 2012, **20**(1), 132–136.
5. A. S. Michael, *J. Propul. Power*, 2012, **14**(6), 981–990.
6. F. T. Chen, H. M. Tan and Y. J. Luo, *Chin. J. Explos. Propellants*, 2000, **23**(3), 19–22.
7. W. Q. Pang, Y. N. Xue and X. Z. Fan, *et al.*, Thermal behavior of tetraethylammonium dodecahydrododecaborates (BHN) and its compatibility with main components of propellant, *Chin. J. Energ. Mater.*, 2012, **20**(3), 280–285.
8. B. S. Igor, I. B. Vladimir and S. Stefan, *Collect. Czech. Chem. Commun.*, 2002, **67**, 679–727.
9. B. S. Igor, V. P. Alexander and N. Daoud, *Collect. Czech. Chem. Commun.*, 2010, **75**, 1149–1199.
10. Y. Nie, H. Y. Chen and J. L. Miao, Advances in the chemistry and applications of the borane anion closo-$[B^{10}H^{10}]^{2-}$ and its derivatives, *Chin. J. Org. Chem.*, 2009, **29**(6), 822–834.
11. E. I. Mutterties, J. H. Balthis, Y. T. Chia, W. H. Knoth and H. C. Miller, *Inorg. Chem.*, 1964, **3**(3), 444–451.
12. N. T. Kuznetsov, G. S. Klimchuk, O. A. Kananaeva and K. A. Solntsev, *Russ. J. Inorg. Chem.*, 1976, **21**(4), 505–508.
13. L. I. Isaenko, K. G. Myakishev, I. S. Posnaya and V. V. Volkov, Izvestiya Sibirskogo Otdeleniya Akademii Nauk SSSR, *Ser. Khim. Nauk*, 1982, **2**, 73–78.
14. T. P. Goddard, *High burning rate propellants with coprecipitated Salts of decahydrodecaboric acid*, US4138282, 1979.
15. P. J. Kaste, *Studies of the Effect of hivelite and other boron compounds on nitramine decomposition by pyrolysis GC-FTIR*, AD-A206802, 1988.
16. Y. N. Xue, Y. N. Li and W. Q. Wang, *et al.*, Synthesis and thermal properties of high burning rate regulator bis (tetraethylammonium decahydrodecaborate), *The 5th Conference on Propellants of Chinese Chemical Society*, LiaoNing, Dalian, 2011, pp. 382–384.
17. X. Yunna, L. Jiaoyi and L. Jian, *et al.*, Synthesis, thermal decomposition mechanism and kinetic equation of the polyhedral boron hydride compound $[(C_2H_5)_4N]_2B_{10}H_{10}$ with high enthalpy of combustion, *Chem. J. Chin. Univ.*, 2015, **36**(2), 375–380.
18. X. Yunna, L. Yani and W. Weiqiang, *et al.*, Synthesis and thermal properties of high burning rate regulator di-tetraethyl ammonium decahydroborate, *Chin. Chem. Soc.*, 2011, 382–384.
19. S. Y. Xu, F. Q. Zhao and J. H. Yi, *et al.*, *Chem. Res. Chin. Univ.*, 2012, **28**(3), 516–519.
20. P. Weiqiang, F. Xuezhong and X. Yunna, *et al.*, Study on the compatibility of tetraethylammonium decahydrodecaborate (BHN) with some energetic components and inert materials, *Propellants, Explos., Pyrotech.*, 2013, **38**, 278–285.

21. J. Fan, W. Xiaofeng and W. Xiaohong, *et al.*, Thermal decomposition characteristics and kinetic parameters of tetraethylammonium decahydrodecaborate, *Chin. J. Explos. Propellants*, 2020, **43**(4), 57–63.
22. D. Zhao, J. Huang and J. R. Song, *et al.*, *Chem. Res. Chin. Univ.*, 2012, **28**(2), 230–233.
23. Y. L. Zhu, Q. J. Jiao, H. Huang and H. Ren, Effect of Aluminum Particle Size on Thermal Decomposition of AP, *Chem. J. Chin. Univ.*, 2013, **34**(3), 662–667.
24. H. E. Kissinger, *Anal. Chem.*, 1957, **29**(11), 1702–1706.
25. A. W. Coats and J. P. Redfern, *Nature*, 1964, **201**(4914), 68–69.
26. Z. R. Liu, *Thermal Analyses for Energetic Materials*, National Defense Industry Press, Beijing, 2008.
27. P. Weiqiang, L. T. DeLuca and A. Gromov, *et al.*, *Innovative Energetic Materials: Properties, Combustion and Application*, Springer Nature, 2020.
28. J. Fan, N. Yulei and P. Yufan, *et al.*, Effect of dodecahydrododecaborate bistetraethylammonium on combustion reaction mechanism of aluminum powder, *Chin. J. Energ. Mater.*, 2021, **29**(10), 937–947.
29. S. Zi-xing, S. Li-li and Y. Rong-jie, Research progress of icosahedronal polyhydroborate $B_{12}H_{12}{}^{2-}$ anion compounds and their application in propellants and explosives, *Chin. J. Explos. Propellants*, 2017, **40**(3), 1–16.
30. H. Gao and J. M. Shreeve, Azole-based energetic salts, *Chem. Rev.*, 2011, **111**(11), 7377–7436.
31. L. Yunlu, X. Mei and W. Jianlong, *et al.*, Advances in the synthesis of poly-nitro furazans, *Chin. J. Org. Chem.*, 2016, **36**(7), 1528–1538.
32. R. P. Singh, R. D. Venna and D. T. Meshri, *et al.*, Energetic nitrogen-rich salts and ionic liquids, *Angew. Chem., Int. Ed.*, 2006, **45**, 3584.
33. M. Hiskey and D. Chavez, Progress in high-nitrogen chemistry in explosives, propellants and pyrotechnics, *Proceedings of the 27th International Pyrotechnics Seminar*, USA: Colorado, 2000, pp. 3–14.
34. T. M. Klapotke, *High energy density materials*, Struct and Bonding, 2007, vol. 125, p. 85.
35. S. Lili, S. Zixing and G. Xiaoyan, *et al.*, Synthesis, characterization and thermal behavior of bis(dialkyl-5-aminotetrazolium) dodecadodecaborates, *Chin. J. Org. Chem.*, 2018, **38**, 2093–2100.
36. S. I. Uspenskaya, K. A. Solntsev and N. T. Kuznetsov, The X-ray diffraction study of alkali metal dodecahydro-closo-dodecaborates, *Zh. Strukt. Khim.*, 1975, **16**(3), 482–484.
37. G. Shoham, D. Schomburg and W. N. Lipscomb, The crystal and molecular structure of $[B_{12}H_{12}[NEt_3H]_2$, *Acta Crystallogr., Sect. C: Cryst. Struct. Commun.*, 1980, **9**(2), 429–434.
38. I. Tiritiris and T. Schleid, Investigations on the crystal structure of lithium dodecahydro-closo-dodecaborate from aqueous solution: $Li_2(H_2O)_7[B_{12}H_{12}]$, *Cheminform*, 2002, **33**(38), 10–15.
39. J. H. Her, W. Zhou and V. Stavila, *et al.*, Role of cation size on the structural behavior of the alkali-metal dodeca-hydro-closo-dodecaborates, *J. Phys. Chem. C*, 2009, **113**(26), 11187–11189.

40. S. Zixing, S. Lili and Y. Rongjie, Research progress of icosahedronal polyhydroborate $B_{12}H_{12}^{2-}$ anion compounds and their application in propellants and explosives, *Chin. J. Explos. Propellants*, 2017, **40**(3), 1–16.
41. B. R. S. Hansen, M. Paskevicius and H. W. Li, *et al.*, Metal boranes: progress and applications, *Coord. Chem. Rev.*, 2015, **323**(4), 60–70.
42. L. Yanhua, S. Zixing and S. Lili, *et al.*, Compatibility of dodecahydrododecaborate with propellants ingredients, *Chin. J. Explos. Propellants*, 2017, **40**(6), 95–100.
43. X. Yunna, W. Weiqiang and L. Jiaoyi, *et al.*, Synthesis, Structure Characterization and Thermal Properties of Novel Cationic Polynitrogen Heterocyclium Decahvdrodecaborates, *Chin. J. Energ. Mater.*, 2016, **24**(3), 274–278.
44. L. Runtao, G. Zemei and W. Xin, *Ring Chemistry—Structure, Reaction, Synthesis and Application*, Chemical Industry Press, Beijing, 2008.
45. V. I. Saldin, M. A. Karpenko and L. N. Ignat'eva, *et al.*, Synthesis and study of hexamethvlenetetrammonium dodecahydro-closo-dedocaborate, *Zh. Neorg. Khim.*, 2009, **54**(1), 15–19.
46. S. Zixing, S. Ming and W. Bingbin, Without crystal water, high nitrogen and high borohyd ride compound: hexamethylenetraammonium dodecahydro-closo-dedocaborate, preparation and thermal properties, *The 6th Conference on Propellants of Chinese Chemical Society*, Jiuquan, 2013.
47. W. Weiqiang, X. Yunna and M. Suning, *et al.*, Synthesis of dihydrazine-1,2,4,5-tetrazine decahydrodecaborate, *Chin. J. Energ. Mater.*, 2014, **22**(3), 428–429.
48. L. Ruibin and L. Zhanxiong, Synthesis and thermalproperty of energetic caged-carborane derivatives, *Chin. J. Synth. Chem.*, 2010, **18**(2), 229–231.
49. N. S. Hosmane, *Boron Science: New Technologies and Applications*, CRC Press, 2012.
50. *Boron-Based Compounds: Potential and Emerging Applications in Medicine*, ed. E. Hey-Hawkins and C. Vinas Teixidor, John Wiley & Sons Ltd., 2018, DOI: 10.1002/9781119275602.
51. I. B. Sivaev, *Russ. J. Inorg. Chem.*, 2021, **66**, 1289.
52. I. N. Klyukin, N. A. Selivanov and A. Y. Bykov, *et al.*, *Russ. J. Inorg. Chem.*, 2020, **65**, 1547.
53. A. V. Nelyubin, N. A. Selivanov and A. Y. Bykov, *et al.*, *Russ. J. Inorg. Chem.*, 2020, **65**, 795.
54. V. V. Avdeeva, E. A. Malinina, K. Y. Zhizhin, and N. T. Kuznetsov, *Russ. J. Coord. Chem.* 2021, **47**, 519.
55. E. A. Kravchenko, A. A. Gippius and N. T. Kuznetsov, *Russ. J. Inorg. Chem.*, 2020, **65**, 546.
56. E. Y. Matveev, I. V. Novikov and A. S. Kubasov, *et al.*, *Russ. J. Inorg. Chem.*, 2021, **66**, 187.
57. E. A. Malinina, S. E. Korolenko and A. P. Zhdanov, *et al.*, *J. Cluster Sci.*, 2020, **32**, 755.
58. V. V. Avdeeva, G. A. Buzanov and E. A. Malinina, *et al.*, *Crystals*, 2020, **10**, 389.

59. S. E. Korolenko, V. V. Avdeeva and E. A. Malinina, *et al.*, *Russ. J. Coord. Chem.*, 2020, **46**, 297.
60. E. Didelot, Y. Sadikin, Z. Łodzian and R. Černya, *Solid State Sci.*, 2019, **90**, 86.
61. N.-D. Van, I. Tiritiris and T. Schleid, *Z. Anorg. Allg. Chem.*, 2004, **630**, 1764.
62. I. Tiritiris, N.-D. Van and T. Schleid, *Z. Anorg. Allg. Chem.*, 2001, **637**, 682.
63. I. Tiritiris and T. Schleid, *Z. Anorg. Allg. Chem.*, 2003, **629**, 1390.
64. E. A. Malinina, K. Y. Zhizhin and I. N. Polyakova, *et al.*, *Russ. J. Inorg. Chem.*, 2002, **47**, 1275.
65. V. V. Drozdova, E. A. Malinina and O. N. Belousova, *et al.*, *Russ. J. Inorg. Chem.*, 2008, **53**, 1024.
66. I. Tiritiris and T. Schleid, *Z. Anorg. Allg. Chem.*, 2003, **629**, 581.
67. T. E. Paxton, M. F. Hawthorne, L. D. Brown and W. N. Lipscomb, *Inorg. Chem.*, 1974, **13**, 2772.
68. E. A. Malinina, V. V. Drozdova, I. N. Polyakova and N. T. Kuznetsov, *Russ. J. Inorg. Chem.*, 2008, **53**, 197.
69. R. D. Dobrott and W. N. Lipscomb, *J. Chem. Phys.*, 1962, **37**, 1779.
70. V. V. Avdeeva, A. S. Kubasov and S. E. Korolenko, *et al.*, Synthesis and structures of copper and thallium (I) coordination compounds $[Cu_2[B_{10}H_{10}]]_n$ and $Tl_2[B_{10}H_{10}]$ with the closo-decaborate anion, *Russ. J. Inorg. Chem.*, 2022, **67**(5), 628–635.
71. K. Hofman and B. Albert, *Z. Kristallogr.*, 2005, **220**, 142.
72. I. B. Sivaev and V. I. Bregadze, in *Organometallic Chemistry Research Perspectives*, ed. R. P. Irwin, Nova Science Pub Inc., 2007.
73. I. B. Sivaev and V. I. Bregadze, *Eur. J. Inorg. Chem.*, 2009, **11**, 1433.
74. Z. J. Leśnikowski, *J. Med. Chem.*, 2016, **59**, 7738.

CHAPTER 3

Interactions of $B_{10}H_{10}[(C_2H_5)_4N]_2$ with Energetic Components and Inert Materials

3.1 Introduction

Tetraethylammonium dodecahydrododecaborate ($B_{10}H_{10}[(C_2H_5)_4N]_2$, BHN), is a new energetic material, which can be used as a main ingredient in cast explosives and solid propellants[1–3] because of its high heat of combustion value. It has the potential for possible use as a high-energy ingredient of fuel-rich solid propellants and explosives from the point of view of the above-mentioned high performance.[4–6] The reactivity or compatibility of $B_{10}H_{10}[(C_2H_5)_4N]_2$ with some energetic components and inert materials is one of the most useful aspects for $B_{10}H_{10}[(C_2H_5)_4N]_2$ in practical application. There are a few reports on the synthesis, thermal decomposition and its development as a combustion catalyst;[7–11] however, the compatibility of $B_{10}H_{10}[(C_2H_5)_4N]_2$ with some energetic components and inert materials used in propellants has seldom been reported. In this chapter, the compatibilities of $B_{10}H_{10}[(C_2H_5)_4N]_2$ with some energetic components and inert materials under non-isothermal conditions investigated by means of DSC[12–16] are reported. In addition, in order to make our conclusion more convincing, an investigation on impact and friction sensitivity properties of the $B_{10}H_{10}[(C_2H_5)_4N]_2$ and its combination with energetic materials was also carried out according to the existing methods.[17,18]

3.2 Experimental

3.2.1 Raw Materials

The tetraethylammonium dodecahydrododecaborate ($B_{10}H_{10}[(C_2H_5)_4N]_2$) used in this work was prepared and purified at Xi'an Modern Chemistry Research

High-energy Combustion Agents of Organic Borohydrides: Synthesis, Characterization and Applications
By XiaoHong Zhang, WeiQiang Pang, Luigi T. DeLuca and Yu Zhao

Published by the Royal Society of Chemistry, www.rsc.org

Institute, and the purity was more than 99.5%.[19] The sample was kept in a vacuum desiccator before use. Glycidyl azide polymer (GAP, 99.8%), cyclo-trimethylenetrinitramine (RDX, 99.6%), cyclotetramethylenete-tranitroamine (HMX, >99.5%), lead 3-nitro-1,2,4-triazol-5-onate (NTO-Pb, >99.3%), hexa-nitrohexaazaisowurtzitane (CL-20, >99.7%), aluminum powder (12.18 mm, >99.9%), magnesium powder (Mg, 44 mm, >99.9%), 3,4-dinitrofurzanfuroxan (DNTF, >99.7%), and *N*-guanylurea-dinitramide (GUDN, >99.7%) were used as energetic components and polyoxy-tetramethylene-*co*-oxyethylene (PET, $\bar{M} = 4000$), polyethylene glycol (PEG, $\bar{M} = 10\,000$), addition product of hexam-ethylene diisocyanate and water (N-100), hydroxyl terminated polybutadiene (HTPB), cupric adipate (AD-Cu, >99.4%), cupric 2,4-dihydroxybenzoate (β-Cu, >99.8%), lead phthalate (φ-Pb, >99.2%), carbon black (C.B., >99.8%), aluminum oxide (Al_2O_3, 99.8%), 1,3-dimethyl-1,3-diphenyl urea (C_2, 99.0%), di-2-ethylhexyl sebacate (DOS, 99.5%) and potassium perchlorate (KP, 99.5%) used as inert materials were industrially procured. Mixtures of $B_{10}H_{10}[(C_2H_5)_4N]_2$ (BHN) and energetic components or inert materials were prepared according to mass ratio 1 : 1.

All the sample systems involved in this investigation, which were purified by different solvents and subsequently dried for 48 h (at 60 °C), were premixed symmetrically and defined as follows (the number in the parentheses is their code designation): BHN (1); GAP (2); 1 : 1-BHN/GAP (3); RDX (4); 1 : 1-BHN/RDX (5); HMX(6); 1 : 1-BHN/HMX (7); NTO-Pb (8); 1 : 1-BHN/NTO-Pb (9); CL-20 (10); 1 : 1-BHN/CL-20 (11); Al (12); 1 : 1-BHN/Al (13); Mg (14); 1 : 1-BHN/Mg (15); DNTF (16); 1 : 1-BHN/DNTF (17); GUDN (18); 1 : 1-BHN/GUDN (19); PET (20); 1 : 1-BHN/PET (21); PEG (22); 1 : 1-BHN/PEG (23); N-100 (24); 1 : 1-BHN/N-100 (25); HTPB (26); 1 : 1-BHN/HTPB (27); AD-Cu (28); 1 : 1-BHN/AD-Cu (29); β-Cu (30); 1 : 1-BHN/β-Cu (31); φ-Pb (32); 1 : 1-BHN/φ-Pb (33); C.B. (34); 1 : 1-BHN/C.B. (35); Al_2O_3 (36); 1 : 1-BHN/Al_2O_3 (37); C_2 (38); 1 : 1-BHN/C_2 (39); DOS (40); 1 : 1-BHN/DOS (41); KP (42); 1 : 1-BHN/KP (43).

3.2.2 Equipment and Experimentation

All the DSC measurements were carried out with a NETZSCH DSC 204 HP in-strument. The conditions of DSC were as follows: sample mass, BHN (about 0.35–0.46 mg) and mixture system (0.50–0.78 mg); heating rate, 10 K min^{-1}; atmosphere, dynamic atmosphere of nitrogen with a flow rate of 50 K min^{-1} and pressure of 0.1 MPa. The BHN, energetic component or inert material or mixture of 1 : 1-BHN/energetic component and 1 : 1-BHN/inert material, was sealed in an aluminum cell, and all the samples were premixed symmetrically.[20,21]

3.2.3 Interactions with Different Methods

3.2.3.1 DSC Method

The evaluated standards of compatibility for explosive and contacted materials[22] are listed in Table 3.1.

Table 3.1 Evaluated standards of compatibility for explosive and contacted materials. Reproduced from ref. 22.

Criteria (ΔT_p/°C)	Rating	
Less than or equal to 2	A	Compatible or good compatibility
3–5	B	Slightly sensitized or moderate compatibility
6–15	C	Sensitized or poor compatibility
15 and above	D	Hazardous or bad compatibility

In Table 3.1, A is safe for use in any explosive design; B is safe for use in testing, when the device will be used in a very short period of time, not to be used as a binder material or when long-term storage is desired; C is not recommended for use with explosive items; D is hazardous, do not use under any conditions.

3.2.3.2 *Vacuum Stability Test*

Sample size: single sample (2.50 ± 0.01) g, mixed sample (5.00 ± 0.01) g, mass ratio of 1 : 1. Under the conditions of constant volume, constant temperature (90.0 ± 0.5) °C and certain vacuum degree, the pressure of gas released within a certain time (40 h) is converted into the gas volume under the standard state, and the compatibility is evaluated by the volume of gas released per unit mass of sample. The volume of gas released by the sample in the standard state is:

$$V_H = 2.69 \times 10^{-3} P(V_O - V_G)/T \tag{3.1}$$

where V_H is the volume of gas released by the sample under the standard state (the amount of gas released), mL; 2.69×10^{-3} is the ratio of temperature and pressure under standard state, kPa; P is the gas pressure released by the sample; V_O is the sum of the volume of reaction vessel and the volume of pressure measured in the connecting pipeline, mL; V_G is the sample volume (mass divided by true density), mL; T is the test room temperature, K. The net outgassing amount is:

$$R = V_C - (V_A + V_B) \tag{3.2}$$

In the formula, R is the net gas discharge of reaction, mL; V_C is the sum of the volume of reaction vessel and the volume of pressure measured in the connecting pipeline, mL; V_A is the outgassing volume of sample a, mL; V_B is the outgassing volume of 8 samples, mL. Each sample is tested three times in parallel, and the average value is taken at the end.

Compatibility evaluation standard: $R < 3.0$ mL, compatibility; $R =$ 3.0–5.0 mL, moderate reaction; 0 mL, incompatible.

3.3 Compatibility of BHN with Propellant Ingredients

3.3.1 DSC Method

3.3.1.1 Compatibility of $B_{10}H_{10}[(C_2H_5)_4N]_2$ with Some Energetic Components

Typical DSC curves of systems (1)–(19) at 0.1 MPa are shown in Figure 3.1, and their maximum exothermic peak temperatures are shown in Table 3.2. In fact, Beach and Canfield gave the test method of compatibility of explosives with polymers in 1971.[17]

The DSC curve of the BHN/GAP (glycidyl azide polymer) mixture shows no endothermic change and two sharp exothermic changes. The exothermic peak temperature of GAP became from 250.3 °C to 246.7 °C for the addition of BHN, thus BHN can increase the decomposition of the GAP binder. The DSC curve of HMX consists of two endothermic peaks and one exothermic peak. The first peak temperature at 247.1 °C is due to the crystal transformation from α to δ; for more detail please see ref. 24. The following endothermic peak at 281.7 °C is the phase change from solid to liquid. The exothermic peak at 283.7 °C is caused by the rapid decomposition reaction. The exothermic peak temperature of the BHN/HMX mixture is lower by 24.5 °C than that of pure BHN and lower by 2.4 °C than that of pure HMX, which suggests that the presence of BHN stabilized HMX effectively.

The maximum exothermic peak temperature difference between BHN and the BHN/GAP mixture (ΔT_p), is 3.6 °C. The value of ΔT_p is 2.4 °C between BHN and the BHN/HMX mixture. According to the standards of compatibility evaluated in Table 3.2, we think that the binary systems BHN/GAP and BHN/HMX have moderate compatibility because ΔT_p is less than 5 °C.

The values of ΔT_p for BHN and BHN/NTO-Pb, BHN/Al and BHN/Mg mixtures are −0.5 °C, −0.5 °C and −0.4 °C, respectively. The value of ΔT_p is 1.9 °C for CL-20 and BHN/CL-20. This phenomenon indicates that there is no chemical reaction between components of the mixture and the mixture has good compatibility.

When BHN and 3,4-dinitrofurzanfuroxan (DNTF) were mixed, the binary eutectic solid system came into being, where the melting point is little higher than that of pure DNTF. The melting point of pure DNTF was 110.0 °C.[25] Because of the large quantity of heat that BHN emitted, it was too late for BHN to melt before decomposition, and there was no endothermic peak before exothermic change. The endothermic peaks of DNTF are caused by the phase change from solid to liquid. The endothermic peaks of the binary system (17) are due to the melting process of DNTF.

The value of ΔT_p for the BHN and BHN/RDX mixture is 25.4 °C, the value of ΔT_p for the BHN and BHN/DNTF mixture is −33.8 °C and the value of ΔT_p for the BHN and BHN/GUDN mixture is 43.8 °C, which is consistent with the destabilization of DNTF and GUDN by the presence of BHN. This means an increase in reactive ability and a decrease in the thermal stability of the

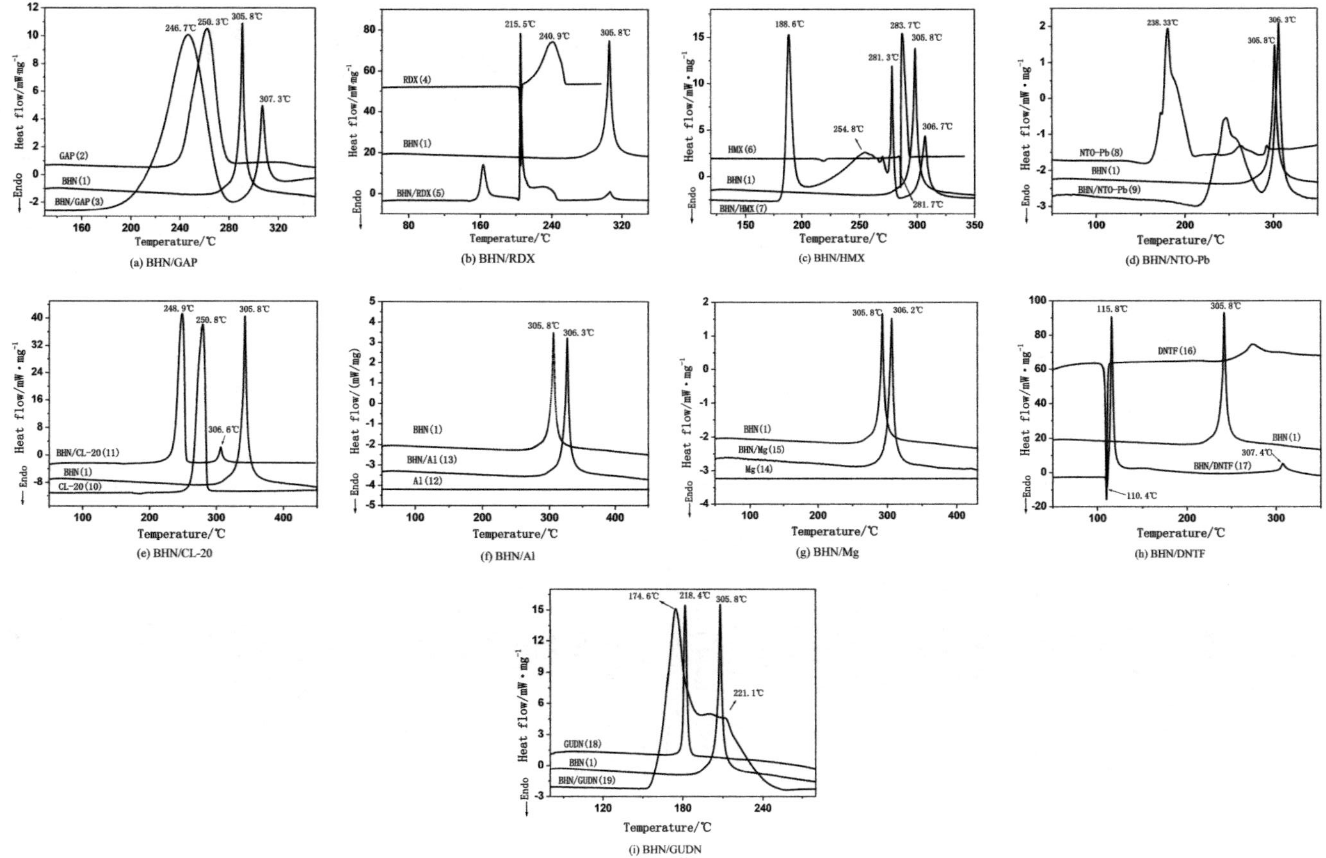

Figure 3.1 The DSC curves of BHN with some energetic components at 0.1 MPa (heating rate: 10 K min^{-1}; sample mass: 0.50–0.78 mg). Reproduced from ref. 23.

Table 3.2 The evaluations of compatibilities for DSC peak temperature (binary system of 1/1-BHN/energetic components).[a] Reproduced from ref. 23.

System		Peak temperatures			
Mixture system	Single system	T_{p1}/°C	T_{p2}/°C	ΔT_p/°C	Rating
1/1-BHN/GAP (3)	GAP	250.3	246.7	3.6	B
1/1-BHN/RDX (5)	RDX	240.9	215.5	25.4	D
1/1-BHN/HMX (7)	HMX	283.7	281.3	2.4	B
1/1-BHN/NTO-Pb (9)	BHN	305.8	306.3	−0.5	A
1/1-BHN/CL-20 (11)	CL-20	250.8	248.9	1.9	A
1/1-BHN/Al (13)	BHN	305.8	306.3	−0.5	A
1/1-BHN/Mg (15)	BHN	305.8	306.2	−0.4	A
1/1-BHN/DNTF (17)	DNTF	273.6	307.4	−33.8	D
1/1-BHN/GUDN (19)	GUDN	218.4	174.6	43.8	D

[a]Mixture system, 1/1-BHN/energetic component binary system; single system, system of the pure energetic component, where exothermic peak temperature is smaller of the two pure components; T_{p1}, the maximum exothermic peak temperature of single system; T_{p2}, the maximum exothermic peak temperature of mixture system; $\Delta T_p = T_{p1} - T_{p2}$.

mixture. RDX, DNTF and GUDN are incompatible with BHN. Because of the large quantity of heat that BHN emitted, there was no endothermic peak before exothermic change for the BHN/RDX, BHN/DNTF and BHN/GUDN mixtures.

The compatibility of the binary system BHN/energetic component decreases in the order: (15) > (9) = (13) > (11) > (7) > (3) > (5) > (17) > (19). The relative thermal stability of the binary system BHN/energetic component decreases in the order: (17) > (9) = (13) > (15) > (7) > (11) > (3) > (5) > (19).

3.3.1.2 *Compatibility of $B_{10}H_{10}[(C_2H_5)_4N]_2$ with Some Inert Materials*

Typical DSC curves of systems of (20)–(43) are shown in Figure 3.2. Their maximum exothermic peak temperatures are shown in Table 3.3. From Figure 3.2 and Table 3.3, the following observations can be made. The DSC curves of BHN/PEG and BHN/C_2 mixtures show one endothermic peak each. The endothermic peaks temperatures 63.8 °C and 122.9 °C are the eutectic points for the systems BHN/PEG and BHN/C_2, respectively. The values of ΔT_p for BHN and BHN/PEG, BHN/PET, BHN/N-100 °C, BHN/HTPB, BHN/AD-Cu, BHN/β-Cu, BHN/φ-Pb, BHN/DOS, BHN/KP, BHN/Al_2O_3, BHN/C_2 and BHN/C.B. mixtures are all less than 2.0 °C, and it was indicated that these binary systems are extremely compatible. To a certain extent, most of the inert materials examined herein can promote the decomposition of BHN. The compatibility of binary system BHN/inert material decrease in the order: (31) = (35) = (37) > (39) = (41) > (43) > (25) > (23) > (21) = (27) = (33) > (29). The relative thermal stability of binary systems BHN/inert material decreases in the order: (29) > (27) > (23) > (25) > (41) > (35) > (31) = (37) > (39) > (43) > (21) = (33).

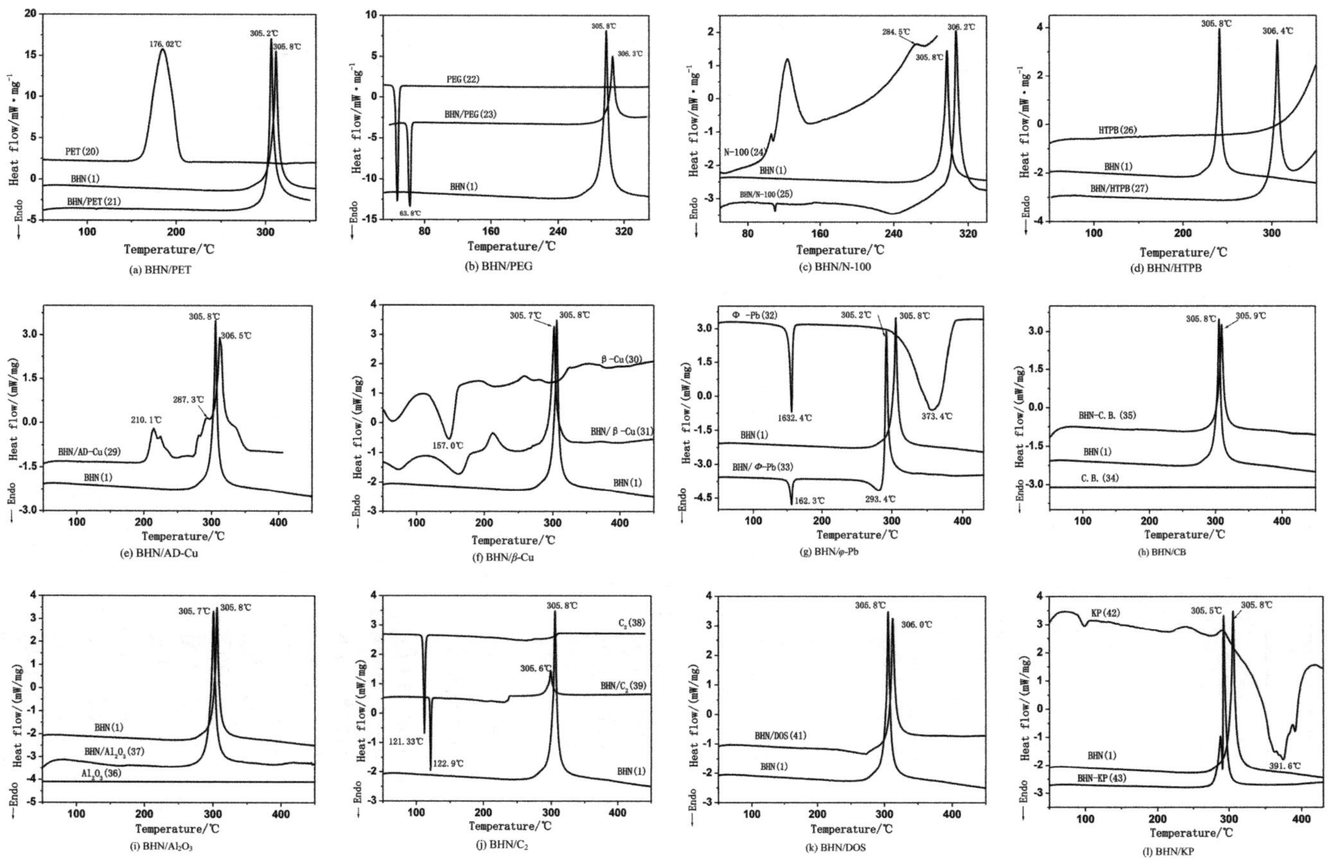

Figure 3.2 The DSC curves of BHN with some inert materials at 0.1 or 1 MPa (heating rate: 10 K min^{-1}; sample mass: 0.50–0.78 mg). Reproduced from ref. 23.

Table 3.3 The evaluations of compatibilities for DSC peak temperature (binary system of 1/1-inert materials/BHN). Reproduced from ref. 23.

System		Peak temperatures			
Mixture system	Single system	T_{p1}/°C	T_{p2}/°C	ΔT_p/°C	Rating
1/1-BHN/PET (21)	BHN	305.8	305.2	0.6	A
1/1-BHN/PEG (23)	BHN	305.8	306.3	−0.5	A
1/1-BHN/N-100 (25)	BHN	305.8	306.2	−0.4	A
1/1-BHN/HTPB (27)	BHN	305.8	306.4	−0.6	A
1/1-BHN/AD-Cu (29)	BHN	305.8	306.5	−0.7	A
1/1-BHN/β-Cu (31)	BHN	305.8	305.7	0.1	A
1/1-BHN/φ-Pb (33)	BHN	305.8	305.2	0.6	A
1/1-BHN/C.B. (35)	BHN	305.8	305.9	−0.1	A
1/1-BHN/Al_2O_3 (37)	BHN	305.8	305.7	0.1	A
1/1-BHN/C_2 (39)	BHN	305.8	305.6	0.2	A
1/1-BHN/DOS (41)	BHN	305.8	306.0	−0.2	A
1/1-BHN/KP (43)	BHN	305.8	305.5	0.3	A

Lastly, one point we must emphasize is that when the system evaluated by DSC is incompatible, we cannot affirm that it is indeed incompatible. By contrast, when the system evaluated by DSC is compatible, we can affirm that it is compatible at ambient pressure. Because the sample was tested under the condition of high temperature, when it is compatible under that condition it will be more compatible at lower temperatures. Whereas, when the system evaluated by PDSC is incompatible, other evaluating methods such as DTA/TG or the vacuum stability test (VST) may be used to confirm it.

3.3.1.3 Compatibility of $B_{12}H_{12}$ Salts with Propellant Ingredients

The interactions between $B_{12}H_{12}$ salts (BH-12-K, BH-12-U, BH-12-G) and propellant ingredients including binders (PET, HTPB), curing agents (TDI, N-100), plasticizers (DOS, A3) and energetic fillers with the mass ratio of 1 : 1, were investigated by means of the DSC technique, and the compatibilities were evaluated; the results are shown in Table 3.4. It can be seen that there is no endothermic and exothermic thermal decomposition at 500 °C for BH-12-K, while there is an exothermic peak at 213.6 °C and 212.8 °C for BH-12-U and BH-12-G, respectively. The decomposition peak temperature of BH-12-K/HTPB mixed system is 0.8 °C higher than that of HTPB. The thermal decomposition temperature of BH-12-K/PET mixed system is greater than 500 °C, and the temperature difference between the mixed system BH-12-U/HTPB and BH-12-U/PET and BH-12-U is $\pm$0.3 °C, indicating that the BH-12-K and BH-12-U with HTPB and PET is safe under general use and the storage temperature. The thermal decomposition peak temperatures of BH-12-G/HTPB and BH-12-G/PET mixed systems are 16.3 °C and 12.8 °C higher than those of sample BH-12-G, respectively, showing that HTPB and PET are helpful for improving the stability of sample BH-12-G.

Meantime, there is no obvious interaction between BH-12-K with N-100 and TDI, but there are obvious exothermic thermal decomposition peak

Table 3.4 The compatibility parameters of dodecahydrododecaborate/ingredients mixed systems. Reproduced from ref. 26.

Sample	T_{p1}/°C	T_{p2}/°C	T_{endo}/°C	T_{pBH}/°C	$T_{pIngredients}$/°C	Rating
BH-12-K	>500	—	>500		—	—
BH-12-U	213.6	—	>500		—	—
BH-12-G	212.8, 230.0, 282.3	—	>500		—	—
PET	>500	—	384.7		—	—
HTPB	363.2	—	454.5		—	—
BH-12-K/HTPB	—	364.0	440.5		0.8	A
BH-12-K/PET	—	>500	410.6		—	A
BH-12-U/HTPB	—	213.3, 360.4	456.4	−0.3	−2.8	B
BH-12-U/PET	—	213.9	>500	0.3	—	A
BH-12-G/HTPB	—	229.1, 349.6	457.2	16.3	−13.6	C
BH-12-G/PET	—	225.6	>500	12.8	—	A
N-100	276.1, 401.0	—	232.8, 344.6, 450.5	—	—	—
TDI	>500	—	96.4, 184.4	—	—	—
BH-12-K/N-100	—	>500	241.1	—	—	A
BH-12-K/TDI	—	>500	>500	—	—	A
BH-12-U/N-100	—	214.4	>500	—	0.8	A
BH-12-U/TDI	—	213.7	149.7	—	0.1	A
BH-12-G/N-100	—	217.8	>500	—	5.0	B

BH-12-G/TDI	—	214.3	149.6	—	1.5	A
A3	>500	—	248.6	—	—	—
DOS	>500	—	233.6, 20 508	—	—	—
BH-12-K/DOS	—	>500	>500	—	—	A
BH-12-K/A3	—	>500	256.8	—	—	A
BH-12-U/DOS	—	214.4	323.2	—	0.8	A
BH-12-U/A3	—	212.6	>500	—	−1.0	A
BH-12-G/DOS	—	217.3	>500	—	4.5	B
BH-12-G/A3	—	221.3	>500	—	8.5	C
AP	243.4	—	283.1, 404.1	—	—	—
RDX	205.5	—	233.4	—	—	—
BH-12-K/RDX	—	213.5	—	—	−19.9	D
BH-12-K/AP	—	283.1	243.2	—	0	—
BH-12-U/RDX	—	195.9	—	−16.9	−37.5	D
BH-12-U/AP	—	210.9	—	−2.7	−72.2	D
BH-12-G/RDX	—	202.1, 220.7	194.4	−10.7	−12.7	C
BH-12-G/AP	—	213.0, 271.4	—	0.2	−11.7	C

temperatures at 214.4 °C, 213.7 °C, 217.8 °C and 214.3 °C for the mixed systems BH-12-U/N-100, BH-12-U/TDI, BH-12-G/N-100 and BH-12-G/TDI, respectively. The exothermic thermal decomposition peak temperature of the mixed system is significantly delayed compared with samples BH-12-U and BH-12-G, which indicates that curing agent N-100 and TDI can improve the stability of samples BH-12-U and BH-12-G.

Moreover, the melting endothermic peak temperature of DOS appears at 233.6 °C, while the melting endothermic peak disappears after mixing with $B_{12}H_{12}$ salts, indicating that the interaction between DOS and $B_{12}H_{12}$ salts is relatively strong. The melting endothermic peak temperature of A3 appeared at 248.6 °C, and the peak temperature advanced after mixing with BH-12-K. After A3 is mixed with BH-12-U, the exothermic peak temperature becomes higher, indicating that its interaction is large and causes exothermic reaction. There is no thermal decomposition exothermic peak for BH-12-K/DOS and BH-12-K/A3 at 500 °C, and while the exothermic peak for mixtures of BH-12-U/DOS moves forward 1.0 °C, the exothermic peak for mixtures of BH-12-U/A3, BH-12-G/A3, BH-12-G/DOS move backward, indicating that the stability of BH-12-U and BH-12-G were improved.

In addition, the melting endothermic peaks temperature of RDX and AP are at 205.5 °C and 243.4 °C, and the exothermic thermal decomposition peak temperatures are at 233.4 °C and 283.1 °C, respectively. The first exothermic peak temperature of BH-12-K/RDX was 19.9 °C ahead, while there was no change for the BH-12-K/AP mixture. There is a large thermal decomposition peak temperature difference between BH-12-U and AP, RDX, indicating that neither of them could mix with RDX and AP.

3.3.2 Vacuum Stability Test (VST) Method

The compatibility of $B_{10}H_{10}[(C_2H_5)_4N]_2$ (BHN) with RDX, HMX, Al powder, Mg powder, φ-Pb, AP, HTPB, DOS, AD-Cu, TDI, GFP and IPDI was studied by using the VST method. The results are shown in Table 3.5. It can be seen that when $B_{10}H_{10}[(C_2H_5)_4N]_2$ is mixed with RDX, HMX, AP, φ-Pb, Al powder, Mg powder, HTPB, DOS and AD-Cu, the amount of outgassing decreases. This may be due to the fact that $B_{10}H_{10}[(C_2H_5)_4N]_2$ is coated with RDX, Al powder, AP and other solid particles after being heated and melted, which increases the stability and reduces the outgassing amount, indicating that $B_{10}H_{10}[(C_2H_5)_4N]_2$ has little interaction with RDX, HMX, AP, φ-Pb, Al and Mg powders and has good compatibility. Furthermore, after mixing $B_{10}H_{10}[(C_2H_5)_4N]_2$ with GFP and IPDI, the outgassing amount increases slightly, while the net outgassing amount is still less than 3 mL, it can be judged according to the judgment standard that $B_{10}H_{10}[(C_2H_5)_4N]_2$ is compatible with GFP and IPDI, and they both can be used in a solid propellant formulation. On the other hand, when $B_{10}H_{10}[(C_2H_5)_4N]_2$ and TDI are mixed, the outgassing amount increases significantly, which indicates that the interaction between $B_{10}H_{10}[(C_2H_5)_4N]_2$ and TDI in the heating process is much more violent. Therefore, TDI and $B_{10}H_{10}[(C_2H_5)_4N]_2$ cannot be used in solid propellant and explosive formulations at the same time.

Table 3.5 Compatibility results of $B_{10}H_{10}[(C_2H_5)_4N]_2$ (BHN) with some energetic materials obtained by means of the VST method. Reproduced from ref. 27.

Sample	Gas evolved/mL	Net gas evolution/mL	Rating
BHN	0.62	—	—
RDX	0.09	−0.49	Compatible
BHN/RDX	0.27		
HMX	0.08	−0.37	Compatible
BHN/HMX	0.52		
Al	0.04	−0.36	Compatible
BHN/Al	0.53		
Mg	0.04	−0.42	Compatible
BHN/Mg	0.41		
NOT-Pb	0.18	−0.35	Compatible
BHN/NTO-Pb	0.58		
AP	0.03	−0.37	Compatible
BHN/AP	0.54		
HTPB	0.18	−0.17	Compatible
BHN/HTPB	0.35		
DOS	0.28	−0.46	Compatible
BHN/DOS	0.35		
AD-Cu	0.21	−0.44	Compatible
BHN/AD-Cu	0.37		
TDI	0.42	3.18	Fair compatibility
BHN/TDI	5.02		
GFP	0.26	0.04	Compatible
BHN/GFP	0.13		
IPDI	0.34	0.14	Compatible
BHN/IPDI	0.12		

3.4 Stability of $B_{10}H_{10}[(C_2H_5)_4N]_2$ Under Impact Load

In recent years, the investigations on $B_{10}H_{10}[(C_2H_5)_4N]_2$ have mainly been focused on thermal decomposition and compatibility with propellant components, and it has been found that $B_{10}H_{10}[(C_2H_5)_4N]_2$ is compatible with the main components of solid propellant and it can be used for solid propellant and explosive formulations.[28] However, the physico-chemical state of the explosives has changed greatly over the detonation temperature range 2000–5000 °C and over the pressure range 10–30 GPa, and its energy release process is very different from that of the low-pressure process.[29,30] Therefore, it is essential to accurately characterize the stability and decomposition process of $B_{10}H_{10}[(C_2H_5)_4N]_2$ under explosion and electric detonation shock wave methods.

3.4.1 Samples and Apparatus

Casting TNT (Φ 40 mm×40 mm, density 1.55 g cm^{-3}); booster JH-14C (Φ 15 mm×15 mm, density 1.68 g cm^{-3}); the purity of $B_{10}H_{10}[(C_2H_5)_4N]_2$ is more than 99.5% prepared in Xi'an Modern Chemistry Research Institute; HMX, purity more than 99.5%, particle size 125 μm. Gansu Yinguang Chemical Industry Group Co., Ltd; gas sampling bag (volume 1 L), Dalian Hyde Technology Co., Ltd.

QP-2010 Plus gas chromatography–mass spectrometry, Shimadzu, Japan; Quanta 600 FEG field emission scanning electron microscope, FEI company, USA; EMPYREAN X-ray powder diffractometer, Netherlands; TENSOR 27 infrared spectrometer, Brooke company, USA.

3.4.2 Stability Tests

3.4.2.1 Single Explosion Impact Test

After $B_{10}H_{10}[(C_2H_5)_4N]_2$ fills the closed chamber, it is compacted (density is 0.90 g cm^{-3}) to ensure that it is not in direct contact with the explosion impact source. The thickness of the impact end cover is 30 mm. A schematic diagram of the device is shown in Figure 3.3. The TNT was fixed in the center of the upper end of the chamber after the fuel chamber was fixed. The booster was JH-14C and was detonated by a detonator. By controlling the thickness of the aluminum partition, the shock wave pressure acting on $B_{10}H_{10}[(C_2H_5)_4N]_2$ was adjusted. Firstly, an aluminum partition, 10 mm thick, was added. After initiation, the gas was collected by a gas collection bag and the mass of $B_{10}H_{10}[(C_2H_5)_4N]_2$ was weighed. Then, the aluminum baffle was removed, the gas after initiation collected, and the mass of $B_{10}H_{10}[(C_2H_5)_4N]_2$ was weighed.

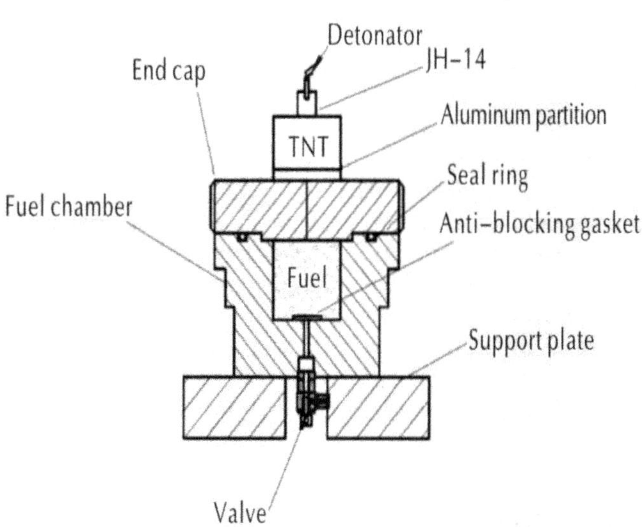

Figure 3.3 Schematic diagram of the explosion shock test device. Reproduced from ref. 28.

3.4.2.2 *Shock Effect of Electric Detonation*

The experimental setup of the electric detonation shock is shown in Figure 3.4. The $B_{10}H_{10}[(C_2H_5)_4N]_2$ fuel was loaded into the ceramic shell, and the Ni–CR wire was used to pass through the center of the fuel, and then compacted and sealed. The wire length diameter ratio and charging voltage were controlled to keep them unchanged to ensure that the wire released the same energy during discharge. The decomposition of $B_{10}H_{10}[(C_2H_5)_4N]_2$ under shock and high temperature was studied by using 20 kV high voltage Ni–CR wire excited in N_2 and air atmosphere. $B_{10}H_{10}[(C_2H_5)_4N]_2$ was initiated in N_2 and air respectively. In order to further improve the impact and thermal effect of the fuel in the initiation process, $B_{10}H_{10}[(C_2H_5)_4N]_2$ and HMX were mixed at a mass ratio of 1 : 1 and then detonated in N_2. The initiation voltage was 20 kV; Φ 0.3 mm×100 mm Ni–CR wire and the mass ratio of Ni to CR was 80 : 20. After the experiment, the gas was collected using a gas collecting bag through a gas valve.

3.4.3 Performance Characterization

The gas products decomposed by $B_{10}H_{10}[(C_2H_5)_4N]_2$ were analyzed by GC–MS to determine the organic gas composition and relative content.

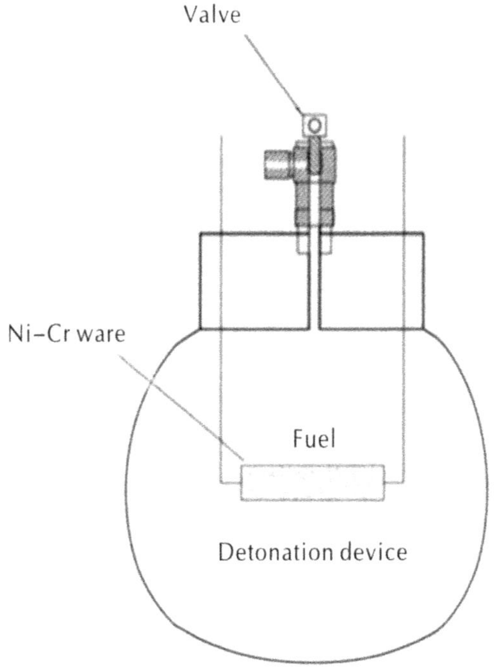

Figure 3.4 Schematic diagram of the electrical detonation shock test device. Reproduced from ref. 28.

The surface morphology of decomposition products was observed by SEM and the composition of solid products was analyzed by EDS.

The crystal structure of the shock decomposition products was analyzed by X-ray diffraction. CuK_α radiation ($\lambda = 0.15406$ nm) was selected.

The working conditions of the X-ray tube were 40 kV and 40 mA, the scanning range was 5–90° and the step speed was 0.2 s^{-1}.

Fourier transform infrared spectrometry was used to test the product, and the composition of the product was analyzed. The test band was 400–4000 cm^{-1}, the resolution was 0.1 cm^{-1}, and the acquisition frequency was 95 file per s.

3.4.4 Stability of $B_{10}H_{10}[(C_2H_5)_4N]_2$ Under a Single Explosion Shock

ANSYS/AUTODYN software was used to calculate the impact force on the fuel medium at the central point of contact between $B_{10}H_{10}[(C_2H_5)_4N]_2$ and the end cover. It is known that the peak value of overpressure of shock wave with adding and removing aluminum partition is 1.25–2.1 GPa. The peak value of overpressure of the shock wave can reach 15.0 GPa by setting a test point at the bottom of TNT explosive. Therefore, the existence of the end cap can reduce the overpressure of the shock wave, reducing the impact force by an order of magnitude, and this means that the high temperature generated by the explosion cannot be quickly transmitted to the combustion agent.

It was found that no gas was collected and the mass of $B_{10}H_{10}[(C_2H_5)_4N]_2$ did not decrease in the two groups of experiments with and without aluminum separators. The closed cage structure of $B_{10}H_{10}[(C_2H_5)_4N]_2$ has good stability against pressure shock, which indicates that a single explosion shock (1.0 GPa) cannot promote the decomposition of $B_{10}H_{10}[(C_2H_5)_4N]_2$. It can be seen that the decomposition of $B_{10}H_{10}[(C_2H_5)_4N]_2$ needs the temperature to reach the corresponding threshold, and the decomposition cannot be completed when only a single pressure of 1.0 GPA is used as the impact source.

3.4.5 Stability of $B_{10}H_{10}[(C_2H_5)_4N]_2$ Under Electric Detonation Impact

3.4.5.1 *Gas Phase Product Analysis*

The gas components of the thermal decomposition products of electric detonation shock were analyzed. The analysis results of three samples are shown in Table 3.6. It was found that $B_{10}H_{10}[(C_2H_5)_4N]_2$ can be decomposed by plasma electric detonation, and its products are mainly organic combustible gases with high calorific value, such as triethylamine, nitriles, alkanes, olefins, alkynes, *etc.* The length of the carbon chain of the gas product is obviously shorter than that of the HMX/$B_{10}H_{10}[(C_2H_5)_4N]_2$ mixture. Except for a small amount of triethylamine, the length of the gas

Table 3.6 Gas products of BHN-10 and BHN/HMX decomposed by electrical detonation shock. Reproduced from ref. 28.

Sample	Products	Chemical formula	Peak area/%
BHN (N_2)	Triethylamine	$C_6H_{15}N$	81.99
	Isoprene monooxide	$C_8H_8O_8$	8.50
	2,2,4,6,6-Pentamethylheptane	$C_{12}H_{26}$	3.50
	Benzene	C_6H_6	3.24
	Butylene	C_4H_2	1.71
	Acetonitrile	C_2H_3N	1.06
BHN (air)	Benzene	C_6H_6	45.49
	Bicyclo[1,1,0]butane	C_4H_6	25.45
	2-Butanone	C_4H_8O	7.03
	Acetonitrile	C_2H_3N	4.97
	Cyclopentadiene	C_5H_6	3.86
	Acrylonitrile	C_3H_3N	2.57
	2-Methyl-1-hexene	C_8H_{16}	2.44
	cis-2-Butene	C_4H_8	2.39
	Triethylamine	$C_6H_{15}N$	1.91
	Toluene	C_7H_8	1.15
BHN/HMX (N_2)	Acetonitrile	C_2H_3N	30.26
	N-Butane	C_4H_{10}	26.14
	Ethyl nitrite	$C_2H_5NO_2$	13.87
	Nitroethane	$C_2H_5NO_2$	13.87
	Nitroethane	$C_2H_5NO_2$	10.06
	Triethylamine	$C_6H_{15}N$	6.76
	Nitromethane	CH_3NO_2	6.61
	Propionitrile	C_3H_5N	4.49
	Acrylonitrile	C_3H_3N	2.26

product was no more than four carbon atoms, indicating that the stability of $B_{10}H_{10}[(C_2H_5)_4N]_2$ is low under the impact of electric detonation, and HMX can improve the decomposition efficiency of $B_{10}H_{10}[(C_2H_5)_4N]_2$ and produce a higher proportion of low carbon chain flammable gas.

3.4.5.2 Analysis of Morphology and Composition of Solid Products

An X-ray diffraction spectrum was used to test the solid powder after initiation, and the results are shown in Figure 3.5. It was found that the main thermal decomposition products of $B_{10}H_{10}[(C_2H_5)_4N]_2$ under the impact of electric detonation in N_2 atmosphere are C_{60} and $B_{12}H_{16}$ with an orthorhombic crystal. In addition to C_{60} and $B_{12}H_{16}$, the characteristic peak of B_2O_3 appears in the solid phase products of $B_{10}H_{10}[(C_2H_5)_4N]_2$ under the impact of electric detonation in air atmosphere. The appearance of the SiO_2 peak is due to the ceramic residue produced after electrical explosion when the sample is filled in the ceramic shell. Moreover, the main composition of BHN/HMX (1 : 1 in mass ratio) products are C, B, and O except for SiO_2 in ceramic residue, which indicates that the product mainly exists in amorphous form after HMX and BHN mix initiation.

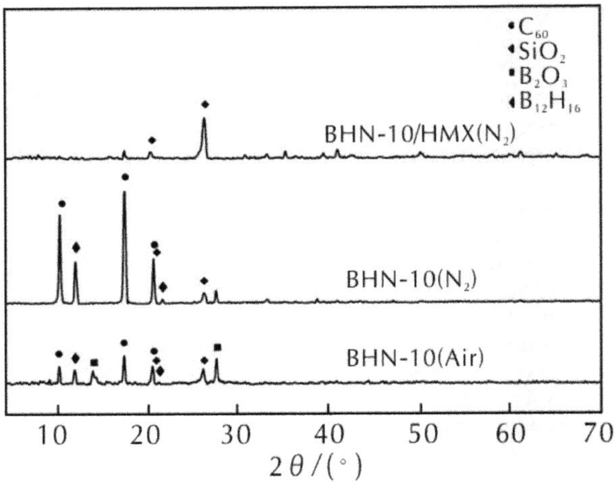

Fig. 3.5 XRD patterns of solid products of BHN and BHN/HMX decomposed by electrical detonation shock. Reproduced from ref. 28.

Scanning electron microscopy (SEM) was used to observe the change in the micro-sized morphology of $B_{10}H_{10}[(C_2H_5)_4N]_2$ powder with an electro detonation product, and energy dispersive spectroscopy (EDS) was used to analyze the element composition of the products. The results are shown in Figure 3.6. It can be seen that in N_2 atmosphere, the structure of $B_{10}H_{10}[(C_2H_5)_4N]_2$ is mainly a polyhedron, which is mostly composed of B.

Because it was in N_2 atmosphere, the B element was not oxidized (Figure 3.6b). No matter whether HMX is added or not, the solid products become flakes with tens of nanometer scale, and the flakes are more evenly dispersed. The main components are C and amorphous B_2O_3. Combined with infrared data analysis, $B_{10}H_{10}[(C_2H_5)_4N]_2$ reacts with oxygen in air to form boron oxide after electro detonation decomposition, and some reacts with water in explosion products to further form boric acid (Figure 3.6d). When HMX is added, the thickness of flake C is further reduced, which is because the existence of HMX increases the pressure and temperature of the electric explosion, and makes the decomposition reaction more complete. Therefore, different from the effect of a single explosion shock on $B_{10}H_{10}[(C_2H_5)_4N]_2$, the high-pressure shock and instantaneous high temperature produced by electric detonation have a better thermal decomposition effect on $B_{10}H_{10}[(C_2H_5)_4N]_2$, and the addition of HMX is conducive to the complete decomposition of $B_{10}H_{10}[(C_2H_5)_4N]_2$ (Figure 3.6f).

Furthermore, EDS was used to analyze the type and content of micro elements in the solid powder. The B and C atomic percentages of pure $B_{10}H_{10}[(C_2H_5)_4N]_2$ in N_2 atmosphere and air atmosphere are 29.3%, 70.7%, 32.2%, and 67.8%, respectively; the atomic percentages of B and C in HMX/BHN (1 : 1 in mass ratio) detonation products in N_2 atmosphere are 23.2% and 76.8%, respectively. Compared with 38.5% and 61.5% of B and C

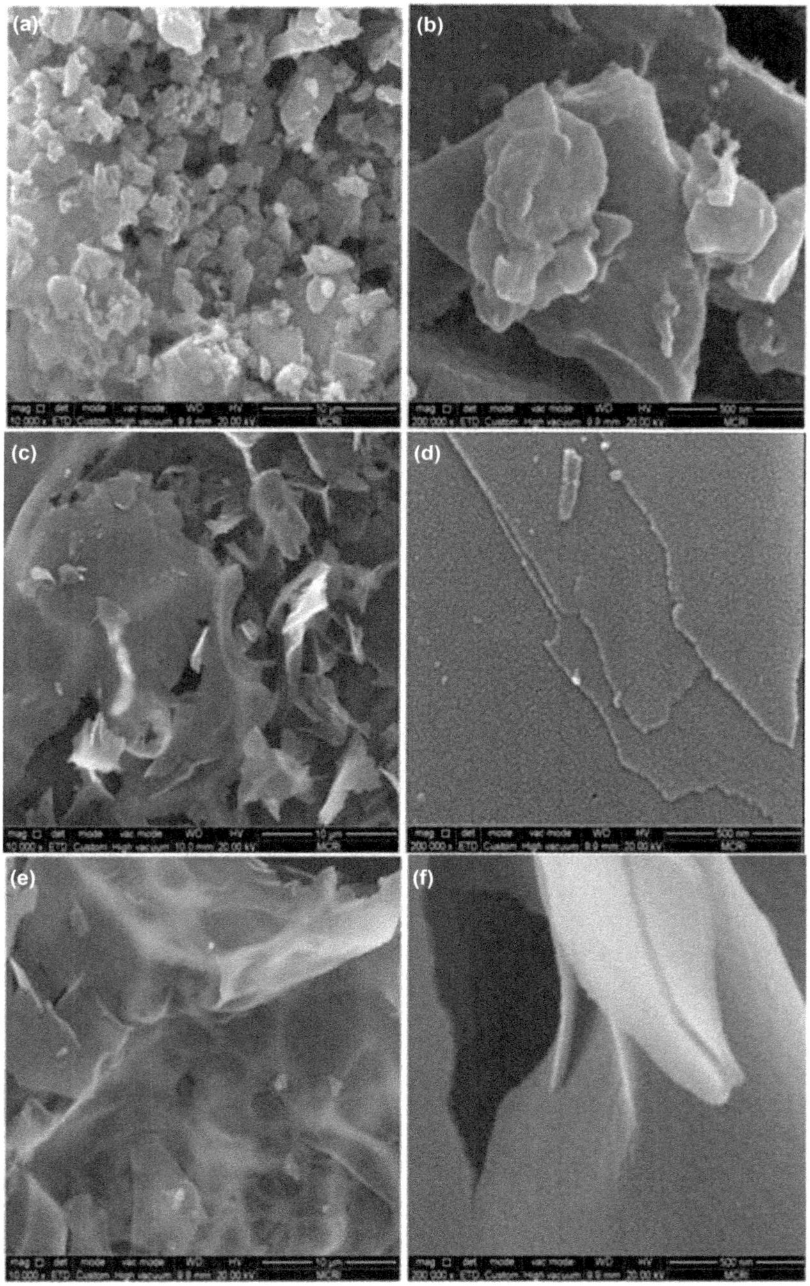

Figure 3.6 Low and high magnification SEM images of solid products decomposed by electrical detonation shock. (a): BHN (N_2), 10 000; (b): BHN (N_2), 200 000; (c): BHN (Air), 10 000; (d): BHN (Air), 200 000; (e): BHN/HMX, 10 000; (f): BHN/HMX, 200 000. Reproduced from ref. 28.

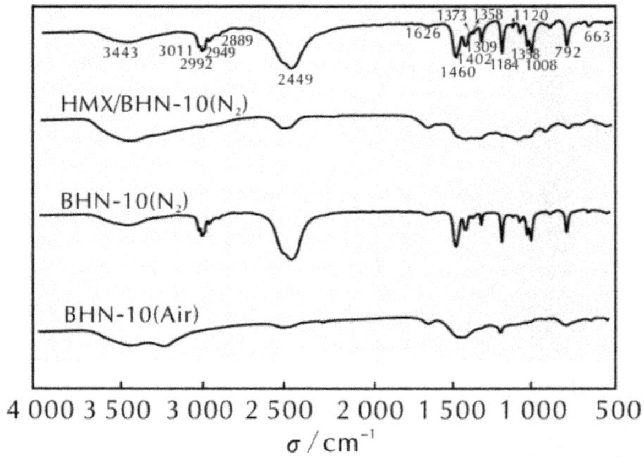

Figure 3.7 Infrared spectrum of solid products decomposed by electrical detonation shock. Reproduced from ref. 28.

atoms in pure $B_{10}H_{10}[(C_2H_5)_4N]_2$, the B and C atom percentages of initiation products are decreased. The reason maybe that the C element transfers to the gas phase in the process of the shock decomposition reaction, while the B element mostly exists in the form of solid phase.

The solid product was tested by infrared spectroscopy and the results are shown in Figure 3.7. It can be seen that 3011 cm^{-1} and 1460 cm^{-1} are the antisymmetric stretching vibration peaks of CH_3; 2949, 1373 and 1358 cm^{-1} are the symmetrical stretching vibration peaks of CH_3; 2992 cm^{-1} and 1402 cm^{-1} are the antisymmetric stretching vibration peaks of CH_2; 2889 cm^{-1} and 1309 cm^{-1} are the symmetrical stretching vibration peaks of CH_2; 1120 cm^{-1}, 1079 cm^{-1} and 792 cm^{-1} are the vibration absorption peaks of C–C. The vibration absorption peak of B–H is at 2449 cm^{-1}. The vibration absorption peak of C–N is at 1184 cm^{-1}. It is found that there are a few residual samples in $B_{10}H_{10}[(C_2H_5)_4N]_2$ under the effect of electro detonation shock decomposition, which indicates that the electro detonation shock cannot decompose $B_{10}H_{10}[(C_2H_5)_4N]_2$ completely. However, the relative peak strength of the standard vibration absorption peak (2449 cm^{-1}) of $B_{10}H_{10}[(C_2H_5)_4N]_2$ decreased significantly with the addition of HMX. The results show that the addition of HMX can further improve the decomposition efficiency of $B_{10}H_{10}[(C_2H_5)_4N]_2$, which is consistent with the results obtained by X-ray diffraction (XRD).

3.5 Performance of $B_{12}H_{12}[(C_2H_5)_4N]_2$ Particles

Among the polyborane ion compounds, dodecahydrododecaboric acid ion $B_{12}H_{12}^{2-}$ compounds have attracted much attention because of their high stability, low toxicity, ease of complete combustion, and convenient preparation advantages.[29,30] Therefore, $B_{12}H_{12}^{2-}$ compound, as one of the combustion agents with many advantages, has been widely used in propellant and explosives.[31]

3.5.1 Materials and Instruments

Sodium borohydride (purity 98%), boron trifluoride ether complex (purity 99%), tetraethyl ammonium chloride (purity 98%), diethylene glycol dimethyl ether (purity greater than 99%), Shanghai Aladdin Biochemical Technology Co., Ltd.

Magna-IR 550 type II Nicolet Fourier transform infrared spectrometer, Thermofisher, United States; D8 advance X-ray diffractometer, Bruker, Germany; Sta 409 PC/QMS 403C thermogravimetric mass spectrometer, Netzsch instruments, Germany.

The laser ignition combustion test can be found in ref. 32.

3.5.2 Preparation of $B_{12}H_{12}[(C_2H_5)_4N]_2$ Samples

Under the protection of nitrogen, sodium borohydride was put into a Schlenk bottle, diethylene glycol dimethyl ether (diethylene glycol dimethyl ether was added with calcium hydride under the protection of high-purity nitrogen before use, stirred for 12 h and filtered for use) was added then stirred and the temperature raised to 100 °C. Then boron trifluoride ether solution was added slowly, and stirring continued for 2 h after dropping. The temperature was slowly raised to 165 °C and the reaction continued for 16 h. In the whole reaction process, the tail gas absorption device was used to absorb the volatile ether and borane gas.

After the reaction, the solvent was evaporated by vacuum distillation to obtain a solid with a light yellow color. Deionized water was slowly added to the solid in nitrogen atmosphere. The solid was dissolved and bubbles released. Then, hydrochloric acid solution was slowly added to release a large amount of gas and white precipitate. Filtrate was obtained by filtration and precipitation. Tetraethyl ammonium chloride was dissolved in deionized water, and the solution dropped into the filtrate to produce white precipitation, then stirred and filtered. The obtained white filter residue was washed with deionized water many times, and finally the product $B_{12}H_{12}[(C_2H_5)_4N]_2$ was obtained.

3.5.3 Composition Structure Analysis

The performance of $B_{12}H_{12}[(C_2H_5)_4N]_2$ was characterized, and the results are shown in Figure 3.8. It can be seen that the large and wide characteristic peak can be observed at 2485–2469 cm^{-1}, corresponding to the B–H stretching vibration peak, and the characteristic peak is in the range of 1467–1458 cm^{-1}, which corresponds to the B–H bending vibration peak. The peak at 2995–2985 cm^{-1} is the stretching vibration peak of C–H (Figure 3.8a). The double peak at chemical shift -14.9 cm^{-1} is the characteristic peak of

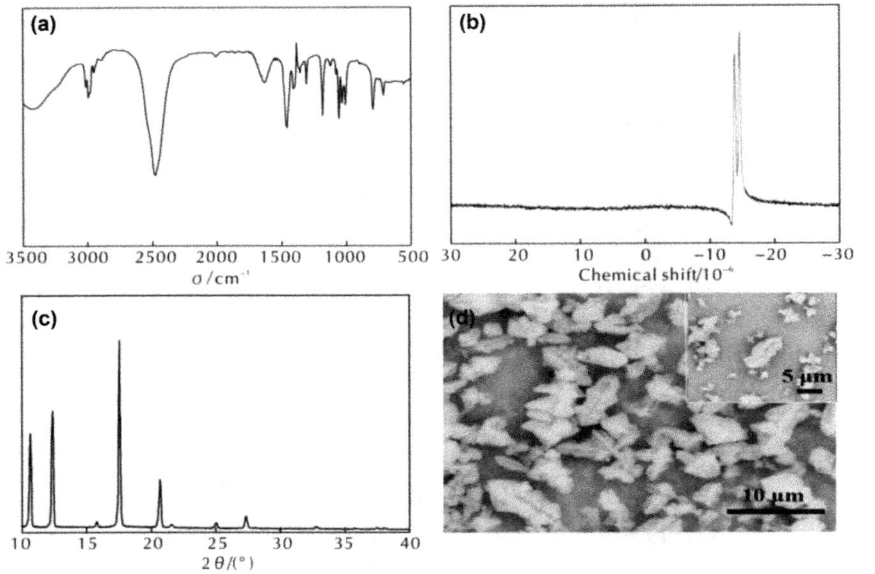

Figure 3.8 Characterization of $B_{12}H_{12}[(C_2H_5)_4N]_2$. (a) IR spectrum; (b) ^{11}B NMR spectrum; (c) XRD pattern; (d) SEM image. Reproduced from ref. 34.

the $B_{12}H_{12}{}^{2-}$ group, which is consistent with the literature.[33] At the same time, there is no other type of B peak, indicating that the purity of this prepared compound is high (Figure 3.8b); the crystallinity is also good. The prepared compound has unique edges and corners, and the particle size is 2–10 μm (Figure 3.8d).

3.5.4 Thermal Decomposition Analysis

The thermal decomposition behavior of $B_{12}H_{12}[(C_2H_5)_4N]_2$ in nitrogen and air were measured by thermogravimetric mass spectrometry (TG–MS), the results are shown in Figure 3.9. It can be seen that $B_{12}H_{12}[(C_2H_5)_4N]_2$ is stable basically when nitrogen is lower than 300 °C. When the temperature is higher than 300 °C, it begins to decompose slowly and lose weight. It undergoes violent decomposition over the temperature range of 300–400 °C, releasing a large amount of volatile substances, and the weight loss rate is 50%. From the TG–MS curves of $B_{12}H_{12}[(C_2H_5)_4N]_2$ in Figure 3.9b, it can be found that $B_{12}H_{12}[(C_2H_5)_4N]_2$ has good thermal stability in air, and there is no obvious mass change below 300 °C.

The mass spectrum monitoring data at this stage showed that hydrogen and ethane were released from the thermal decomposition of the material, so it was speculated that the tetraethylammonium cation was cracked. The material undergoes the second stage of decomposition, and the weight loss rate is 16.0% over the temperature range of 400–600 °C. The hydrogen

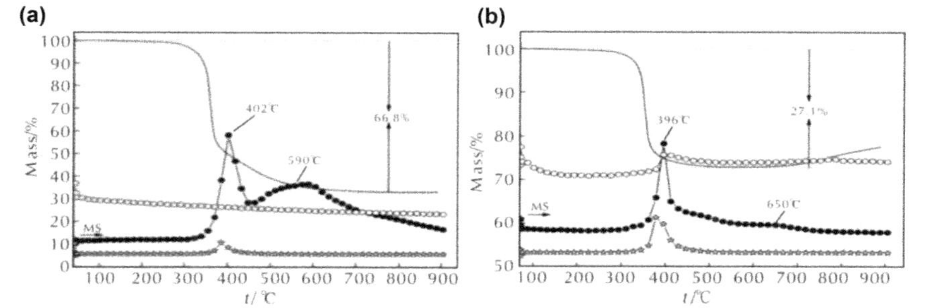

Figure 3.9 TG–MS results of $B_{12}H_{12}[(C_2H_5)_4N]_2$ in different atmospheres at the heating rate of 20 °C min^{-1}. (a) In nitrogen; (b) in air. Reproduced from ref. 34.

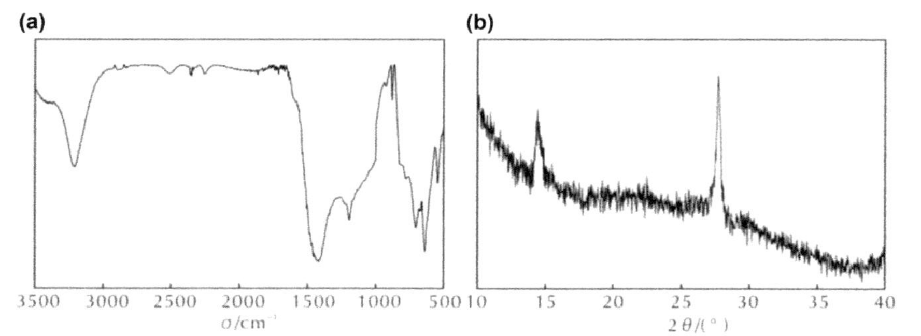

Figure 3.10 Infrared (IR) spectrum and XRD patterns of the thermal decomposition product of $B_{12}H_{12}[(C_2H_5)_4N]_2$ after pyrolysis in nitrogen. (a) IR spectrum; (b) XRD pattern. Reproduced from ref. 34.

release was detected in this stage by mass spectrometry, but there was no ethane release. The hydrogen release may be caused by the further cracking of the boron cage. Above 600 °C, the material basically has no mass loss, and the total weight loss of the material is 66.8% at 900 °C. During the whole heating decomposition process, there is no released ethylborane gas detected, confirming that the boron cage would not be cracked in nitrogen to form small molecular borane.

In addition, the solid product of pyrolysis in nitrogen is black powder, and its infrared spectrum and XRD spectrum are shown in Figure 3.10. We can see that the B–H stretching vibration peak of the raw material at 2485–2469 cm^{-1} basically disappeared, and the C–H stretching vibration peak at 2995–2985 cm^{-1} completely disappeared after pyrolysis, which confirmed that these chemical valence bonds were involved in the pyrolysis reaction (Figure 3.10a). The characteristic diffraction peak of $B_{12}H_{12}[(C_2H_5)_4N]_2$ disappeared completely, and no unknown crystalline substances appeared in the product after pyrolysis in nitrogen atmosphere (Figure 3.10b).

Comparing the thermogravimetric test results of $B_{12}H_{12}[(C_2H_5)_4N]_2$ in nitrogen and air, it can be found that $B_{12}H_{12}[(C_2H_5)_4N]_2$ has good thermal stability and oxidation resistance. When $B_{12}H_{12}[(C_2H_5)_4N]_2$ was heated to more than 300 °C, it began to crack to generate hydrogen and hydrocarbon gaseous products. The residual solid of cracking is easy to oxidize, and the generated oxide has a certain degree of stability. It would continue to oxidize at higher temperature, resulting in obvious weight gain of the material.

3.5.5 Thermal Decomposition Kinetics

In order to further calculate the kinetic parameters and thermal decomposition mechanism of $B_{12}H_{12}[(C_2H_5)_4N]_2$, the thermal decomposition kinetics were investigated at different heating rates (2.5 °C min^{-1}, 5.0 °C min^{-1}, 10.0 °C min^{-1}, and 20.0 °C min^{-1}). The DSC and TG–DTG curves of $B_{12}H_{12}[(C_2H_5)_4N]_2$ are shown in Figure 3.11. It can be seen that the first exothermic peak of $B_{12}H_{12}[(C_2H_5)_4N]_2$ is at 319.3–341.2 °C (the peak temperatures at 2.5, 5, 10, and 20 °C min^{-1} are 315.3, 324.9, 326.2, and 341.2 °C, respectively), the second exothermic peak at 319.1–365.6 °C (319.1, 328.6, 334.2, and 365.6 °C, respectively), and the third exothermic peak at 321.8–406.2 °C (321.8, 331.6, 341.8, and 406.2 °C, respectively). Two endothermic peaks are superimposed in the middle of the decomposition exothermic peak, of which the first endothermic peak is 318.2–349.8 °C (318.2, 327.6, 330.7, and 349.8 °C, respectively), and the second endothermic peak is 320.0–389.3 °C (320.0, 329.3, 336.9, and 389.3 °C, respectively) (Figure 3.11a).

Under these four heating rates, the total mass loss in the thermal decomposition process is in the range from 32.3% to 33.9%. The remaining mass of the sample is due to other solid components of condensed matter. The endothermic peak in the middle of the $B_{12}H_{12}[(C_2H_5)_4N]_2$ exothermic peak becomes more obvious, and will appear when the heating rate is low. There is an endothermic peak between the two exothermic peaks in the DSC curve, but no mass loss is found on the corresponding TG curve, indicating that the thermal decomposition process of $B_{12}H_{12}[(C_2H_5)_4N]_2$ is similar to

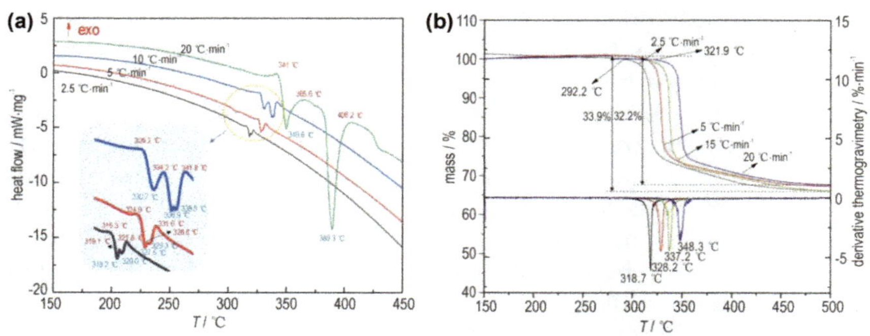

Figure 3.11 DSC and TG–DTG curves of $B_{12}H_{12}[(C_2H_5)_4N]_2$ at different heating rates. (a) DSC curves; (b) TG–DTG curves. Reproduced from ref. 38.

Table 3.7 Kinetic parameters for the thermal decomposition reaction of $B_{12}H_{12}[(C_2H_5)_4N]_2$. Reproduced from ref. 38.

		E/kJ mol^{-1}		r		
β/K min^{-1}	T_p/K	Kissinger	Ozawa	Kissinger	Ozawa	lg(Ak s^{-1}) Kissinger
2.5	591.9	206.4	205.8	0.9987	0.9983	17.5
5	601.4					
10	610.4					
20	621.5					

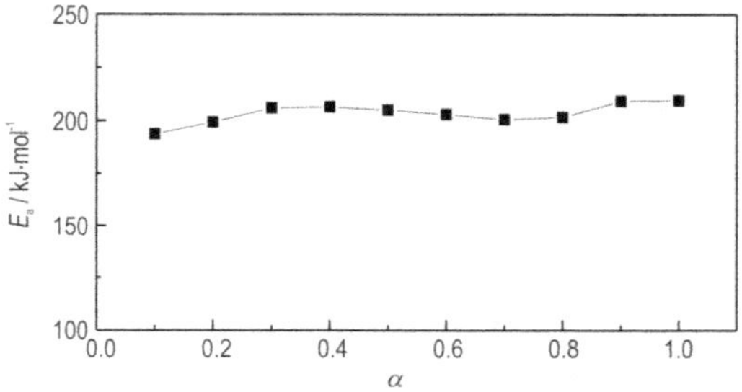

Figure 3.12 Apparent activation energy varies with the decomposition extent of $B_{12}H_{12}[(C_2H_5)_4N]_2$. Reproduced from ref. 38.

$B_{10}H_{10}[(C_2H_5)_4N]_2$, and a melt decomposition type intermediate product is generated in the thermal decomposition process.[35–37] The mass loss of $B_{12}H_{12}[(C_2H_5)_4N]_2$ begins at about 292.2–321.9 °C. With the increase in heating rate, the process of mass loss will shift to a higher temperature. Table 3.7 and Figure 3.12 show the apparent activation energy of $B_{12}H_{12}[(C_2H_5)_4N]_2$ calculated by the Kissinger and Ozawa methods. The whole thermal decomposition process of $B_{12}H_{12}[(C_2H_5)_4N]_2$ is different from that of $B_{10}H_{10}[(C_2H_5)_4N]_2$, and there is no obvious two-step thermal decomposition reaction.

3.5.6 Thermal Decomposition Gas Product Analysis of $B_{12}H_{12}[(C_2H_5)_4N]_2$

Figure 3.13 shows MS spectra of $B_{12}H_{12}[(C_2H_5)_4N]_2$ gas products at different temperatures. It can be seen that when the temperature exceeds 310 °C, $B_{12}H_{12}[(C_2H_5)_4N]_2$ gradually begins to decompose, and the infrared characteristic absorption intensity of the gas continues to increase. When the thermal decomposition reaction occurs, there are absorption peaks of ethylene (950 cm^{-1} and 2990–3166 cm^{-1}),[39] ethane (2950 cm^{-1} and 2972 cm^{-1}), and ammonia (967 cm^{-1} and 932 cm^{-1}).

The gas products corresponding to the ionic fragment m/z is H_2 ($m/z = 2$); ethylene ($m/z = 14$, 25, 26, 27, 28); NH_3 ($m/z = 14$, 15, 16, 17); ethane ($m/z = 14$, 15, 16, 25, 26, 27, 28, 29, 30). The ion current intensity began to

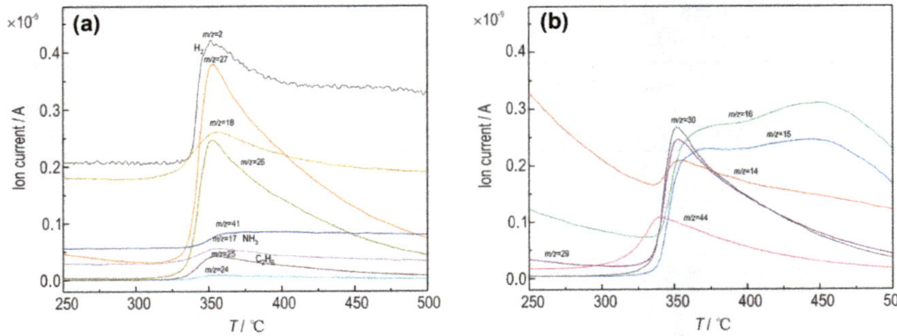

Figure 3.13 MS spectra of $B_{12}H_{12}[(C_2H_5)_4N]_2$ gas products at different temperatures. Reproduced from ref. 38.

rise at 302.2 °C and reached a peak at 352.7 °C. Apart from H_2, which, as a non-polar gas, could not be detected in gas infrared, the results of mass spectrometry were consistent with those of infrared.

In the decomposition process of $B_{10}H_{10}[(C_2H_5)_4N]_2$, the increase in H_2 particle flow is significantly ahead of the increase in NH_3 and C_2H_6 ion flow, which is considered to be the first dehydrogenation process of the boron cage.[40] However, in the thermal decomposition process of $B_{12}H_{12}[(C_2H_5)_4N]_2$, the increase in H_2 ion flow and NH_3, C_2H_6 ion flow intensity occurred simultaneously, which indicates that the dehydrogenation reaction of $B_{12}H_{12}[(C_2H_5)_4N]_2$ did not occur in advance during the decomposition process, and the dehydrogenation of the boron cage on its anion occurred simultaneously with the decomposition of tetraethyl ammonium cation.

3.5.7 Thermal Decomposition Condensed Product Analysis of $B_{12}H_{12}[(C_2H_5)_4N]_2$

The infrared *in situ* cell is used to analyze the condensed phase products in the thermal decomposition process of $B_{12}H_{12}[(C_2H_5)_4N]_2$, and the three-dimensional curve of the absorbance of the condensed phase products changing with temperature can be obtained as shown in Figure 3.14. It can be seen that the functional groups of $B_{12}H_{12}[(C_2H_5)_4N]_2$ at the initial reaction stage are the same as those of $B_{12}H_{12}[(C_2H_5)_4N]_2$ at normal temperature, and the main functional groups include $\nu_{as}(CH_3) = 3011$, 1460 cm^{-1}; $\nu_s(CH_3) = 2981$, 1373, 1358 cm^{-1}; $\nu_{as}(CH_2) = 2992$, 1402 cm^{-1}; $\nu_s(CH_2) = 2889$, 1309 cm^{-1}; ν(C–C) = 1120, 1079, 792 cm^{-1}; ν(B–H) = 2449 cm^{-1}; ν(C–N) = 1184 cm^{-1}. Among them, the vibration peak of the B–H bond shifted to 2481 cm^{-1}, and the characteristic functional groups appeared at 2007 cm^{-1} and 711 cm^{-1}. When the temperature reaches 300 °C, the absorbance of $B_{12}H_{12}[(C_2H_5)_4N]_2$ decreases, indicating that $B_{12}H_{12}[(C_2H_5)_4N]_2$ begins to decompose at this temperature.

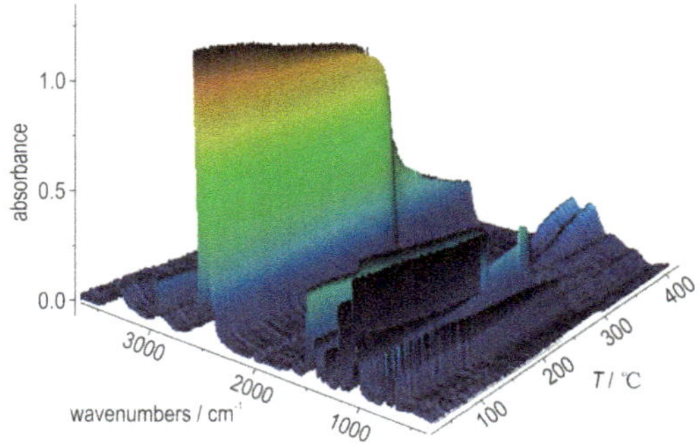

Figure 3.14 Infrared absorbance of condensed phase of $B_{12}H_{12}[(C_2H_5)_4N]_2$ at different temperatures. Reproduced from ref. 38.

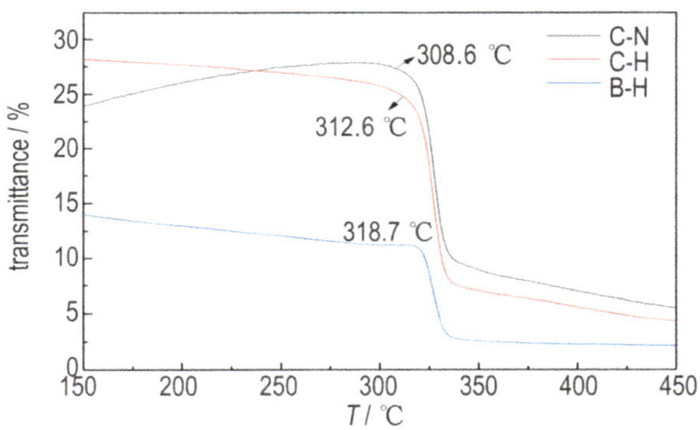

Figure 3.15 Infrared transmittance of the condensed phase of $B_{12}H_{12}[(C_2H_5)_4N]_2$ at different temperatures. Reproduced from ref. 38.

$B_{12}H_{12}[(C_2H_5)_4N]_2$ is composed of cage-like negative ions of dodecyl borate and positive ions of tetraethyl ammonium. The change in IR characteristic absorption intensity of main functional groups with heating time is shown in Figure 3.15.

The infrared peak of $B_{12}H_{12}[(C_2H_5)_4N]_2$ is composed of a closed cage structure B–H bond ($\nu_{B\text{-}H} = 2481$ cm^{-1}) and C–H bond of alkyl in a cation ($\nu_{C\text{-}H} = 1407$ cm^{-1}) and C–N key ($\nu_{C\text{-}N} = 1184$ cm^{-1}). According to Figure 3.14, the decomposition of the B–H bond is 318.7 °C, about 12 °C higher than that of the C–N bond and the C–H bond, which indicates that the decomposition of $B_{12}H_{12}[(C_2H_5)_4N]_2$ is consistent with the results of gas mass spectrometry.

The thermal decomposition mechanisms of $B_{12}H_{12}[(C_2H_5)_4N]_2$ and $B_{10}H_{10}[(C_2H_5)_4N]_2$ are different. For the thermal decomposition of $B_{12}H_{12}[(C_2H_5)_4N]_2$, it is not the dehydrogenation process of the boron cage, and the dehydrogenation process lags behind the decomposition process of

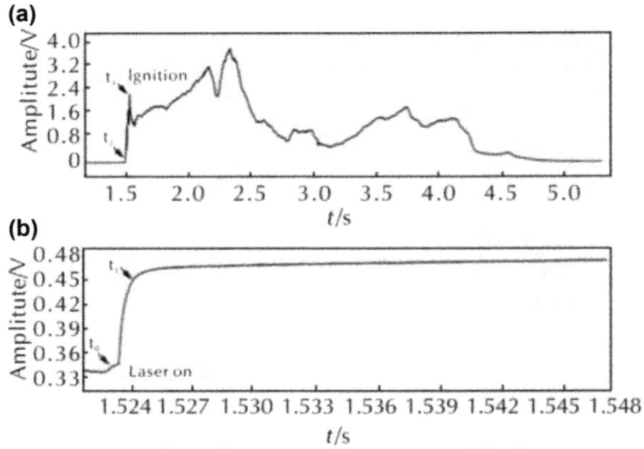

Figure 3.16 Time history of the ignition of $B_{12}H_{12}[(C_2H_5)_4N]_2$ with a diameter of 1.1 mm. (a) PMT response signal for burning; (b) Si detector response signal for laser. Reproduced from ref. 34.

a cation. There are three endothermic peaks and two exothermic peaks in the thermal decomposition process of $B_{12}H_{12}[(C_2H_5)_4N]_2$. The first peak in the thermal decomposition process of $B_{12}H_{12}[(C_2H_5)_4N]_2$ is the melting decomposition product peak, the second peak is the thermal decomposition of the tetraethyl ammonium cation peak, and the third one is the exothermic peak formed by the dehydrogenation of the boron cage.

3.5.8 Ignition Performance Detection

In order to quantitatively determine its ignition and combustion characteristics, the synthesized $B_{12}H_{12}[(C_2H_5)_4N]_2$ powder is placed in the burner, the sample is ignited by continuous laser irradiation, and the ignition and combustion process of the sample is monitored, as shown in Figure 3.16. The equivalent diameter of the sample to be tested is 1.1 mm. As calculated according to eqn (3.1), the ignition energy density and the ignition delay time are 4.72×10^6 W m^{-2} and 45.8 ms, respectively.

Figure 3.17 shows the maximum surface temperature of $B_{12}H_{12}[(C_2H_5)_4N]_2$ particles during the ignition and combustion process. It can be seen that the surface temperature of particles increases rapidly after laser heating. When the surface temperature reaches 800–1000 °C, the particles are ignited. The heating rate range from "heating" to "ignition" is 2.49×10^6 to 3.11×10^6 K s^{-1}, and the particles begin self-sustaining combustion after ignition. The particles maintain a high surface temperature during the combustion process, in which the peak temperature is 1800 °C. Meantime, in the process of stable combustion, the surface temperature is accompanied by small vibrations, which are presumably caused by the release of gas in the combustion process. After the combustion is extinguished, the surface temperature drops rapidly, and then reaches a "platform area" with a slight

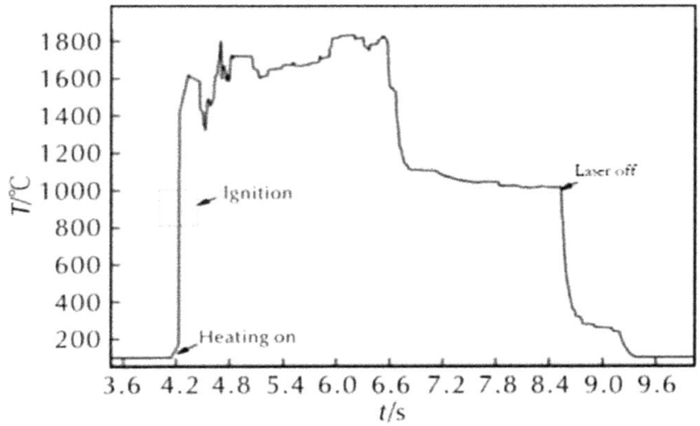

Figure 3.17 Surface temperature of $B_{12}H_{12}[(C_2H_5)_4N]_2$ during its combustion process in air. Reproduced from ref. 34.

temperature change. When the laser is turned off, the surface temperature drops further rapidly to room temperature.[35]

3.5.9 Combustion Flame Structure Characteristics

The combustion flame structure evolution process of $B_{12}H_{12}[(C_2H_5)_4N]_2$ in air from 45.8 ms to 842.5 ms is shown in Figure 3.18. It can be seen that the combustion flame structure is near the sample surface, and the flame appears at 45.8 ms for the first time, indicating that the material is ignited. At 61.8 ms, it can be seen that many small flames are produced by combustion, and the end of the flame is yellow green. In addition, obvious flue gas traces appeared after the combustion time was 359.1 ms, and the flue gas trace gradually weakened and then disappeared after the combustion time was 677.8 ms. The total combustion time of $B_{12}H_{12}[(C_2H_5)_4N]_2$ particles was 2.647 s. During the combustion process, no particle melting and sample foaming were observed.

The black powder is obtained after $B_{12}H_{12}[(C_2H_5)_4N]_2$ particles are burned in air. The product was analyzed by infrared and XRD, as shown in Figure 3.19. As can be seen from Figure 3.19a, the B–H characteristic peak of $B_{12}H_{12}[(C_2H_5)_4N]_2$ at 2485–2469 cm^{-1} is significantly weakened, the C–H stretching vibration peak at 2995–2985 cm^{-1} basically disappears, and the characteristic wide peak of B–O is generated near 1350 cm^{-1} after air combustion. As can be seen from Figure 3.19b, the characteristic diffraction peak of $B_{12}H_{12}[(C_2H_5)_4N]_2$ disappears completely after combustion, and the combustion product contains unknown crystalline substances. This may be due to the fact that both B–H and tetraethyl cations of $B_{12}H_{12}[(C_2H_5)_4N]_2$ participate in the combustion reaction and produce boroxides during combustion in air.

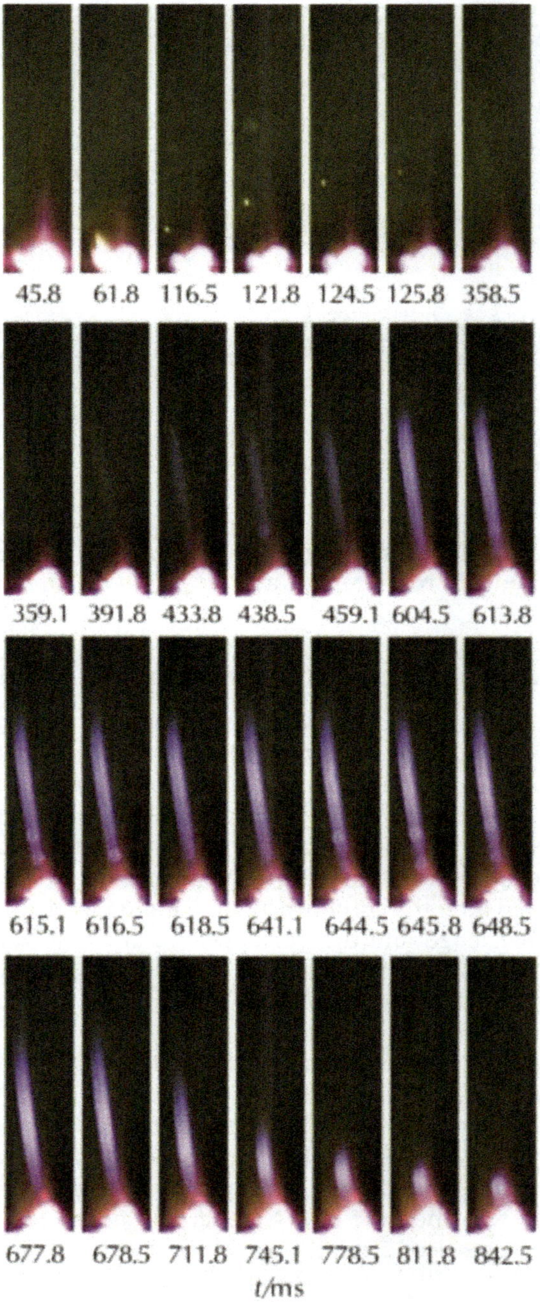

Figure 3.18 Combustion process of $B_{12}H_{12}[(C_2H_5)_4N]_2$ in air from 45.8 ms to 842.5 ms. Reproduced from ref. 34.

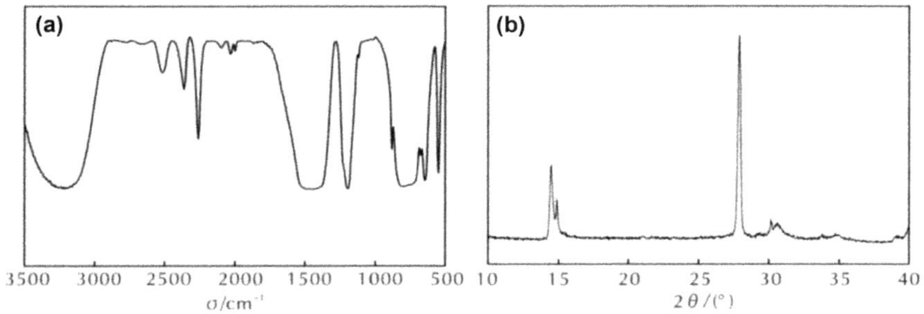

Figure 3.19 Infrared (IR) spectrum and XRD pattern of the combustion products of $B_{12}H_{12}[(C_2H_5)_4N]_2$. (a) IR spectrum; (b) XRD pattern. Reproduced from ref. 34.

3.6 Reaction of $B_{10}H_{10}[(C_2H_5)_4N]_2$ with Halogenated Aromatic Nitrogen Heterocyclic Salts

Decahydrodecaborate anion ($B_{10}H_{10}{}^{2-}$) is a kind of boron hydrogen anion with a highly symmetrical hexagonal cage structure, which is composed of an equilateral triangle with a boron atom as the apex. Most of its salts have excellent thermal stability and chemical stability.

Eight new aromatic nitrogen heterocyclic ammonium salts of decahydrodecaborate were synthesized by the reaction of tetraethyl ammonium decahydrodecaborate with amino substituted *N*-methyl (or ethyl) pyridinium halide, *N*-methyl (or ethyl) imidazolium halide, benzimidazolium halide, benzothiazole halide and benzoselenide halide. The experimental results show that the alkyl group attached to the heterocyclic nitrogen atom in halogenated aromatic nitrogen heterocyclic ammonium salts plays a decisive role in the reaction behavior. When the heterocyclic quaternary nitrogen atom is attached to methyl, the double replacement product bis(*N*-methyl heterocyclic ammonium) decahydrodecaborate is often obtained. When ethyl is attached to the heterocyclic quaternary nitrogen atom, the mono replacement product (*N*-ethyl heterocyclic ammonium) tetraethyl ammonium decahydrodecaborate is obtained. The eight new compounds were all stable solids, and their molecular structures were confirmed by elemental analysis and IR spectra.

Bis(*N*-methyl) aromatic nitrogen heterocyclic ammonium salt of decahydrodecaborate:

(1) Bis(*N*-methyl) aromatic nitrogen heterocyclic ammonium salt of decahydrodecaborate: bis(*N*-methyl-2-aminopyridine) decahydrodecaborate, white solid, yield 97%, analysis: measured value (%): B, 32.27; N_2, 16.48; C, 42.29; H, 8.47; $C_{12}H_{28}B_{10}N_4$ calculated value (%): B, 32.13; N, 16.66; C, 42.83; H, 8.39, IR (cm^{-1}): ν_{N-H}, 3360 vs, 3300 vs, 3260 vs; ν_{B-H}, 2420 vs, 2350 vs, sh; δ_{N-H}, 1650 vvs; $\nu_{Aromatic\ ring\ skeleton}$, 1620 s, 1584 vvs, 1515 s; $\nu_{boron\ cage}$, 1070 w, 1010 s, br, 668 vs.

(2) Bis(1-3 methyl-3-aminopyridine) decahydrodecaborate, white flake crystal, yield 80%, analysis: measured value (%): B, 32.11; N, 16.56; $C_{12}H_{28}B_{10}N_4$ calculated value (%): B, 32.13; N, 16.6, IR (cm^{-1}): ν_{N-H}, 3430 vs, 3350 vs; ν_{B-H}, 2430 vvs, 2375 sh, δ_{N-H}, 1627 vs; $\nu_{Aromatic\ ring\ skeleton}$, 1590 vs, 1563 sh, 1515 vs; $\nu_{boron\ cage}$, 1071 w, 1032 m, 1019 s, 686 ms.

(3) Bis(1,2-dimethyl-3-ethyl-5,6-dichlorobenzimidazole) decahydrodecaborate, yellowish solid, yield 57%, analysis: measured value (%): B, 18.14; N, 8.82; C, 40.31; H, 5.98, $C_{22}H_{36}B_{10}C_{14}N_4$ calculated value (%): B, 17.83; N, 9.24; C, 43.57; H, 6.03, IR (cm^{-1}): ν_{B-H}, 2475 sh, 2445 vs, 2400 vs, 2360 sh; $\nu_{Aromatic\ ring\ skeleton}$, 1600 w, 1537 sh, 1528 s, 1510 sh; $\nu_{boron\ cage}$, 1063 m, 1030 sh, 1010 ms, 667 mw, ν_{C-Cl}, 659 m.

(4) Bis(1,2-dimethyl-6-methoxybenzoselenide) decahydrodecaborate, beige solid, yield 77%, analysis: measured value (%): B, 17.69; N, 4.34; $C_{18}H_{28}B_{10}N_2O_2Se_2$ calculated value (%): B, 18.00; N, 4.67, IR (cm^{-1}): $\nu_{C-H(OCH_3)}$, 2830 mw; ν_{B-H}, 2450 sh, 2400 vvs, 2350 sh; $\nu_{Aromatic\ ring\ skeleton}$, 1600 vs, 1580 ms; $\nu_{boron\ cage}$, 1086 vw, 1070 vw, 1020 sh, 1010 s, 676 w.

(5) Bis(1,2-dimethyl-5-chlorobenzothiazole) decahydrodecaborate, yellowish solid, yield 60%, analysis: measured value (%): B, 20.80; N, 5.32; calculation of $C_{18}H_{28}B_{10}Cl_2N_2S_2$ calculated value (%): B, 20.97; N, 5.44, IR (cm^{-1}): ν_{B-H}, 2445 s, sh, 2400 vs, 2350 sh; $\nu_{Aromatic\ ring\ skeleton}$, 1600 m, 1578 ms; $\nu_{boron\ cage}$, 1070 w, 1023 sh, 1010 s, 675 w.

Tetraethyl ammonium salt of aromatic nitrogen heterocyclic ammonium decahydrodecaborate:

(1) Tetraethyl ammonium decahydrodecaborate (1-ethyl-2-aminopyridine), white solid, yield 48%, analysis: measured value (%): B, 29.00; N, 11.52; C, 48.08; H, 11.56; $C_{15}H_{41}B_{10}N_3$ calculated value (%): B, 29.09; N, 11.31; C, 48.48; H, 11.12; IR (cm^{-1}): ν_{N-H}, 3380 vvs, 3310 vvs, 3270 s; δ_{B-H}, 2530 vs, sh, 2430 vvs, br, 2380 vs, sh; ν_{N-H}, 1660 vvs, br, $\nu_{Aromatic\ ring\ skeleton}$, 1628 vs, 1590 vs, 1522 s; $\nu_{boron\ cage}$, 1088 m, 1071 m, 1060 m, 1020 vs, 1000 s, sh, 676 mw.

(2) Tetraethyl ammonium decahydrodecaborate (1-ethyl-3-aminopyridine), white (yellowish) solid, yield 90%, analysis: measured value (%): B, 29.26; N, 10.86; C, 48.11; H, 11.10; $C_{15}H_{41}B_{10}N_3$ calculated value (%): B, 29.09; N, 11.31; C, 48.48; H, 11.12; IR (cm^{-1}): ν_{N-H}, 3440 s, 3345 s; ν_{B-H}, 2450 vs, br, 2370 vs, sh; $\delta_{N-H} + \nu_{Aromatic\ ring\ skeleton}$, 1632 s, 1620 m, sh, 1590 s, 1517 vs; $\nu_{boron\ cage}$, 1080 w, 1060 w, 1037 s, 1018 vs, 691 s.

(3) Tetraethyl ammonium decahydrodecaborate (1-ethylimidazole), white crystal, yield 93%, analysis: measured value (%): B, 30.96; N, 12.04; $C_{13}H_{39}B_{10}N_3$ calculated value (%): B, 31.28; N, 12.16; IR (cm^{-1}): $\nu_{N-H} + \gamma = C\text{–}H$, 3100 ms, br; ν_{B-H}, 2440 vvs, 2350 sh, $\nu_{Imidazole\ ring\ skeleton}$, 3100 ms, br, 1575 mw; $\nu_{boron\ cage}$, 1093 w, 1072 w, 1015 s, 670 w.

3.7 Effect of $B_{12}H_{12}[(C_2H_5)_4N]_2$ Particles on Combustion Performance of Al Simulation

The chemical equation and kinetic parameters of the reaction can be obtained by analyzing the products and reaction process of the thermal decomposition reaction of $B_{12}H_{12}[(C_2H_5)_4N]_2$ (BHN-12). On this basis, the chemical reaction process was studied by hydrodynamic method to explore the influence mechanism of $B_{12}H_{12}[(C_2H_5)_4N]_2$ thermal decomposition products on the reaction process of Al.

3.7.1 Theory and Simulation Method

3.7.1.1 Simulation Model Construction and Mesh Generation

A two-dimensional physical model is established by using Solidworks software.[41] The inlet is a quarter circular boundary with a radius of 0.5 m, and the outlet is a circular boundary with a radius of 6 m. The entrance radius of 0.5 m is selected because the pressure time history data calculated by AUTO-DYN shows that this point is less affected by the wall reflection, and the distance is moderate. It can be used as the boundary between near-field and long-range. A physical model diagram is shown in Figure 3.20.

The MESHING module of ANSYS is used to mesh the model. The tangential and circumferential directions are divided into 60 parts with equal distance. The radial direction adopts 120 parts of grids with a growth rate of 1.02 from the center to the periphery, a total of 7200 grids. Figure 3.21 shows the division of grids.

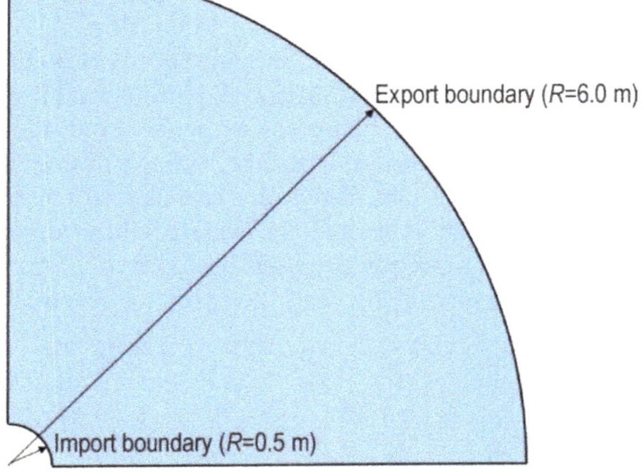

Figure 3.20 Physical model of the after-burning reaction. Reproduced from ref. 38.

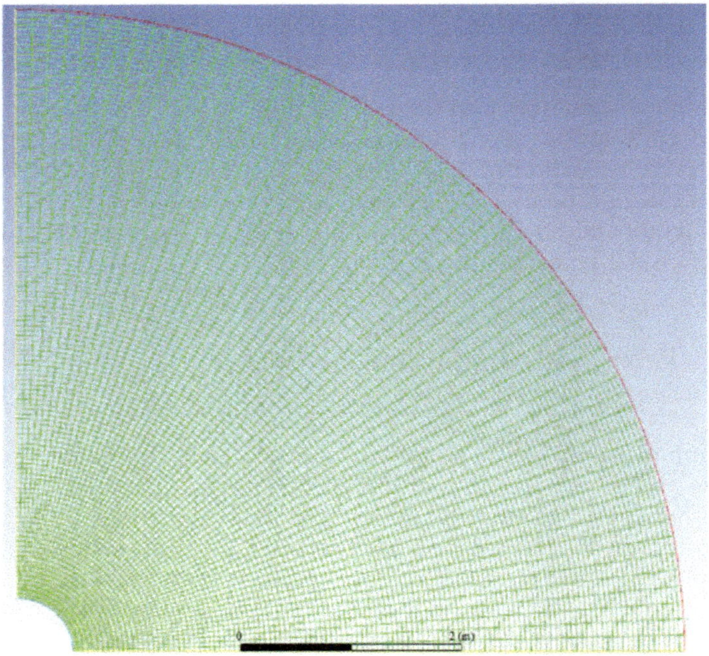

Figure 3.21 Grid of the model fluid calculation domain. Reproduced from ref. 38.

3.7.1.2 Boundary Conditions and Solution Control

It is difficult to simulate the chemical reaction process of aluminum powder in the flow field by using the conventional explosion simulation software. Therefore, in order to restore the explosion process of explosives, the pressure data at 0.5 m near the AUTO-DYN explosion point is used as the boundary condition of the inlet pressure. This distance is also used as the boundary position between the near field (explosion field) and the far field (flow field). The ground boundary is set as the wall. The pressure outlet is set to pressure out, the outlet atmospheric pressure is set to 101.325 kPa, the user-defined pressure function (UDF) is used for import, and the inlet pressure action time is 5.2 ms. The inlet boundary is set as the (pressure inlet) interface, and the direction is the normal phase of the curve. Due to the rapid pressure rise caused by the explosion reaction, the fitting accuracy of the explosion pressure field cannot be achieved by using a simple function. Therefore, the pressure data at 0.5 m is divided into four segments for polynomial fitting, and the pressure expression is:

$$p=\begin{cases} 101.332 & 0\leq t\leq 0.12269 \text{ ms} \\ b_1X^8+b_2X^7+\ldots+b_8X+b_9 & 0.12269<t\leq 0.13573 \text{ ms} \\ c_1X^8+c_2X^7+\ldots+c_8X+c_9 & 0.13573<t\leq 0.30147 \text{ ms} \\ d_1X^9+d_2X^8+\ldots+d_8X+d_9 & 0.30147<t\leq 5.2 \text{ ms} \end{cases} \tag{3.3}$$

where, the values of fitting polynomials are shown in Table 3.8.

Table 3.8 Piecewise fitting of the coefficients of explosion pressure. Reproduced from ref. 38.

b_i (i = 1–9)	b_1	b_2	b_3	b_4	b_5
	-6.10443×10^{21}	6.30105×10^{21}	-2.84487×10^{21}	7.33804×10^{20}	-1.18272×10^{20}
	b_6	b_7	b_8	b_9	
	1.21975×10^{19}	-7.86036×10^{17}	2.89389×10^{16}	-4.66022×10^{14}	
c_i (i = 1–9)	c_1	c_2	c_3	c_4	c_5
	-5.16363×10^{12}	9.64971×10^{12}	-7.82198×10^{12}	3.58995×10^{12}	-1.01967×10^{12}
	c_6	c_7	c_8	c_9	
	1.83399×10^{11}	-2.03806×10^{2}	1.27808×10^{10}	-3.45784×10^{7}	
d_i (i = 1–9)	d_1	d_2	d_3	d_4	d_5
	1.36653×10^{8}	−31.68522	304.14845	1560.62231	4614.04934
	d_6	d_7	d_8	d_9	
	−7904.41808	7498.09833	−3494.28755	620.4646	

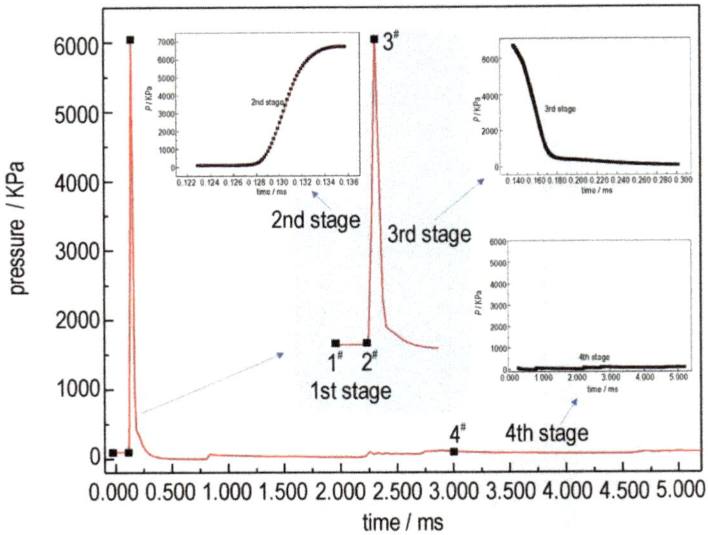

Figure 3.22 Explosion pressure–time curves and its polynomial fitting results. Reproduced from ref. 38.

Figure 3.22 shows the explosion pressure time history curve and fitted function curves. The first section is atmospheric pressure, so no fitting is made for this section. It can be seen that the function curve fits well with the explosion pressure field data, which can meet the requirements for calculation accuracy.

The DPM model and air phase coupling calculation method are adopted, and the solid aluminum powder particles in the fuel as discrete phase are calculated. The average particle size of $B_{12}H_{12}[(C_2H_5)_4N]_2$ fuel is set as 57.8 μm, the density is 930 kg m^3, the heat capacity is 1000 J K^{-1}, the mass diffusion coefficient is 5×10^{-12}. The simple algorithm is used to solve the discretized momentum equation, and the iteration time step is 5×10^{-6} s, the total number of iteration time steps is 20 000, each time step is iterated 10 times, and the dispersion process of fuel within 20 ms is calculated.

3.7.1.3 Turbulence Model and Discrete Phase Model Setup

For the k–ε proposed by Spalding and Lauder,[42] the model has high stability, economy and high calculation accuracy. The SST k–ω model is selected, the transverse dissipation derivative term is added to the model, and the transport process of turbulent shear stress is considered in the definition of turbulent viscosity. Import the grid into Fluent software, set it as transient, add C_2H_6, C_2H_4, H_2, NH_3 to the fluid components, and add Al, Al_2O_3, $B_{12}H_{12}[(C_2H_5)_4N]_2$ to the air and solid components (the components are determined by the decomposition products later). The reaction type of particles is set to the particle surface type.

3.7.1.4 Selection of Physical Parameters and Determination of the Reaction Equation

The reaction between aluminum powder and air is simulated by the total package reaction model $Al + 0.75O_2 \rightarrow 0.5Al_2O_3$. The eddy-dissipation model is selected to simulate the interaction between chemical reactions in turbulent flow. According to the analysis results of the $B_{12}H_{12}[(C_2H_5)_4N]_2$ thermal decomposition mechanism, the $B_{12}H_{12}[(C_2H_5)_4N]_2$ thermal decomposition reaction can be expressed as a one-step reaction, as shown in formula (3.4):

$$B_{12}H_{12}\left[(C_2H_5)_4N\right]_2 \rightarrow 13H_2\uparrow + 2C_2H_6\uparrow + 2C_2H_4\uparrow + 2NH_3\uparrow + 8C + 12B \tag{3.4}$$

Due to the high reaction temperature of C and B, the reaction between C and B and oxygen can be ignored here. Assuming that the gas produced by $B_{12}H_{12}[(C_2H_5)_4N]_2$ in the post combustion reaction can completely react with oxygen, the following reactions will occur in the post combustion reaction process in addition to the combustion of Al fuel:

$$H_2 + 0.5O_2 \rightarrow H_2O \tag{3.5}$$

$$C_2H_4 + 3O_2 \rightarrow 2CO_2 + 2H_2O \tag{3.6}$$

$$C_2H_6 + 3.5O_2 \rightarrow 2CO_2 + 3H_2O \tag{3.7}$$

$$2NH_3 + 1.5O_2 \rightarrow N_2 + 3H_2O \tag{3.8}$$

Add reactants from the material database provided by Fluent software. The types of gas phase include O_2, N_2, CO_2, Co, CH_4, NH_3, H_2, C_2H_6, C_2H_4. The types of solid phases include Al_2O_3, C, B, $B_{12}H_{12}[(C_2H_5)_4N]_2$, Al. The pre-exponential factor and activation energy can be calculated from the kinetic parameters, and the mass diffusion coefficient and other reaction parameters are input. The component model is volumetric volume reaction, and the mixed material is methane–air-2step. Add chemical reactions (3.4)–(3.8). Set the injection mass flow of $B_{12}H_{12}[(C_2H_5)_4N]_2$ and Al as 10 $kg\,s^{-1}$, the molar mass of Al particles as 27 $kg\,kmol^{-1}$, the density as 2719 $kg\,m^{-3}$, the specific heat as 871 $J\,Kg^{-1}\,k^{-1}$, and the standard enthalpy as 1.84474×10^7 $J\,kmol^{-1}$, and thermal conductivity as 202.4 $W\,m^{-1}\,K^{-1}$. The reaction type is chemical reaction on the particle surface, the mass diffusion coefficient is 5×10^{-12}, the pre-exponential factor is 6.7 s^{-1}, and the activation energy is 1.138×10^8 $J\,kg\,mol^{-1}$.[43,44] The molar mass of the combustion product of Al_2O_3 is 102 $kg\,kmol^{-1}$, the density is 2000 $kg\,m^{-3}$, and the standard enthalpy is 1.6757×10^{12} $J\,kmol^{-1}$. The granularity of $B_{12}H_{12}[(C_2H_5)_4N]_2$ is set to 57.8 μm. The density is 960 $kg\,m^{-3}$, the heat capacity is 1000 $kg\,m^{-3}$, and the mass diffusion coefficient is 5×10^{-12}.

Based on the above selected calculation model, the turbulence equation, the conservation equation and the user-defined parameter equation are solved by the SIMPLE algorithm.[45] The time step is set to 0.0001 s and the number of calculation steps is 10 000.

3.7.2 Analysis of Pressure Field and Temperature Field

The pressure *vs.* time curves of the measuring point at the position 1–5 m away from the explosion center are shown in Figure 3.23. It can be seen that the pressure peak decreases gradually with the increase of the distance from the explosion center, and the peak width of the pressure peak increases significantly, indicating that the afterburning effect occurred. This is mainly because the aerobic chemical reaction of aluminum powder forms a continuous expansion wave. At the same time, borohydride fuel undergoes a decomposition reaction in the post combustion reaction to produce a large number of combustible gases. The generation of gases can reduce the pressure attenuation process of the system in the post combustion reaction, further reduce the attenuation degree of the secondary pressure peak, and prolong the damage time of the shock wave of thermobaric explosives.

Figure 3.24 shows the cloud chart of pressure *vs.* time at different times. It can be seen that, with the extension of time, the reaction boundary gradually diffuses, and the pressure in the central area of the explosion reaction process forms a negative pressure (blue cloud chart). This is because the combustion chemical reaction in the flow field is occurring, and the borohydride fuel is decomposed in the process of dispersion. Their decomposition can produce a large number of combustible gases to react with oxygen, and the air is compressed under high temperature and high pressure to produce shock waves. With the passage of time, the shock wave gradually attenuates in the propagation process. When it is reduced to atmospheric pressure, the inertia of air particles continues to decrease to a negative value. At this time, the intensity of chemical reaction decreases.

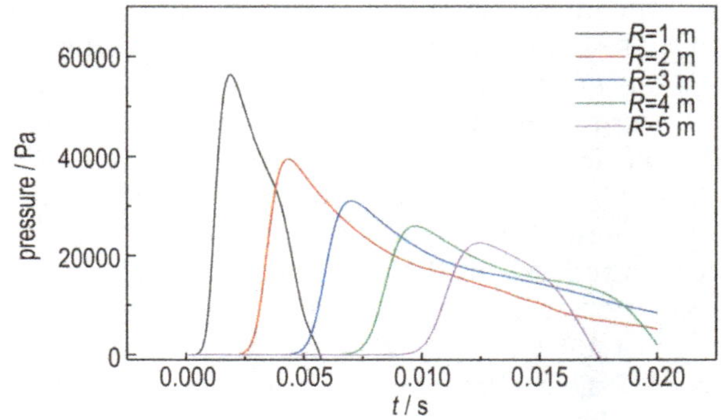

Figure 3.23 Time *vs.* pressure curves of explosion field at different positions. Reproduced from ref. 38.

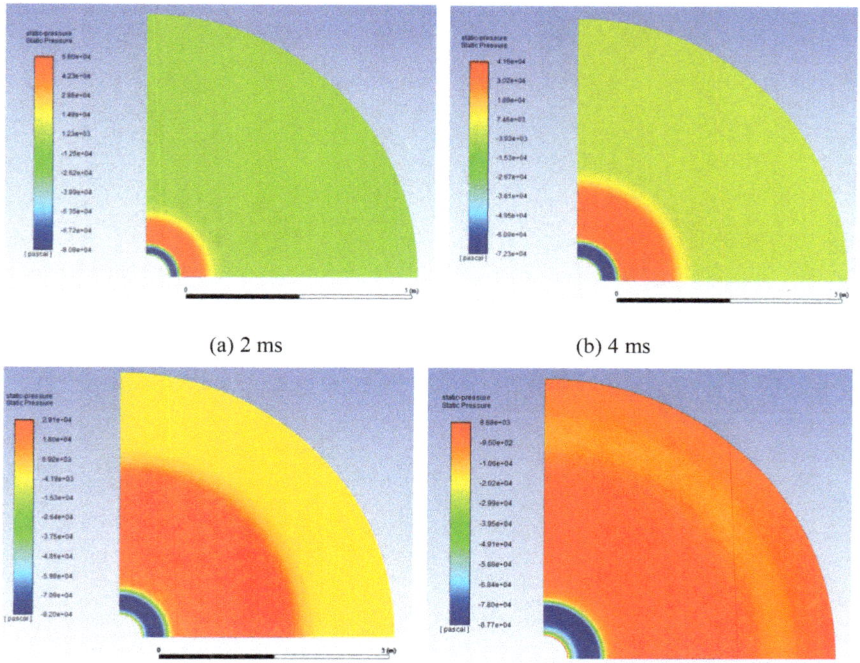

(a) 2 ms (b) 4 ms

(c) 8 ms (d) 20 ms

Figure 3.24 Cloud chart of pressure–time at different times. Reproduced from ref. 38.

It can be seen from Figure 3.24d that, with the passage of time, the gas outside the flow field is replenished again due to pressure, and eventually tends to stabilize near the atmospheric pressure.

The cloud diagrams of the temperature distribution at different times are shown in Figure 3.25. It can be seen from Figure 3.25a that aluminum powder begins to react at the initial time. This is due to the formation of high-temperature products after the explosion of the main explosive. At the position 0.5 m away from the explosion point, high-temperature and high-pressure products are generated at the position of the inlet boundary conditions calculated by fluent software. Aluminum powder reacts with the air in the surrounding environment to produce an oxidation–reduction reaction under the heat transfer of high-temperature products. In addition, $B_{12}H_{12}[(C_2H_5)_4N]_2$ decomposes under high temperature to produce combustible gases, which will also react with the surrounding air under high temperature to generate a large amount of chemical heat, thus raising the temperature of the air medium inside and around the fireball. When the afterburning reaction occurs, the high-temperature fireball continues to expand radially under the action of the shock wave, forming a large high-temperature reaction field. According to Figure 3.25d, the temperature of the

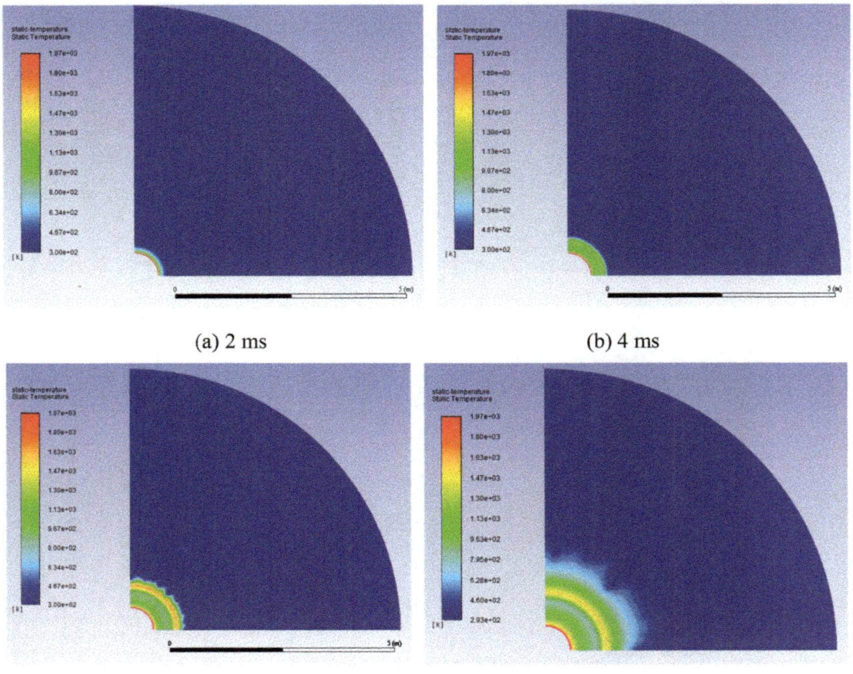

(a) 2 ms (b) 4 ms

(c) 8 ms (d) 20 ms

Figure 3.25 Cloud chart of temperature versus time. Reproduced from ref. 38.

reaction center is about 1800 °C, which is about 300 °C higher than the surrounding temperature. This is the result of the reaction between the gas products decomposed by $B_{12}H_{12}[(C_2H_5)_4N]_2$ and the surrounding oxygen. Compared with conventional high explosives, the fireball temperature of the warm pressure explosion has a relatively high temperature inside the fireball in addition to the explosion center, and the decomposition of $B_{12}H_{12}[(C_2H_5)_4N]_2$ can further promote the after-burning of Al; thus, the action range of the fireball is further enlarged and the reaction time is prolonged, which has potential application prospects.

3.7.3 Concentration Analysis of Reactants and Products

By tracing the product components, the concentration information of fuel particles in the post combustion reaction process can be obtained, and the generation time and distribution of gas products can be obtained. The reaction model can better simulate the post combustion reaction of $B_{12}H_{12}[(C_2H_5)_4N]_2$ fuel in the explosion flow field. The concentration distribution of reactants and products in the post combustion reaction is shown in Figure 3.26. It can be seen from Figure 3.26a that the mass concentration of $B_{12}H_{12}[(C_2H_5)_4N]_2$ in the dispersion process is 0.072–0.179 kg m^{-3}, and the

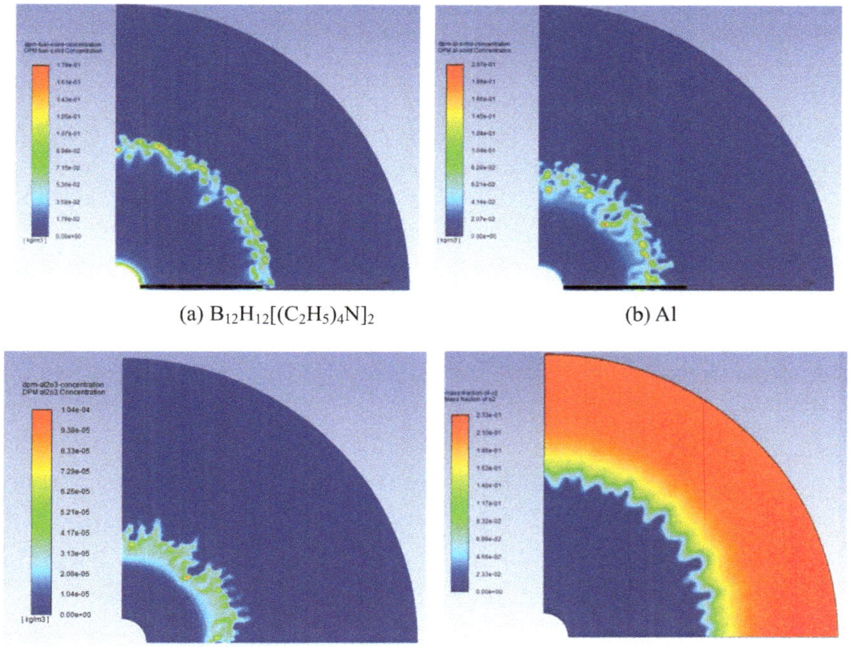

(a) $B_{12}H_{12}[(C_2H_5)_4N]_2$ (b) Al

(c) Al_2O_3 (d) O_2

Figure 3.26 Concentration cloud chart of $B_{12}H_{12}[(C_2H_5)_4N]_2$, Al, Al_2O_3, and O_2 at 20 ms. Reproduced from ref. 38.

product is dispersed outward in an arc. By comparing the dispersion results of $B_{12}H_{12}[(C_2H_5)_4N]_2$ and Al in Figure 3.26a and b, it can be seen that the dispersion speed of $B_{12}H_{12}[(C_2H_5)_4N]_2$ is higher than that of Al. The average weight of $B_{12}H_{12}[(C_2H_5)_4N]_2$ particles set in the calculation model is about 1.09×10^{-10} g ($d = 57.8$ μm, $\rho = 930$ kg m^{-3}), which is higher than the average weight of Al particles by 1.14×10^{-11} g ($d = 10$ μm, $\rho = 2719$ kg m^{-3}), which is consistent with ref. 42 that the particle dispersion velocity increases with an increase in particle mass fraction, indicating that the calculation results are relatively accurate. At 20 ms, the dispersion radius of Al is 2.5 m, and that of $B_{12}H_{12}[(C_2H_5)_4N]_2$ is 3 m. It can be seen from Figures 3.26c and d that Al_2O_3 was generated by a chemical reaction during the dispersion of Al, and the oxygen depletion zone was formed in the central region due to the reaction.

Since the decomposition gas products of $B_{12}H_{12}[(C_2H_5)_4N]_2$ include H_2, CH_4, C_2H_4, and NH_3, the cloud chart results of the gas products in the post combustion reaction are relatively similar, so only H_2 as a representative of gas products is discussed. The cloud diagram of H_2 distribution at different times is shown in Figure 3.27. It can be seen that when the time of post combustion reaction reaches 4 ms, the gas product appears, indicating that $B_{12}H_{12}[(C_2H_5)_4N]_2$ has undergone a decomposition reaction in 4 ms to produce

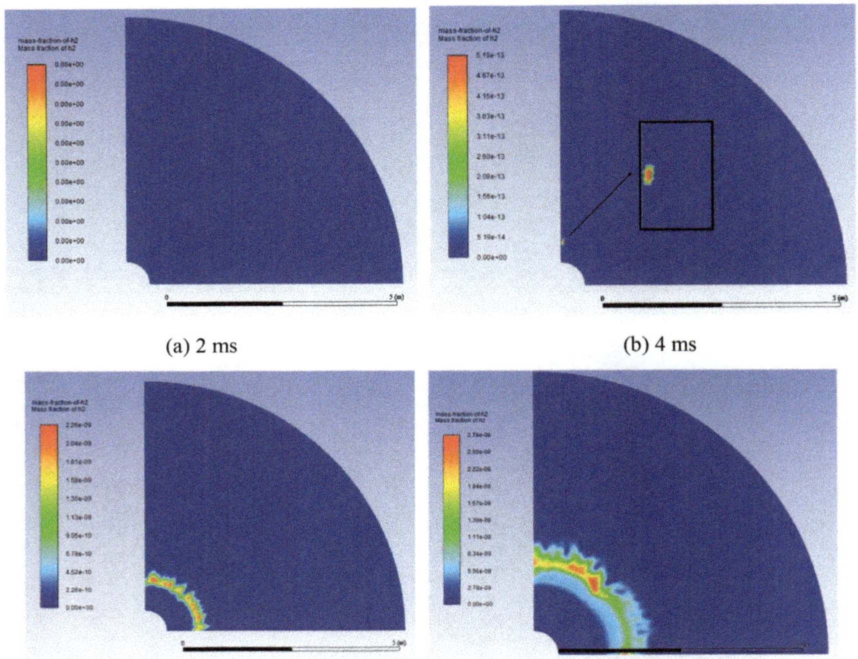

(a) 2 ms (b) 4 ms

(c) 8 ms (d) 20 ms

Figure 3.27 Cloud chart of H_2 concentration *vs.* time. Reproduced from ref. 38.

combustible gas. The gas product can further react with oxygen under high temperature conditions to improve the reaction temperature of the system.

3.8 Ignition and Combustion Features of Tetramethylammonium Octadecahydro-eicosaborate ($B_{20}H_{18}[(CH_3)_4N]_2$, $(Me_4N)_2B_{20}H_{18}$)

The closed boron cage ionic compounds in the ionic borohydride compounds, such as $B_{10}H_{10}{}^{2-}$ and $B_{12}H_{12}{}^{2-}$, have attracted the attention of researchers due to their advantages of good water vapor stability, thermal stability, low toxicity, *etc.*[27,34] It has been reported that these compounds can be oxidized and coupled to form polymers. For example, $(B_{10}H_{10})^{2-}$ can be coupled to obtain dimer anions $B_{20}H_{18}{}^{2-}$ and $B_{20}H_{19}{}^{-}$, and they have prospects in high-energy fuel applications.[46–48]

3.8.1 Characterization and Analysis of Compositions

The compound of $(Me_4N)_2B_{20}H_{18}$ can be synthesized by aqueous-phase oxidation from tetraethyl ammonium decaborate $(Et_4N)_2B_{10}H_{10}$ and ferric chloride

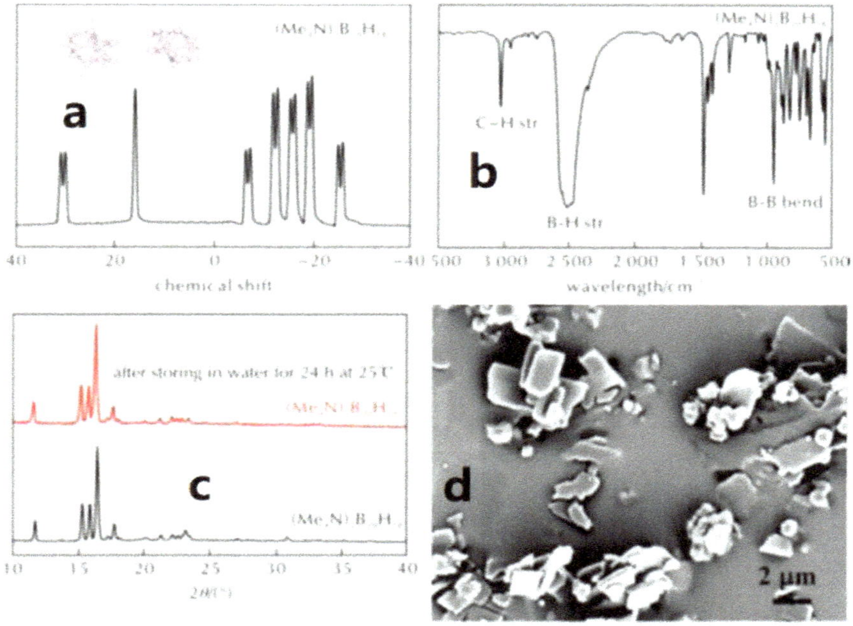

Figure 3.28 ^{11}B NMR spectrum, IR spectrum, XRD patterns and SEM image of $(Me_4N)_2B_{20}H_{18}$. Reproduced from ref. 50.

($FeCl_3$) as the raw materials; a detailed synthesis process can be found in ref. 49. Figure 3.28 shows the ^{11}B NMR spectrum, IR spectrum, XRD patterns and SEM image of $(Me_4N)_2B_{20}H_{18}$. It can be found that the ^{11}B NMR of $(Me_4N)_2B_{20}H_{18}$ prepared has 31.07 (d, 2B), 16.03 (s, 2B), − 6.75 (d, 2B), − 12.46 (d, 4B), − 15.81 (d, 4B), − 19.25 (d, 4B), − 25.49 ppm (d, 2B) characteristic peaks, and there are no other obvious impurity peaks in the ^{11}B NMR spectrum (Figure 3.28a). The large and wide characteristic peak observed at 2500 cm^{-1} is the stretching vibration peak of the B–H bond. The characteristic peaks of 1020 cm^{-1} and 700 cm^{-1} correspond to the vibration absorption peak of B–B in the boron cage. In addition, there is a stretching vibration peak of the C–H bond near 3000 cm^{-1} (Figure 3.28b). The prepared sample has good crystallinity. After placing the sample in water at room temperature for one day, the measured XRD spectrum is consistent with the original sample, proving that $(Me_4N)_2B_{20}H_{18}$ has good water stability (Figure 3.28c). The prepared powder is of flaky structure with particle size of 2–5 μm. The density of the prepared $(Me_4N)_2B_{20}H_{18}$ is 0.96 g cm^{-3} (Figure 3.28d).

3.8.2 Thermal Decomposition Analysis

The thermogravimetric test results of $(Me_4N)_2B_{20}H_{18}$ in nitrogen and air atmosphere are shown in Figure 3.29. It can be found that the sample prepared has good thermal stability in nitrogen atmosphere at the heating rate

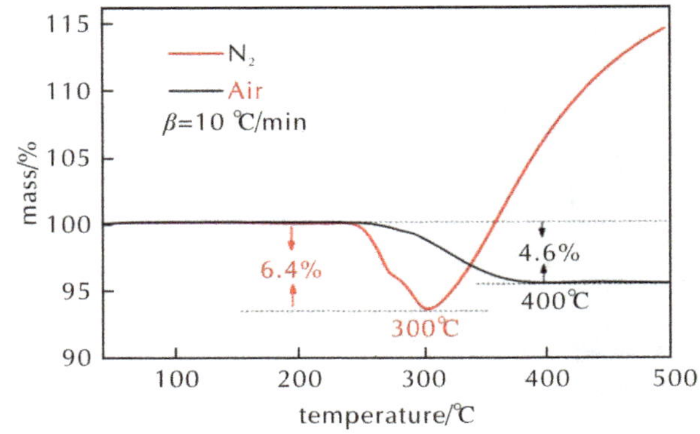

Figure 3.29 TG curves of $(Me_4N)_2B_{20}H_{18}$ in N_2 and air at the heating rate of 10 °C min^{-1}. Reproduced from ref. 50.

of 10 °C min^{-1}. There is no obvious mass change before 250 °C, and it begins to slowly thermal decompose and lose weight when the temperature is higher than 250 °C, and its weight loss is 4.6% over the temperature range 250–400 °C. Above 400 °C, the sample has no weight loss. In addition, the pyrolysis solid product obtained after heating to 500 °C in nitrogen is black powder. $(Me_4N)_2B_{20}H_{18}$ also has good thermal stability in air. There is no obvious mass change when it is lower than 240 °C, and its weight decrease over the temperature range 240–300 °C is 6.4%. When the temperature is higher than 300 °C, the sample weight continues to increase, showing a different behavior from that in N_2.

Figure 3.30 shows the IR spectrum and XRD pattern for the thermal decomposition of $(Me_4N)_2B_{20}H_{18}$ in N_2. It can be shown that the C–H stretching vibration peak at 3000 cm^{-1} of the raw material completely disappeared, and the intensity of the B–H stretching vibration peak at 2500 cm^{-1} decreased significantly after pyrolysis (Figure 3.30a). After pyrolysis in N_2 the characteristic diffraction peak of $(Me_4N)_2B_{20}H_{18}$ completely disappeared and the product was amorphous (Figure 3.30b).

3.8.3 Ignition and Combustion Characteristics

It was seen that the measured mass calorific value of $(Me_4N)_2B_{20}H_{18}$ is 55.5 MJ kg^{-1} by using an oxygen bomb calorimeter, and white products were generated after combustion. At the same time, the substance was easily ignited by an open fire in air.[51] In order to quantitatively determine its ignition and combustion characteristics, hot air at 400 °C was used to ignite the sample. The combustion image and the surface temperature of the sample during combustion were taken by high-speed camera and infrared thermal imager, respectively. Figure 3.31 shows the combustion flame images of $(Me_4N)_2B_{20}H_{18}$ in air. It can be seen that the sample can be successfully ignited with 400 °C hot air. When the sample ignited, the flame expanded rapidly, indicating that the substance had high activity. During the

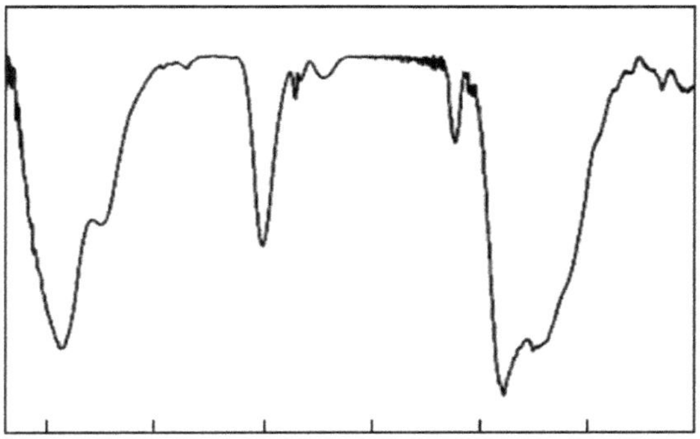

(a) IR spectrum

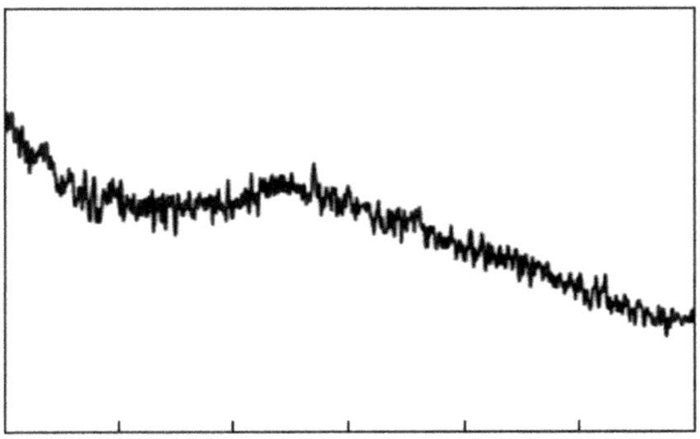

(b) XRD pattern

Figure 3.30 IR spectrum and XRD pattern for the thermal decomposition of $(Me_4N)_2B_{20}H_{18}$ in N_2. Reproduced from ref. 50.

combustion process, the combustion flame is green, which is the characteristic combustion flame color of boron. At the same time, there were bright yellow particles floating up, maybe because the sample powder was blown up. In addition, the sample has a high burning rate. After 0.6 s, the main body of the sample burned and then the combustion flame gradually disappeared; the whole process took about 1.1 s.

Figure 3.32 shows the maximum surface temperature of $(Me_4N)_2B_{20}H_{18}$ particles during ignition and combustion process. It can be found that, under the action of hot air, the thermal oxidation surface temperature of the sample rises after a short induction period. After a period of time, the sample body ignites and burns, and the surface temperature rises rapidly to the maximum temperature of 1100 °C. During the combustion process, the surface temperature of the sample was basically stable, indicating that the

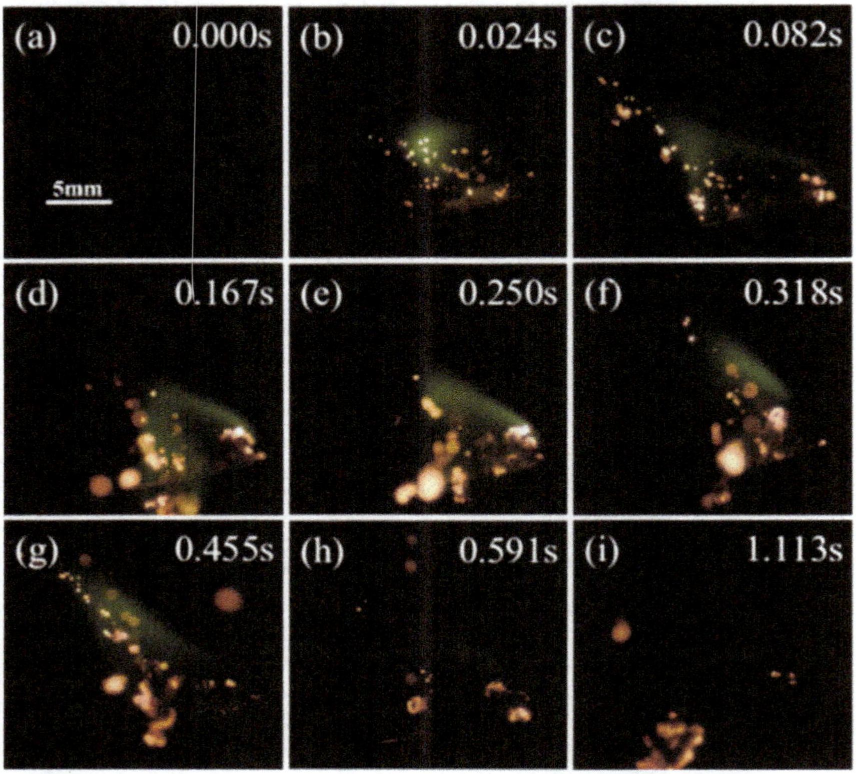

Figure 3.31 The combustion flame images of $(Me_4N)_2B_{20}H_{18}$ in air. Reproduced from ref. 50.

combustion of the sample is relatively stable. The main combustion lasted for 1–2 s, which was basically consistent with the observation results of the combustion flame evolution process. In addition, after the sample was completely burned, its surface temperature decreased rapidly.

Unlike the large amount of white product produced by the combustion of ab oxygen bomb in a pure oxygen environment, $(Me_4N)_2B_{20}H_{18}$ is burnt in air to obtain black powder, which may be caused by the insufficient combustion of the sample in air. Figure 3.33 shows the IR spectrum and XRD pattern for the combustion products of $(Me_4N)_2B_{20}H_{18}$ in air. After sample combustion in air, the B–H characteristic peak of $(Me_4N)_2B_{20}H_{18}$ at 2500 cm^{-1} was significantly weakened, but did not completely disappear. The C–H stretching vibration peak at 3000 cm^{-1} basically disappeared. At the same time, the characteristic broad peak of B–O was generated near 1350 cm^{-1} (Figure 3.33a). After sample combustion, the characteristic diffraction peak of $(Me_4N)_2B_{20}H_{18}$ completely disappeared, and the combustion product contained unknown crystalline substances, and it was speculated that the B–H and tetramethyl cations of

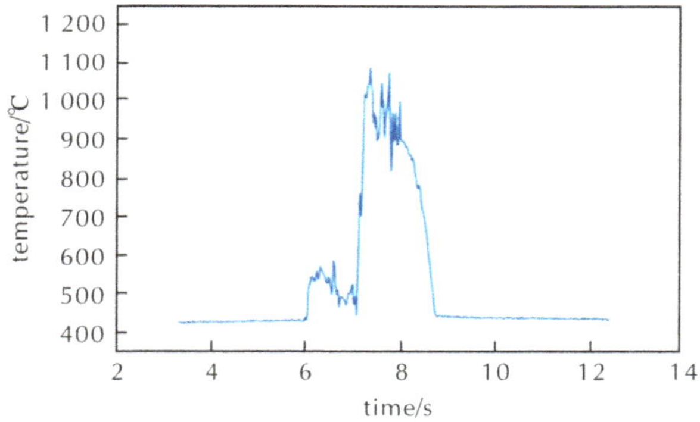

Figure 3.32 The surface temperature for the combustion process of $(Me_4N)_2B_{20}H_{18}$ in air. Reproduced from ref. 50.

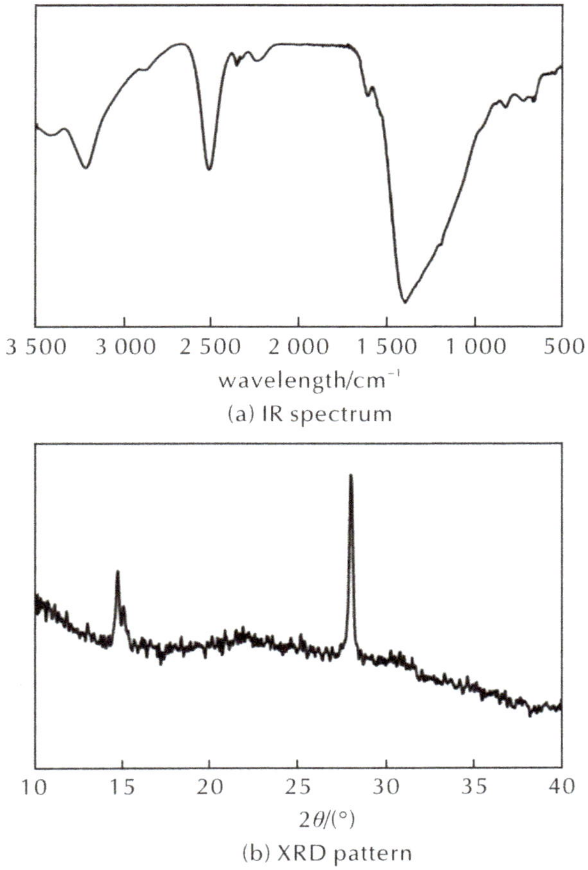

Figure 3.33 IR spectrum and XRD pattern for the combustion products of $(Me_4N)_2B_{20}H_{18}$ in air. Reproduced from ref. 50.

$(Me_4N)_2B_{20}H_{18}$ participated in the combustion reaction and generated boron oxides during the combustion process in air (Figure 3.33b).

To summarize, when $(Me_4N)_2B_{20}H_{18}$ is ignited and burned at high temperature in air, the pyrolysis occurs first, the solid products of pyrolysis oxidize rapidly, and the surface temperature rises. When the temperature rises to a certain degree, the sample main body can be ignited. The gas and solid thermal decomposed products participate in the combustion, and the combustion of borohydride group produces a characteristic green combustion flame during the sample combustion process.

3.9 Conclusions

(1) The systems of BHN/NTO-Pb, BHN/CL-20, BHN/Mg, and BHN/Al powder binary mixtures have good compatibility, the BHN/GAP and BHN/HMX binary mixtures have moderate compatibility, and the BHN/RDX, BHN/DNTF and BHN/GUDN binary mixtures have bad compatibility according to the evaluated standard of compatibility.
(2) BHN/PET, BHN/PEG, BHN/N-100, BHN/HTPB, BHN/AD-Cu, BHN/β-Cu, BHN/φ-Pb, BHN/C.B., BHN/Al_2O_3, BHN/C_2, BHN/DOS, and BHN/KP mixtures are compatible.
(3) There was no consequential affiliation between sensitivity and compatibility, whereas, if the energetic materials can react with each other when they are mixed, they must be incompatible and very sensitive.

References

1. W. P. C. de Klerk, M. A. Schrader and A. C. van der Steen, Compatibility testing of energetic materials, which techniques?, *J. Therm. Anal. Calorim.*, 1999, **56**, 1123–1131.
2. S. Q. Tang and H. X. Ding, Boron hydrides as solid propellant burning rate modifiers of the latest progress, *J. Propul. Tech.*, 1983, **4**, 35.
3. M. Golfier, H. Graindorge, Y. Longevialle and H. Mace, New energetic molecules and their applications in energetic materials, 29th Int. Annual Conference of ICT, Karlsruhe, Germany, June 30–July 3, 1998, pp. 3/1.
4. Y. Y. Zheng and M. Z. Wang, High burning rate modifiers-polyhedral boron hydride, *Chin. J. Explos. Propellants*, 1989, **12**, 32 (in Chinese).
5. L. Jizhen, F. Xuezhong and F. Xiping, *et al.*, Compatibility study of 1,3,3-trinitroazetidine with some energetic components and inert materials, *J. Therm. Anal. Calorim.*, 2006, **85**, 779–784.
6. Q.-L. Yan, X.-J. Li and L.-Y. Zhang, *et al.*, Compatibility study of trans-1,4,5,8-Tetranitro-1,4,5,8-tetraazadecalin (TNAD) with some energetic components and inert materials, *J. Hazard. Mater.*, 2008, **34**, 529–534.
7. Z. R. Liu, C. M. Yin, C. H. Y. Wu and M. N. Chang, The characteristic temperature method to estimate kinetic parameters from DTA curves and evaluate the compatibility of explosives, *Propellants, Explos., Pyrotech.*, 1986, **11**, 10–15.

8. S. Zeman, Some predictions in the field of the physical thermal stability of nitramines, *Thermochim. Acta*, 1997, **302**, 11–16.
9. S. Zeman, On the physical thermal stability of some cyclic nitramines. 21st International Pyrotechnics Seminar, Moscow, Russia, September 11–15, 1995, pp. 940–947.
10. D. Power and T. R. Spalding, Some Comments on the Mechanism of the Preparation of $(Et_4N)_2B_{10}H_{10}$ by Thermolysis of Et_4NBH_4, *Polyhedron*, 1985, **4**, 1329.
11. J. M. Makhlouf and G. T. Hefferman, *Method of preparing tetraethylammonium decahydrodecaborates*, US3373202, Callery Chemical Company, Pittsburgh, PA, USA, 1968.
12. E. L. M. Krabbendam-La Haye, W. P. C. de Klerk, M. Miszczak and J. Szymanowski, Compatibility testing of energetic materials at TNO-PML and MIAT, *J. Therm. Anal. Calorim.*, 2003, **72**, 931–936.
13. J. A. F. F. Rocco, J. E. S. Lima and A. G. Frutuoso, *et al.*, Thermal degradation of a composite solid propellant examined by DSC, *J. Therm. Anal. Calorim.*, 2004, **75**, 551–557.
14. P. R. Patil, V. N. Krishnamurthy and S. S. Joshi, Differential scanning calorimetric study of HTPB-based composite propellants in presence of nano-ferric oxide, *Propellants, Explos., Pyrotech.*, 2006, **31**, 442–446.
15. R. Sivabalan, M. B. Talawar and N. Senthilkumar, *et al.*, Studies on azotetrazolate-based high nitrogen content high energy materials potential additives for rocket propellants, *J. Therm. Anal. Calorim.*, 2004, **78**, 781–786.
16. W. P. C. de Klerk, A. E. D. M. van der Heijden and W. H. M. Veltmans, Thermal analysis of the high energetic material HNF, *J. Therm. Anal. Calorim.*, 2001, **64**, 973–985.
17. N. E. Beach and V. K. Canfield, *Compatibility of Explosives with Polymers (III)*, Plastics Technical Evaluation Center Picatinny Arsenal, NJ, USA, 1971, DOI: AD-721004.
18. H. H. Cady, *Evaluation of Ferrocyanide/Nitrate Explosive Hazard*, Los Alamos National Laboratory, Los Alamos, NM, USA, LA-12589-MS, UC-900 and UC-906, 1992, pp. 4–9.
19. Y. N. Xue, Y. N. Li and W. Q. Wang, *et al.*, Synthesis and Thermal Properties of High Burning Rate Modifiers (Bis-)Tetraethylammonium Decahydrodecaborate, Fifth National Conference on Chemical Propellants. Chinese Chemical Society, Dalian, P. R. China, September 8–9, 2011.
20. S.-Y. Xu, F.-Q. Zhao and S.-W. Li, *et al.*, Impact and friction sensitivity of composite modified double base propellant containing hexanitrohexaazaisowurtzitane (CL-20), *J. Propul. Tech.*, 2006, **27**, 182–186.
21. W. Q. Pang, Y. N. Xue and X. Z. Fan, *et al.*, Thermal behavior of tetraethylammonium dodecahydrododecaborates (BHN) and its compatibility with main components of propellant, *Chin. J. Energ. Mater.*, 2012, **20**(3), 280–285.
22. Z. R. Liu, *Thermal decomposition of energetic materials*, National Defense Industry Press, Beijing, 2008.

23. P. Weiqiang, X. Huixiang and L. Linquan, *et al.*, Compatibility of high energetic boron-hydrogenincendiary agent with some common components in solid propellant by means of DSC method, *J. Solid Rocket Technol.*, 2013, **36**(1), 67–73.
24. M. Farber and R. D. Srivastava, *Thermal decomposition of HMX*, 16th JANNAF Combustion Meeting, 308, CPIA Pub., 1979, pp. 723–729.
25. L. Wei, Q.-L. Wang and S.-W. Liu, Application of High Energy Density Compounds CL-20, DNTF and ADN in High Energy Propellant, *Chin. J. Explos. Propellants*, 2009, **32**, 17–20 (in Chinese).
26. L. Yanhua, S. Zixing and S. Lili, *et al.*, Compatibility of dodecahydrododecaborate with propellants ingredients, *Chin. J. Explos. Propellants*, 2017, **40**(6), 95–100.
27. P. Weiqiang, X. Yunna and F. Xuezhong, *et al.*, Thermal behavior of tetraethylammonium dodecahydrododecaborates (BHN) and its compatibility with main components of propellant, *Chin. J. Energ. Mater.*, 2012, **20**(3), 280–285.
28. J. Fan, W. Xiaofeng and N. Yulei, *et al.*, Stability of BHN-10 under explosion and electroc detonation shock waves, *Chin. J. Explos. Propellants*, 2019, **42**(6), 631–636.
29. E. L. Muetterties, J. H. Balthis and Y. T. Chia, *et al.*, Chemistry of boranes. VIII. salts and acids of $B_{10}H_{10}{}^{2-}$ and $B_{12}H_{12}{}^{2-}$, *Inorganic Chemistry*, 1964, **3**(3), 444–451.
30. P. Mingjing, M. Genwang and W. Wane, *et al.*, On energy releasing efficiency of thermobaric fuel containing aluminum in dispersing with explosion, *Chin. J. Univ. Sci. Technol. China*, 2007, **37**(3), 276–283.
31. S. Zixing, S. Lili and Y. Rongjie, Research progress of icosahedronal polyhedroborate $B_{12}H_{12}{}^{2-}$ anion compounds and their application in propellants and explosives, *Chin. J. Explos. Propellants*, 2017, **40**(3), 1–16.
32. M. H. Rao and K. Muralidharank, Closo-dodecaborate $(B_{12}H_{12})^{2-}$ salts with nitrogen based cations and their energetic properties, *Polyhedron*, 2016, **115**, 105–110.
33. I. A. Ellis, R. Schaeffer and D. F. Gaines, A convenient preparation of $(B_{12}H_{12})^{2-}$ salts, *J. Am. Chem. Soc.*, 1963, **85**(23), 3885–3886.
34. P. Xinxin, H. Xuefeng and L. Shengji, *et al.*, Ignition and Combustion Characteristics of Tetraethylammonium Dodecahydrododecaborates, *Chin. J. Explos. Propellants*, 2017, **40**(3), 1–16.
35. S.-Q. Tang and H.-X. Ding, New Evolution of boron hydrogen compounds as high burning rate additives for solid propellant, *Winged Missile*, 1972, **5**, 35–51.
36. E.-X. Tian, A new type of organic boron compound - carbon boron alkyl, *Sci. Tong Bao*, 1964, **10**, 22–28.
37. X. Huang, S. Li and X. Zheng, *et al.*, Combustion mechanism of a novel energetic fuel candidate based on amine metal borohydrides, *Energy Fuels*, 2016, **30**(2), 1383–1389.

38. J. Fan, N. Yulei and P. Yufan, *et al.*, Effect of dodecahydrododecaborate bistetraethylammonium on combustion reaction mechanism of aluminum powder, *Chin. J. Energ. Mater.*, 2021, **29**(10), 937–947.
39. M. Kaltchev, A. W. Thompson and W. T. Tysoe, Reflection-absorption infrared spectroscopy of ethylene on palladium (111) at high pressure, *Surf. Sci.*, 1997, **391**(1–3), 145–149.
40. J. Fan, W. Xiaofeng and W. Xiaohong, *et al.*, Thermal decomposition characteristics and kinetic parameters of tetraethylammonium decahydrodecaborate, *Chin. J. Explos. Propellants*, 2020, **43**(1), 57–63.
41. I. A. Olaru, CFD analysis in solidworks flow simulation for two mixing fluids with different temperatures in nozzles, *J. Eng. Stud. Res.*, 2020, **26**(1), 41–46.
42. A. Moen, L. Mauri and V. D. Narasimhamurthy, Comparison of k–ε models in gaseous release and dispersion simulations using the CFD code FLACS, *Process Saf. Environ. Prot.*, 2019, **130**, 306–316.
43. A. J. Tulis, *Criteria for the detonation of aluminum particles dispersed in air*, Illinois Institute of Technology, Chicago, 1982.
44. M. L. Chan and G. W. Meyers, *Advanced thermobaric explosive compositions*, US6955732B1, 2005.
45. Y. Liang and M. Beckstead, *Numerical simulation of quasi-steady, single aluminum particle combustion in air*//36th AIAA Aerospace Sciences Meeting and Exhibit, 1998, p. 254.
46. D. L. Frost, S. Goroshin and F. Zhang, Jet formation during explosive particle dispersal, *Proceedings of the 21st Military Aspects of Blast and Shock*, Jerusalem, Israel, 2010.
47. A. Kaczmarczyk, R. D. Dobrott and W. N. Lipscomb, Reactions of ${B_{10}H_{10}}^{2-}$ ion, *Proc. Natl. Acad. Sci. U. S. A.*, 1962, **48**, 729–733.
48. M. Y. Jin, Z. Yang and S. Cheng, *et al.*, Fast sodium-ion conduction in a novel comjuncto-hydroborate of $Na_4B_{20}H_{18}$, *ACS Appl. Energy Mater.*, 2022, **5**(12), 15578–15585.
49. M. F. Hawthorne, K. Shelly and F. Li, The versatile chemistry of the $[B_{20}H_{18}]^{2-}$ ions: novel reactions and structural motifs, *Chem. Commun.*, 2002, **6**, 547–554.
50. J. Mengyuan, Z. Jiankai and L. Yutong, *et al.*, Ignition and combustion characteristics of tetramethylammonium octadecahydro-eicosaborate, *Chin. J. Explos. Propellants*, 2022, **45**(6), 856–861.
51. B. L. Chamberland and E. L. Muetterties, Chemistry of boranes. XVIII. Oxidation of B10H102- and its derivatives, *Inorg. Chem.*, 1964, **3**, 1450–1456.

CHAPTER 4

Performance of Solid Propellants Containing $B_{10}H_{10}[(C_2H_5)_4N]_2$ and $B_{12}H_{12}[(C_2H_5)_4N]_2$

4.1 Introduction

Improving the performance of solid propellants is always an important aspect for researchers, especially the combustion performance of solid propellants.[1,2] Nano-sized particles, due to their small particle size, large surface area, many surface atoms complex microstructures and defects of grain, have high catalytic activity; therefore, replacing conventional catalysts in solid propellant by nano-sized catalysts has become a key research hot topic to improve the combustion performance of propellants.[3] However, the addition of nano-sized particles to the propellant formulation may worsen the processibility of solid propellant slurry because of its large specific surface area. The highly energetic combustion agent, tetraethylammonium decahydrodecaborate ($B_{10}H_{10}[(C_2H_5)_4N]_2$), a new energetic material with a high heat of combustion (49.5 MJ kg^{-1}) and low mechanical sensitivity ($H_{50} > 128$ cm, $P = 0\%$), can be used as a main ingredient in cast explosives and solid propellants;[4] $B_{12}H_{12}[(C_2H_5)_4N]_2$ has similar properties to $B_{10}H_{10}[(C_2H_5)_4N]_2$.

4.2 Properties of $B_{10}H_{10}[(C_2H_5)_4N]_2$/nAl (Nano Aluminum Powder) Composite

Even though $B_{10}H_{10}[(C_2H_5)_4N]_2$ is a kind of ionic borohydride with cage like polyhedron structure, it has high combustion efficiency and good burning rate regulation function. It can be used as gas generator, fuel additive and burning rate regulator of solid propellant. By introducing nano-sized aluminum (nAl) into the combustion agent, the energy and energy release

High-energy Combustion Agents of Organic Borohydrides: Synthesis, Characterization and Applications
By XiaoHong Zhang, WeiQiang Pang, Luigi T. DeLuca and Yu Zhao

Published by the Royal Society of Chemistry, www.rsc.org

efficiency of the energetic material system can be greatly improved, and the composite also has good ignition performance. For instance, Dillier *et al.*[5] studied nano boron and aluminum composite particles and found that adding 0.6% boron/aluminum (B/Al) composite particles can effectively improve the combustion efficiency of an AP/HTPB composite propellant system. Belal *et al.*[6] studied the effect of a Mg/Al system on the combustion performance of an HTPB based composite solid propellant. The results show that the Mg/Al composite particles with mechanical activation have higher reaction completeness and combustion efficiency. Zakiyya *et al.*[7] assembled two-dimensional nano MoO_3 and nAl particles to form a new type of nano energetic composite, which increased the burning rate of the system to 1730 $m\,s^{-1}$ and greatly increased the peak pressure. Bagalkote *et al.*[8] synthesized nAl and Fe_2O_3 xerogel composite particles by the sol–gel method, and the burning rate of the solid propellant system increased by 36%. In addition, it was found that graphene/nAl composites can greatly improve the mechanical strength and energy of the material system, which has potential application prospects in the field of solid propellant.[9–12] In order to improve the ignition and combustion features of $B_{10}H_{10}[(C_2H_5)_4N]_2$, $B_{10}H_{10}[(C_2H_5)_4N]_2$ was compounded with nAl to improve the combustion efficiency of $B_{10}H_{10}[(C_2H_5)_4N]_2$ and solve the agglomeration problem of nAl powder.

4.2.1 Experimental

$B_{10}H_{10}[(C_2H_5)_4N]_2$ was prepared by Xi'an Institute of Modern Chemistry with a purity of more than 99.5%; nano-sized aluminum (nAl) was prepared by the electro detonation method, with medium particle size of 80 nm and active aluminum content of 75%; ethyl acetate, Chengdu Cologne Chemicals Co., Ltd, purity AR; acetonitrile, Chengdu Cologne Chemicals Co., Ltd, purity AR.

FS-600n ultrasonic processor of Shanghai Shengxi Ultrasonic Instrument Co., Ltd, with a frequency of 20 kHz; LICHEN DF-101S type collector constant temperature magnetic stirrer constant temperature water bath pot.

4.2.2 Characterization Methods

Quanta 600 FEG field emission scanning electron microscope, acceleration voltage 20 kV; ASAP2020 specific surface area analyzer, micromeritics, USA; using nitrogen adsorbent, the purity of adsorbent is 99.99%, vacuum rate is 10 °C min^{-1}, adsorption temperature is 77 K, equilibrium time is 60 min, degassing time is 500 min^{-1}. The model 449 TG-DSC synchronous thermal analyzer of Netzsch is used for synchronous thermal analysis. The temperature range of measurement is 30–500 °C, the sample volume is 1.0 mg, the heating rate is 10.0 °C min^{-1}, and the flow rate of AR is 25 mL min^{-1}. The infrared spectrometer is the 5700 infrared spectrometer of Nicolet company of USA; the spectral range is 650–4000 cm^{-1}, the detector is DTGS, the resolution $>$0.09 cm^{-1}, the purge gas flow rate is 50 mL min^{-1}, the protection gas flow rate is 25 mL min^{-1}, and the gas *in situ* cell temperature is room temperature ~450 °C. The mass spectrometer is a QMS 403 quadrupole mass spectrometer of Netzsch; the temperature of the connecting tube between thermal analysis instrument and mass spectrometry is 190 °C, the interface temperature is 200 °C, the test mass range is 1–300.0 amu, the resolution is <0.5 amu, and the detection limit $>1\times10^{-6}$.

4.2.3 The Preparation Process of $B_{10}H_{10}[(C_2H_5)_4N]_2$/nAl Composite

The $B_{10}H_{10}[(C_2H_5)_4N]_2$/nAl composites were prepared by a solvent–non-solvent method. Through screening a variety of solvent–non-solvent systems, acetonitrile was selected as the solvent phase and ethyl acetate as the non-solvent phase.

Firstly, 5.0 g $B_{10}H_{10}[(C_2H_5)_4N]_2$ was weighed and put into a 200 mL beaker, acetonitrile was added and continuously stirred to make $B_{10}H_{10}[(C_2H_5)_4N]_2$ fully dissolved in the solvent phase. Then, the beaker was put into a magnon and placed in a constant temperature water pot, and the heating temperature was set at 80 °C and the rotating speed was set at 2000 rad min^{-1}. After the temperature reached 80 °C, the beaker was kept for 10 min. In another beaker, 5.0 g nAl powder was added to ethyl acetate, which was fully stirred to disperse in the non-solvent phase. Then, nAl powder dispersion was added into the beaker placed in a constant temperature water bath, and the ultrasonic rod was fixed on the tripod and placed in the solution for 1 h. After the reaction, it was filtered with filter paper and dried in a vacuum oven at 80 °C for 6 h to obtain a gray powder sample. At the same time, the control sample was a mixture of $B_{10}H_{10}[(C_2H_5)_4N]_2$ and nAl prepared by dispersing and drying with ethyl acetate. A flow chart of the preparation of the $B_{10}H_{10}[(C_2H_5)_4N]_2$/nAl composite is shown in Figure 4.1.

4.2.4 Results and Discussion

4.2.4.1 SEM Analysis

Figure 4.2 shows SEM photos of the $B_{10}H_{10}[(C_2H_5)_4N]_2$/nAl composite at low and high resolutions and the $B_{10}H_{10}[(C_2H_5)_4N]_2$/nAl mixture. It can be seen that $B_{10}H_{10}[(C_2H_5)_4N]_2$ is a polyhedral structure with size of 1–2 μm, showing that the nAl particles can be uniformly composite on the surface of $B_{10}H_{10}[(C_2H_5)_4N]_2$ by the solvent–non-solvent method (Figure 4.2a). Even at low magnification, the agglomeration of nano aluminum is not obvious, which indicates that the composite $B_{10}H_{10}[(C_2H_5)_4N]_2$/nAl sample has good dispersion and overcomes the disadvantages of the agglomeration of single component nAl powder (Figure 4.2b). Moreover, the agglomeration of nAl and $B_{10}H_{10}[(C_2H_5)_4N]_2$ in the sample is serious; the structure of polyhedron has been destroyed, indicating that the solvent evaporation method can synthesize

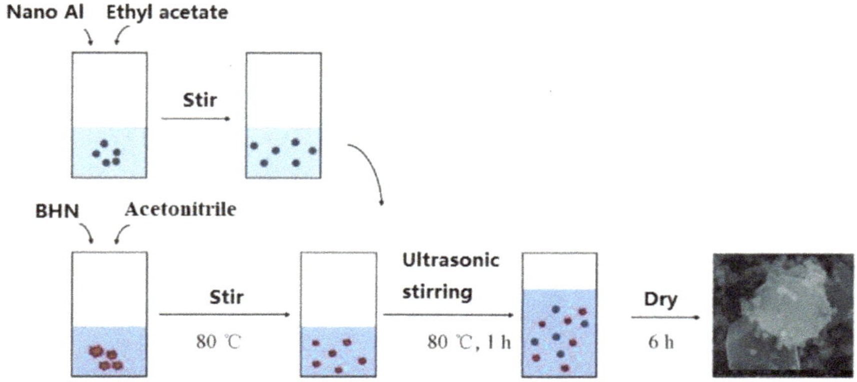

Figure 4.1 Flow chart of the preparation of $B_{10}H_{10}[(C_2H_5)_4N]_2$/nAl composite. Reproduced from ref. 13.

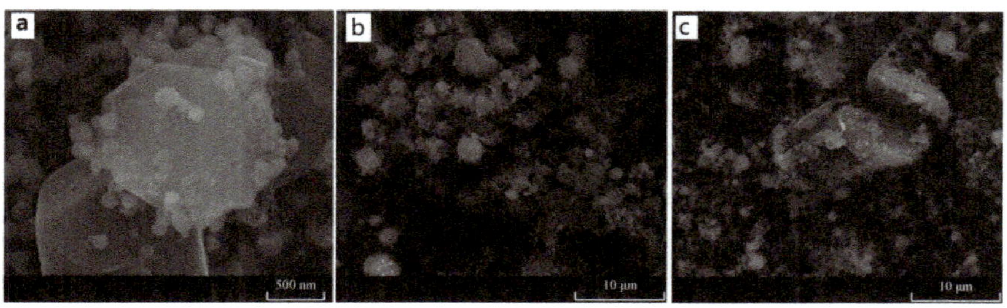

Figure 4.2 SEM images of $B_{10}H_{10}[(C_2H_5)_4N]_2$/nAl composite and $B_{10}H_{10}[(C_2H_5)_4N]_2$/nAl mixture. (a) $B_{10}H_{10}[(C_2H_5)_4N]_2$/nAl composite, high resolution; (b) $B_{10}H_{10}[(C_2H_5)_4N]_2$/nAl composite, low resolution; (c) $B_{10}H_{10}[(C_2H_5)_4N]_2$/nAl mixture. Reproduced from ref. 13.

$B_{10}H_{10}[(C_2H_5)_4N]_2$/nAl composite with good dispersion, and the method can significantly regulate the crystal morphology of the composite (Figure 4.2c).

4.2.4.2 Fourier Transform Infrared Spectroscopy Analysis

The infrared spectrum of the $B_{10}H_{10}[(C_2H_5)_4N]_2$/nAl composite is shown in Figure 4.3. The infrared spectrum of the $B_{10}H_{10}[(C_2H_5)_4N]_2$/nAl composite is consistent with the standard infrared spectrum of $B_{10}H_{10}[(C_2H_5)_4N]_2$. $\nu_{as}(CH_3) = 3011$, 1460 cm^{-1}, $\nu_s(CH_3) = 2950$, 1373, 1358 cm^{-1}; $\nu_{as}(CH_2) = 2992$, 1401 cm^{-1}; $\nu_s(CH_2) = 2889$, 1309 cm^{-1}; $\nu_{C\text{-}C} = 1120$, 1079, 1034, 1008, 792 cm^{-1}; $\nu_{B\text{-}H} = 2449$ cm^{-1}; $\nu_{C\text{-}N} = 118.4$ cm^{-1}, indicating that no new organic compounds and other products were formed after the formation of the complex.

4.2.4.3 Nitrogen Adsorption Analysis (BET)

In order to further explore the change in the surface morphology of the sample after the formation of the composite, the specific surface area of the

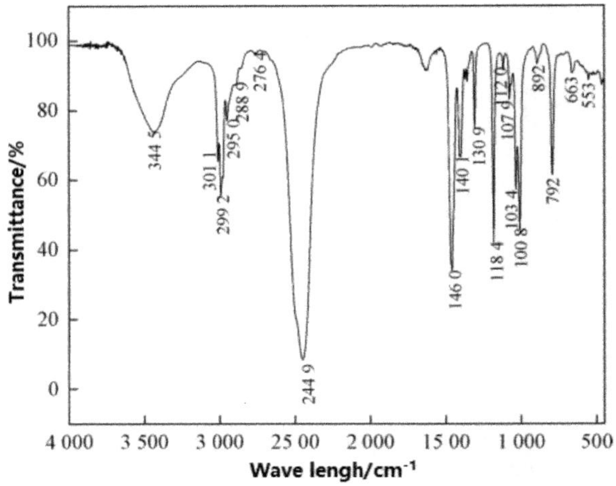

Figure 4.3 Infrared spectrum of the $B_{10}H_{10}[(C_2H_5)_4N]_2$/nAl composite. Reproduced from ref. 13.

sample was analyzed by using a specific surface area analyzer. The nitrogen adsorption capacities of the obtained $B_{10}H_{10}[(C_2H_5)_4N]_2$/nAl composite and the control sample are shown in Figure 4.4. It can be seen that the nitrogen adsorption capacity of the sample increases gradually with the increase of specific pressure, and finally reaches saturation. The specific surface area of the $B_{10}H_{10}[(C_2H_5)_4N]_2$/nAl composite is 7.06 $m^2\,g^{-1}$, while the specific surface area of the mixture of $B_{10}H_{10}[(C_2H_5)_4N]_2$ and nAl as the control sample is 0.20 $m^2\,g^{-1}$, indicating that the specific surface area of the $B_{10}H_{10}[(C_2H_5)_4N]_2$/nAl composite prepared by the solvent–non-solvent method is much larger than that of the ordinary mixture. The reason may be that, after the solvent volatilization of the ordinary mixture, the nAl powder has a serious agglomeration phenomenon, which greatly reduces the specific surface area of the system. The surface of $B_{10}H_{10}[(C_2H_5)_4N]_2$ polyhedron can provide the attachment site for nAl, reduce the agglomeration of nAl, and the larger specific surface area of the composite will also be conducive to the improvement of ignition performance and combustion efficiency of composite fuel during combustion.

4.2.4.4 *Thermal Performance Analysis*

The high-energy combustion agent $B_{10}H_{10}[(C_2H_5)_4N]_2$ has much influence on the thermal decomposition temperature of solid rocket propellant. The thermal decomposition including DSC and TG–DTG of $B_{10}H_{10}[(C_2H_5)_4N]_2$ was carried out, and the results are shown in Figure 4.5. It can be seen that the thermal decomposition process of $B_{10}H_{10}[(C_2H_5)_4N]_2$ is the main decomposition process. When it was heated, there was a large amount of heat released. The thermal decomposition peak temperature is 305.8 °C

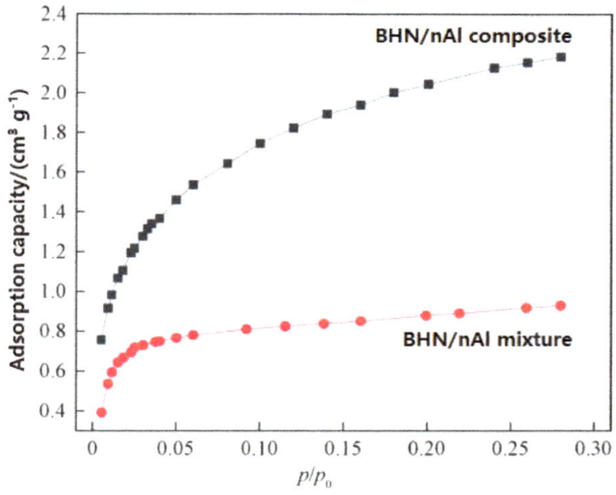

Figure 4.4 N_2 adsorption diagram of the $B_{10}H_{10}[(C_2H_5)_4N]_2$/nAl composite and mixture. Reproduced from ref. 13.

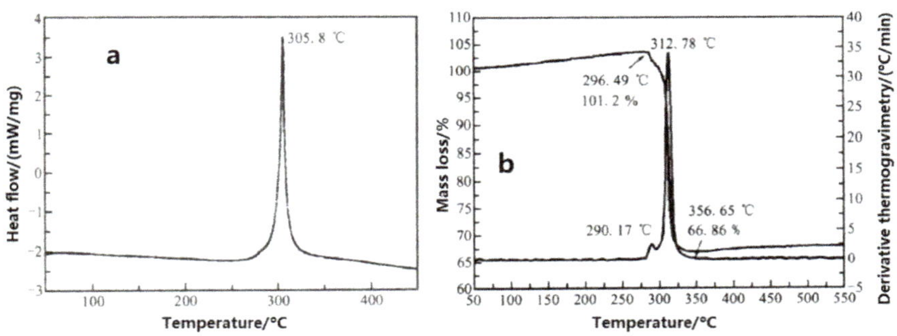

Figure 4.5 DSC and TG–DTG curves of $B_{10}H_{10}[(C_2H_5)_4N]_2$. Reproduced from ref. 14.

(Figure 4.5a). The initial temperature of thermal decomposition is 296.49 °C, and the end temperature is 356.65 °C. The mass loss in this stage is 34.34% (Figure 4.5b). After thermal decomposition, there is about 66.8% residue, which may be due to the absence of oxygen elements in the molecular structure of $B_{10}H_{10}[(C_2H_5)_4N]_2$. The residue after thermal decomposition in nitrogen atmosphere is mainly the carbon skeleton after decomposition, and after multiple sampling and X-ray fluorescence diffraction analysis, the thermal composition of the residue is basically carbon. Compared with the thermal decomposition of nitrocellulose (under the same conditions, the first thermal decomposition peak temperature of nitrocellulose is 202.6 °C), the thermal decomposition peak temperature of $B_{10}H_{10}[(C_2H_5)_4N]_2$ is 104.2 °C higher, indicating that the thermal stability of $B_{10}H_{10}[(C_2H_5)_4N]_2$ is better than that of nitrocellulose.

In order to explore the influence of nAl on the reaction characteristics of $B_{10}H_{10}[(C_2H_5)_4N]_2$ in the $B_{10}H_{10}[(C_2H_5)_4N]_2$/nAl composite, the thermal decomposition properties of $B_{10}H_{10}[(C_2H_5)_4N]_2$ and the $B_{10}H_{10}[(C_2H_5)_4N]_2$/nAl composite samples were studied. The heating rate β was 10 °C min^{-1}. The DSC (differential scanning calorimeter) and TG (thermal gravimetric analysis) curves at 10 °C min^{-1} are shown in Figure 4.6. It can be seen that the thermal decomposition process of $B_{10}H_{10}[(C_2H_5)_4N]_2$ is divided into two stages: the first exothermic peak is wider than the second, and the shoulder peak appears at 285.1 °C. The peak temperature of the second exothermic peak is 304.4 °C. The formation of $B_{10}H_{10}[(C_2H_5)_4N]_2$/nAl composite can advance the decomposition temperature of $B_{10}H_{10}[(C_2H_5)_4N]_2$ to 262.0 °C and 298.8 °C, respectively. There is also an endothermic peak between the two stages of thermal decomposition of the $B_{10}H_{10}[(C_2H_5)_4N]_2$/nAl composite, which may be due to the formation of an intermediate product of melting decomposition type during the decomposition process of $B_{10}H_{10}[(C_2H_5)_4N]_2$. Generally, the melting endothermic peak of melting decomposition type material is not obvious when the heating rate is low. With the increase in heating rate, the melting endothermic peak gradually becomes sharp, showing that the addition of nAl promotes the formation of an endothermic peak during the heating process of $B_{10}H_{10}[(C_2H_5)_4N]_2$, and promotes the formation and decomposition of $B_{10}H_{10}[(C_2H_5)_4N]_2$ intermediates (Figure 4.6a).

Furthermore, the mass loss process of $B_{10}H_{10}[(C_2H_5)_4N]_2$ and $B_{10}H_{10}[(C_2H_5)_4N]_2$/nAl composite can be divided into two stages, which is consistent with the DSC curve, and no mass loss corresponding to the endothermic process in the DSC curve was found in the TG curve, which also confirmed the formation of $B_{10}H_{10}[(C_2H_5)_4N]_2$ melt decomposition intermediate. The total mass loss of the $B_{10}H_{10}[(C_2H_5)_4N]_2$/nAl composite is 42.5%. Because the mass ratio of $B_{10}H_{10}[(C_2H_5)_4N]_2$ to nAl in the $B_{10}H_{10}[(C_2H_5)_4N]_2$/nAl composite is 1 : 1, the highest temperature in the whole heating process does not exceed the reaction temperature of nAl, so the total mass loss of $B_{10}H_{10}[(C_2H_5)_4N]_2$/nAl composite is 21.2%.

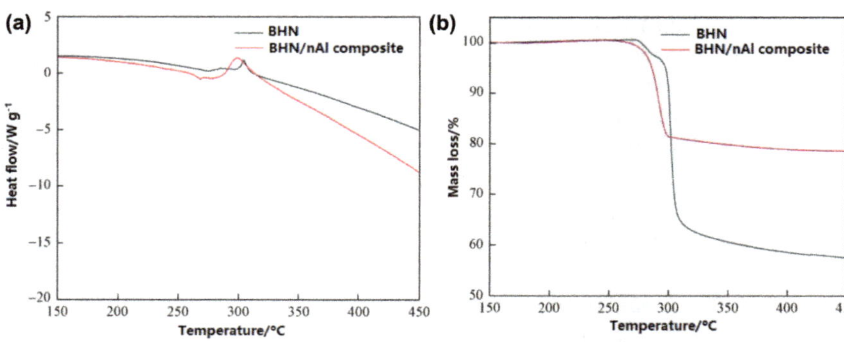

Figure 4.6 DSC and TG curves of $B_{10}H_{10}[(C_2H_5)_4N]_2$ and $B_{10}H_{10}[(C_2H_5)_4N]_2$/nAl composite. (a) DSC; (b) TG. Reproduced from ref. 13.

The thermal decomposition ratio of $B_{10}H_{10}[(C_2H_5)_4N]_2$ in the sample is consistent with that of single $B_{10}H_{10}[(C_2H_5)_4N]_2$. This indicates that nAl powder can promote the reaction of $B_{10}H_{10}[(C_2H_5)_4N]_2$, but it will not take part in the reaction and change the reaction mechanism at room temperature ~500 °C.

Figure 4.7 is a temperature programmed analysis of $B_{10}H_{10}[(C_2H_5)_4N]_2$/nAl composite at the heating rate of 10 °C min^{-1}. Infrared spectrograms of the gas produced by thermal decomposition experiments at different temperatures were taken. The change in infrared absorption intensity with temperature can reflect the change in gas products in the process of $B_{10}H_{10}[(C_2H_5)_4N]_2$ heating. It can be seen from the results of the TG and DSC curves that the formation of the composite can advance the decomposition of $B_{10}H_{10}[(C_2H_5)_4N]_2$. With the reaction going on, the infrared characteristic absorption intensity of $B_{10}H_{10}[(C_2H_5)_4N]_2$ gas increases continuously, producing gas. The characteristic absorption peak of ethylene appears at 950 cm^{-1} and 2990–3166 cm^{-1}, the main characteristic absorption peak of ethane appears at 2950 cm^{-1} and 2972 cm^{-1}, and the weak characteristic absorption peak of NH_3 appears at 967 cm^{-1} and 932 cm^{-1}.

The temperature dependence of the gas mass spectrum ion current of small molecule gas products released from the decomposition of the $B_{10}H_{10}[(C_2H_5)_4N]_2$/nAl composite was observed by a mass spectrometry detector at the heating rate of 10 °C min^{-1}, and the results are shown in Figure 4.8. It can be seen that the mass composition ratios of ion fragments captured by the mass spectrometry detector are 2, 13, 14, 15, 16, 17, 18, 24, 25, 26, 27, 28, 29, 30. The corresponding gas products are hydrogen ($m/z = 2$), ethane ($m/z = 14, 15, 16, 25, 26, 27, 28, 29, 30$), ethylene ($m/z = 14, 25, 26, 27, 28$) and NH_3 ($m/z = 14, 15, 16, 17$). The intensity of the ion current begins to increase at about 270 °C and reaches a peak at about 310 °C. As H_2 is a non-polar gas, there is no obvious signal in the infrared absorption process; the rest of the detected gases are consistent with the infrared spectrum results. According to the results of infrared radiation (IR) and mass spectrum (MS), the thermal decomposition gas

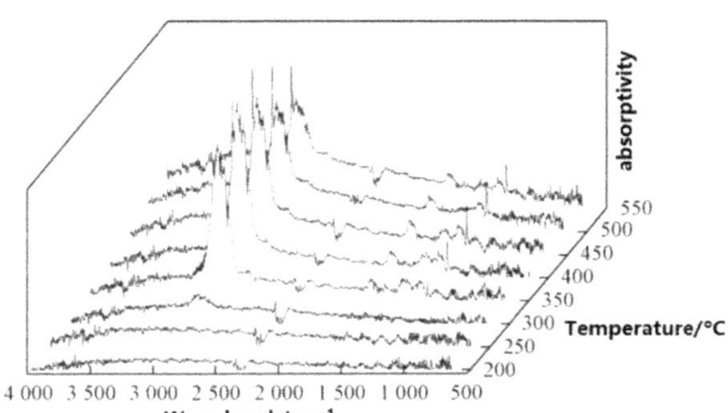

Figure 4.7 Infrared spectrum *vs.* temperature of gases decomposed from $B_{10}H_{10}[(C_2H_5)_4N]_2$/nAl composite. Reproduced from ref. 13.

products are mainly small molecule gases, such as ethylene, ethane, ammonia and hydrogen, which are combustible gases with high heat of combustion.

4.3 Effect of $B_{10}H_{10}[(C_2H_5)_4N]_2$ on the Properties of HTPB Based Propellants

There are a few reports on the synthesis and evolution of combustion catalysts in solid propellants and explosives.[15,16] The most commonly used, in a large range of applications, are obtained by co-precipitation, of decahydrodecaborate salts, together with some oxidizers such as $K_2B_{10}H_{10}$, $Cs_2B_{10}H_{10}$, $(NH_4)_2B_{10}H_{10}$, $Cs_2B_{10}H_{10}{\cdot}CsNO_3$, $Cs_2B_{10}H_{10}/KNO_3$, *etc.*[17–20] Pang *et al.*[14] studied the effect of tetraethylammonium decahydro-decaborate ($B_{10}H_{10}[(C_2H_5)_4N]_2$) on the properties of fuel-rich solid propellant and it was found that the addition of $B_{10}H_{10}[(C_2H_5)_4N]_2$ to the fuel-rich propellant can increase the mass and volume heat of combustion of solid propellant, and the pressure exponent of fuel-rich solid propellants decrease with an increase in the mass fraction of $B_{10}H_{10}[(C_2H_5)_4N]_2$ in the solid propellant formulation. Moreover, the effects of $B_{12}H_{12}[(C_2H_5)_4N]_2$ on the combustion properties of nitrate ester plasticized polyether (NEPE) propellants were reported.[21] Their results showed that it is not an effective catalyst for the decomposition of ammonium perchlorate (AP), and although it can increase the decomposition of nitramines, the burning rate of NEPE propellants can be increased by the addition of this compound, and also that a "platform" appears over the high-pressure range 7–11 MPa. The use of a new type of burning rate catalyst with high catalytic activity is one of the important ways to improve the combustion properties of solid propellant. Thus, from the point of view of the high performance mentioned above, it has the potential for possible use as an energetic ingredient in solid propellants.[22,23]

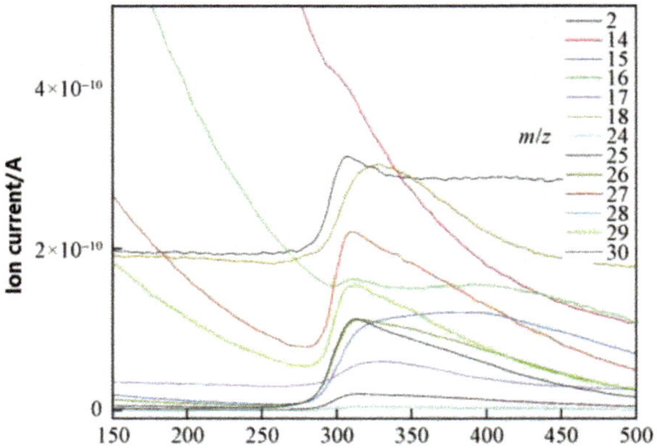

Figure 4.8 Mass spectrum of gaseous products decomposed from $B_{10}H_{10}[(C_2H_5)_4N]_2$/nAl composite at different temperatures. Reproduced from ref. 13.

However, there are few reports on the combustion properties of composite solid propellants containing hydroborate ammonium ($B_{10}H_{10}[(C_2H_5)_4N]_2$). In this chapter, the characteristics of the $B_{10}H_{10}[(C_2H_5)_4N]_2$ particles were analyzed by using the diagnostic techniques of scanning electron microscopy (SEM) and grain size distribution. Different mass fractions of $B_{10}H_{10}[(C_2H_5)_4N]_2$ particles were added to the formulations and three different propellant compositions with and without $B_{10}H_{10}[(C_2H_5)_4N]_2$ were produced. The focus of this chapter is on how $B_{10}H_{10}[(C_2H_5)_4N]_2$ affects the combustion properties (burning rate and pressure exponent) of fuel-rich solid propellant and composite solid propellant, placing emphasis on the thermal decomposition, combustion flame structure and combustion residue performance of the solid propellants, which could be used for solid rocket motor applications.

4.3.1 Experimental

4.3.1.1 Raw Materials

The binder system consists of hydroxyl-terminated polybutadiene (HTPB, $E_{OH} = 7.8 \times 10^{-4}$ mol g^{-1}) plasticized with di-2-etylhexyl sebacate (DOS) to increase the processibility, which is cured by 2,4-toluene diisocyanate (TDI). The additives isophthaloyl-bis-(2-methylaziridine) (HX-752) and bi-(2-methy-1-aziridinyl) phosphine oxide (MAPO) are used as a cross-linker and a bonding agent, respectively. Aluminum powders with an average particle size of 5 μm were used in the formulations (purchased from Nanjing Chemical Industry Co., Ltd), and parts of AP were replaced by hydroborate ammonium salt ($B_{10}H_{10}[(C_2H_5)_4N]_2$). The involved propellant formulations also contain catocene (GFP) and chromite copper (CC) as a burning rate modifier. Ammonium perchlorate (AP), as oxidizer, was always used and kept tri-modal. The first and the second modal of AP consisted of pure research grade AP (>99% pure) with an average particle sizes of 0.178–0.250 mm and 0.105–0.150 mm, respectively. The third modal of AP was made by grinding AP (>99% purity) in a fluid energy mill to an average particle size of around 1–3 μm.

4.3.1.2 Tested Ingredients and Formulations

The composite solid propellant used as the baseline, is an HTPB/AP/Al/catalyst formulation with mass ratios 11.5/65/18/5.5 for all involved samples. Different mass fractions of $B_{10}H_{10}[(C_2H_5)_4N]_2$ additives were used as catalyst additive by replacing AP particles with 0.178–0.250 mm diameter. Except where otherwise stated, all propellants were manufactured, processed, and tested at Xi'an Modern Chemistry Research Institute under identical conditions and using identical procedures. The samples FRP-1, FRP-2, FRP-3, FRP-4 are fuel-rich solid propellants containing 0%, 3%, 6%, 10% of $B_{10}H_{10}[(C_2H_5)_4N]_2$ additives, respectively. The samples SPN-1, SPN-2, SPN-3 are THPB-based composite solid propellants containing 0%, 3%, 6% of $B_{10}H_{10}[(C_2H_5)_4N]_2$ additives, respectively.

4.3.1.3 *Processing*

Propellant formulations were mixed in 500 g batches using a vertical planetary mixer of 2 L capacity. All batches were mixed and cast under a vacuum by a slurry cast technique.[24] The propellant was cured at 50 °C for 3 days in a water jacketed oven to solidify the propellant samples.

4.3.2 Characterization Methods

4.3.2.1 *BET and SEM Tests*

The specific surface area measurement was computed from the nitrogen adsorption isotherm, obtained by static volumetric measurement at a liquid nitrogen boiling temperature (77 K).[25] Samples were out-gassed at 100 °C for at least 4 h at an absolute pressure of less than 10^{-3} torr. All measurements were carried out on a completely automated instrument (ASAP 2010 from Micromeritics, USA), leading to the final value of the specific surface areas expressed in $m^2\,g^{-1}$. Electron microscopy was used to study the shape, size, morphology, and defects of the powders. The morphologies of solid particles were examined using scanning electron microscope (SEM) technology.

4.3.2.2 *Particle Size Distribution Experiments*

Particle size and size distribution of samples were performed through laser scattering (Malvern Master sizer 2000) using a dry dispersion unit (Scirocco). Tests were performed with the same feed rate (45%) and varying feeding pressure in order to achieve sensitivity on this parameter. The quantity of material per test was about 0.07–0.09 grams, fed through the "Micro Volume" tray. Obscuration filtering was switched on and set to values within the range of 0.5–10%.

4.3.2.3 *Density Test*

The density (ρ, g cm^{-3}) measurement of solid propellants was carried out on a Model AG 104 Mettler Toledo balance with samples of 30 mm × 30 mm × 10 mm rectangular sticks in steeping medium of liquid paraffin at the temperature of (20 ± 2) °C.

4.3.2.4 *Hazardous Properties Test*

The impact sensitivity test was carried out on the drop weight impact machine at the Xi'an Modern Chemistry Research Institute. The drop-weight impact machine used is based on a design that the Explosives Research Laboratory at Bruceton, Pennsylvania (USA), developed during World War II.[26] The sample is placed on a fixed roughened anvil. A hammer of known contact area is positioned above the sample and the weight is raised to a predetermined height and dropped. The conditions of this test were as follows: temperature, 18 °C; free-falling height, 25 cm; hammer weight,

2.0 kg; sample mass, about 40 mg; experiment repeat, 25 times for probability calculations to obtain H_{50} values (The H_{50} value represents the height from which dropping a 5 kg weight results in an explosive event in 50% of trials).

The friction sensitivity test was carried out with the method of ABL (above base line) in Xi'an Modern Chemistry Research Institute.[27] In the ABL friction test, the sample is placed on an anvil, and a known force is applied hydraulically through a stationary wheel. A pendulum is used to propel the sliding anvil at any of several standard velocities perpendicular to the force vector. Sample initiation is detected by visual means (spark or flame). The conditions of this test were as follows: temperature, 18 °C; pendulum angle, 90°; sample mass, about 20 mg; experiment repeat, 25 times for probability calculations and an average value of explosive probability P (%) was obtained with 5 peering tests.

4.3.2.5 TG–DTG Analysis

The thermal decomposition processes of the propellants were measured by differential thermal analysis (DTA) and thermogravimetry (TG). The equipment was operated under the condition of the flowing nitrogen at atmospheric pressure. TG and DTA tests were carried out with a heating rate of 10 K min^{-1}, while 3.0 mg samples were heated in Al_2O_3 pans without cover.

4.3.2.6 Burning Rate Test Method

The strand burning rates of the propellants were determined in the pressure range of 1–12 MPa by means of a fuse-wire technique.[28] The methodology involved the combustion of strands (ignited by means of a nichrome wire) with the dimensions of 150 mm × 5 mm × 5 mm in a nitrogen pressurized steel bomb. For the trial conducted, the burning rates were computed from the time that was recorded at each pressure level for each sample.

4.3.2.7 Combustion Flame Structure Tests

The combustion flame structure tests were carried out in a transparent combustion bomb with four sides. The combustion behavior of the composite propellant, containing different aluminum samples with the dimensions of 1.5 mm × 5 mm × 15 mm, were performed at the scheduled pressure under N_2 atmosphere, and ignited by chromium–nickel wires from top to bottom.

4.3.3 Results and Discussion

4.3.3.1 SEM and Grain Size Distribution Analysis

Detailed information on particles, concerning morphology and particle size distribution, was collected by running a series of advanced diagnostic techniques, including SEM, BET, and grain size distribution. The morphology is one of the most important parameters for characterizing the quality of the particles. Specific material properties such as ability to flow, reactivity as well as compressibility, and its hardening potential, are determined by the size distribution. The solid particle size distribution is a key factor in its atomization processes and other chemical processes such as combustion. The surface photographs and particle size distribution of the $B_{10}H_{10}[(C_2H_5)_4N]_2$ particles were analyzed by means of SEM and a laser particle analyzer, and the results are shown in Figure 4.9. The relevant parameters are listed in Table 4.1.

It can be seen that the microstructure of the $B_{10}H_{10}[(C_2H_5)_4N]_2$ particles is irregular in shape, and the corresponding diameters of the $B_{10}H_{10}[(C_2H_5)_4N]_2$ powder $d_{50} = 94.8$ μm. Corresponding to the large value of d_{50}, the specific surface area and the span of the particles is 0.08 $m^2 g^{-1}$ and 1.536 $m^2 g^{-1}$, respectively. Thus, it can be deduced that in order to improve the processing properties of the solid propellants, the prepared particles need to be coated or prilled to have a rounded surface.[29,30]

4.3.3.2 *Effects of $B_{10}H_{10}[(C_2H_5)_4N]_2$ on the Energetic Properties of Fuel-rich Propellants*

$B_{10}H_{10}[(C_2H_5)_4N]_2$, as one of the high combustion agents, provides much improvement to the energetic properties of solid propellants. The energetic properties (density, mass heat of combustion and volume heat of combustion) were measured and calculated and the data are shown in Table 4.2.

The date in Table 4.2 indicates that the mass heat of combustion and volume heat of combustion of fuel-rich propellants increase with increasing mass fraction of $B_{10}H_{10}[(C_2H_5)_4N]_2$ particles in the propellant formulation, whereas the density decreases. Compared with the reference formulation (sample FRP-1), the mass heat of combustion and volume heat of combustion of the fuel-rich solid propellant containing 10% mass fraction of $B_{10}H_{10}[(C_2H_5)_4N]_2$ (sample FRP-4) are increased by 9.0% and 2.1%, respectively, whilst the density is decreased by 6.3%, which would have a large influence on an engine's propellant loading, with its finite volume.

At the same time, the heat of explosion of composite propellant containing $B_{10}H_{10}[(C_2H_5)_4N]_2$ was measured, and the results listed in Table 4.3. It can be seen that the heat of explosion and the heat of explosion efficiency decrease with increasing mass fraction of $B_{10}H_{10}[(C_2H_5)_4N]_2$ in the compositions.

4.3.3.3 *The Hazardous Properties of $B_{10}H_{10}[(C_2H_5)_4N]_2$/Energetic Materials and Propellants*

4.3.3.3.1 The Impact and Friction Sensitivity of $B_{10}H_{10}[(C_2H_5)_4N]_2$ with Some Energetic Materials. There are many uses for knowing the sensitivity of a material to impact and friction stimuli. These can purpose evaluation of material sensitivity over time. The sensitivity of the systems of

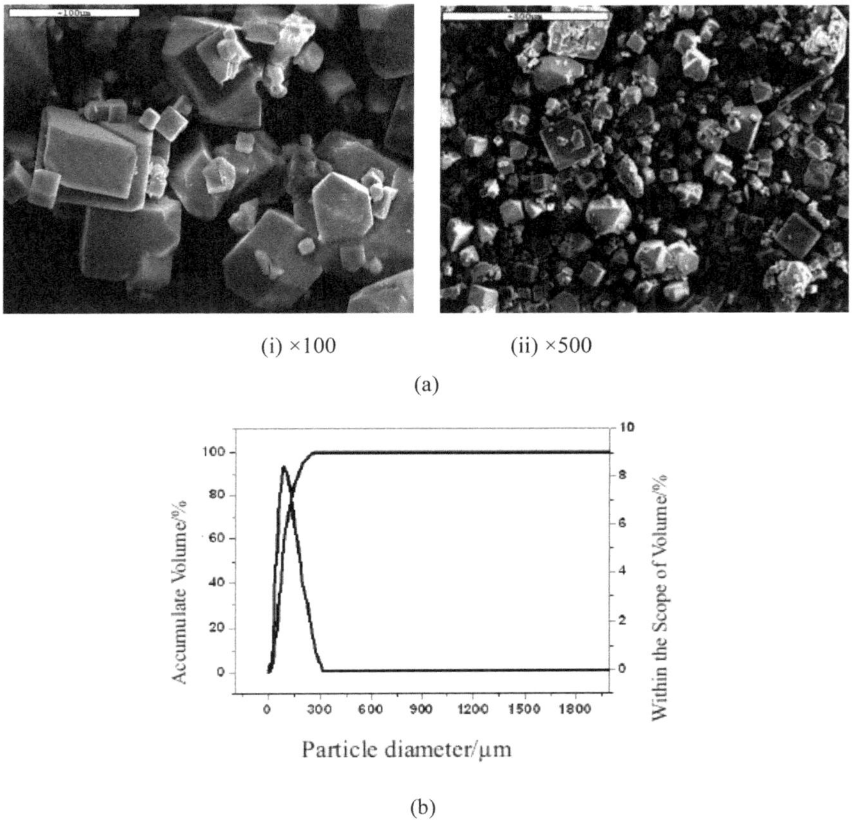

Figure 4.9 SEM photographs (a) and particle size distribution (b) of the $B_{10}H_{10}[(C_2H_5)_4N]_2$ particles. Reproduced from ref. 14.

Table 4.1 Particle size parameters of BHN. Reproduced from ref. 14.[a]

Sample	$D_{[3,2]}$, μm	$D_{[4,3]}$, μm	d_{10}, μm	d_{50}, μm	d_{90}, μm	Span	SSA, $m^2\ g^{-1}$	ρ, $g\ cm^{-3}$
$B_{10}H_{10}[(C_2H_5)_4N]_2$	73.1	106.7	43.9	94.8	189.4	1.536	0.08	0.92

[a] SSA: specific surface area.

$B_{10}H_{10}[(C_2H_5)_4N]_2$ with some energetic materials can be confirmed this way. Comparison of sensitivity of two materials can determine if one material could replace another in a particular process.

According to ref. 31, $B_{10}H_{10}[(C_2H_5)_4N]_2$ is insensitive, and it was approved by our test. From the results of the impact and friction sensitivity of $B_{10}H_{10}[(C_2H_5)_4N]_2$ with some energetic materials (Table 4.4), the following conclusions can be made.

Table 4.2 The energetic properties of fuel-rich propellants with and without $B_{10}H_{10}[(C_2H_5)_4N]_2$ particles. Reproduced from ref. 14.

Samples	Measured density/g cm^{-3}	Measured weight heat of combustion/MJ kg^{-1}	Volume heat of combustion/ MJ cm^{-3}
$B_{10}H_{10}[(C_2H_5)_4N]_2$	0.94	49.50	46.53
FRP-1	1.63	21.81	35.55
FRP-2	1.59	22.62	36.06
FRP-3	1.56	23.29	36.28
FRP-4	1.53	23.77	36.30

Table 4.3 Comparison of the tested and theoretical heat of explosion of composite propellants with and without $B_{10}H_{10}[(C_2H_5)_4N]_2$.

Samples	Mass fraction of $B_{10}H_{10}[(C_2H_5)_4N]_2$/%	Heat of explosion/kJ kg^{-1} Theoretical value	Heat of explosion/ kJ kg^{-1} Measured value	Heat of explosion efficiency/%
SPN-1	0	6987	6858	98.15
SPN-2	3	6767	6286	92.89
SPN-3	6	6410	5621	87.69

It can be known from the results in Table 4.4 that the $B_{10}H_{10}[(C_2H_5)_4N]_2$ in combination with some energetic materials are insensitive except for (7) and (11). The HMX and CL-20 in systems (7) and (11) are very sensitive to mechanical stimuli, and it makes the system more sensitive. Contrasting the results with compatibility experimental data, there was no consequential affiliation between sensitivity and compatibility. Systems (17) and (19) are incompatible while their impact and friction sensitivities are very low. Binary system (11) is compatible while its impact and friction sensitivity is very high. However, if the energetic materials can react with each other when they are mixed, they must be incompatible and very sensitive. Binary systems (3), (9), (13), and (15) are of great compatibility and low sensitivity.

4.3.3.3.2 Effects of $B_{10}H_{10}[(C_2H_5)_4N]_2$ Particles on the Hazard Properties of Fuel-rich Propellants. The mechanical sensitivity of a solid propellant reflects the degree of difficulty of initiation by an external mechanical action, which is one of the most important parameters to assess the safety of solid propellants. Therefore, it was necessary for us to investigate the mechanical sensitivity of fuel-rich solid propellants containing $B_{10}H_{10}[(C_2H_5)_4N]_2$ particles. Table 4.5 shows the results of fuel-rich propellants with and without $B_{10}H_{10}[(C_2H_5)_4N]_2$ particles.

It can be seen from the data in Table 4.5 that the impact and friction sensitivity of the prepared $B_{10}H_{10}[(C_2H_5)_4N]_2$ samples are rather low, especially the friction sensitivity which is 0% at 250 N, and indicates that it is safe with respect to mechanical stimulation and it is feasible for application in fuel-rich solid propellants. The impact sensitivity increases significantly (from 21.99 to 2.47 N m) with increases in the $B_{10}H_{10}[(C_2H_5)_4N]_2$ mass fraction in the propellant composition, whereas the friction sensitivity changes little, with only marginal increases. Compared with the data for the sample FRP-1, the higher the impact sensitivity value, the safer the impact sensitivity of the propellant sample, which may be attributed to the crystal structure of the $B_{10}H_{10}[(C_2H_5)_4N]_2$ and the active groups on the surface of the $B_{10}H_{10}[(C_2H_5)_4N]_2$ particles, which can react with some of the ingredients added in the fuel-rich solid propellant formulations.

4.3.3.4 *Effects of $B_{10}H_{10}[(C_2H_5)_4N]_2$ on the Thermal Decomposition of Fuel-rich Propellants*

The thermal decomposition (DSC and TG–DTG) of fuel-rich propellant containing different mass fractions of $B_{10}H_{10}[(C_2H_5)_4N]_2$ with a heat rate of 10 °C min^{-1} at 0.1 MPa was investigated; the curves are shown in Figures 4.10 and 4.11.

It can be seen from Figure 4.10 that there are three main mass loss stages (stages I, II and III) of fuel-rich propellants with different mass fractions of $B_{10}H_{10}[(C_2H_5)_4N]_2$. The first thermal decomposition stage begins at about 172.3–195.2 °C and ends at 318.3–326.1 °C, accompanied by 21.32–24.78% mass loss, indicating there is significant exothermic peak over this temperature range and the thermal decomposition of this process plays a leading role. The second thermal decomposition stage follows the I stage, and ends at about 365.08–397.2 °C, the corresponding peak temperature is

Table 4.4 Results of the impact and friction sensitivity of $B_{10}H_{10}[(C_2H_5)_4N]_2$ (BHN) with some energetic materials. Reproduced from ref. 32 with permission from John Wiley & Sons, Copyright © 2013 Wiley-VCH Verlag GmbH & Co. KGaA, Weinheim.[a]

Single systems	Impact/%	Friction/%	Mixture systems	Impact/%	Friction/%
BHN (1)	>125.8 cm	0	1/1-BHN/GAP (3)	0	0
GAP (2)	0	0	1/1-BHN/RDX (5)	0	3
RDX (4)	0	0	1/1-BHN/HMX (7)	71	82
HMX (6)	100	100	1/1-BHN/NTO-Pb (9)	0	0
NTO-Pb (8)	0	0	1/1-BHN/CL-20 (11)	76	84
CL-20 (10)	96	100	1/1-BHN/Al (13)	0	0
Al ① (12)	0	0	1/1-BHN/Mg (15)	0	0
Mg ② (14)	0	0	1/1-BHN/DNTF (17)	8	7
DNTF (16)	100	90	1/1-BHN/GUDN (19)	5	10
GUDN (18)	>159 cm	0			

[a]The numbers in parentheses are the same as in Section 3.2.1; ①, ② angle of pendulum = 90°; falling weight = 5.0 kg; total weight of testing sample = 50 mg. The others: angle of pendulum = 66°; falling weight = 2.0 kg; rotal weight of testing sample = 30 mg.

Table 4.5 The hazardous properties of BHN and fuel-rich propellants with and without BHN particles. Reproduced from ref. 14.

	Impact sensitivity		Friction sensitivity	
Sample	H_{50}/cm	Standard deviation S (log value)	P/%	Level of confidence 95%
BHN	>125.9	—	0	(0%, 14%)
FRP-1	112.2	0.30	92	(76%, 99%)
FRP-2	32.3	0.37	88	(69%, 98%)
FRP-3	19.5	0.23	96	(80%, 100%)
FRP-4	12.6	0.10	88	(69%, 98%)

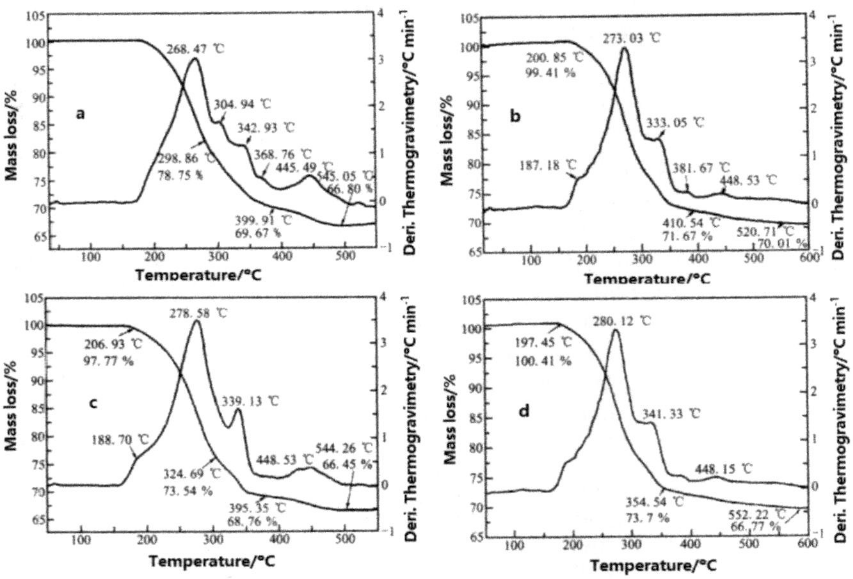

Figure 4.10 TG–DTG curves of fuel-rich propellants containing different mass fractions of $B_{10}H_{10}[(C_2H_5)_4N]_2$. Reproduced from ref. 14.

333.1–342.9 °C, accompanied by 4.89–5.13% mass loss, attributed to the thermal decomposition of GFP in the formulation. The peak temperature of the third thermal decomposition stage is in the range 445.5–448.5 °C, accompanied by 1.58–2.87% mass loss, which maybe attribute to the addition of additives in the propellant formulations. Moreover, there are small amounts of residue after thermal decomposition, maybe attributable to the thermal decomposition residue of HTPB and $B_{10}H_{10}[(C_2H_5)_4N]_2$ in nitrogen atmosphere being mainly a cracked carbon skeleton. It also can be found that the exothermic peak temperature in the first thermal decomposition stage increases with increasing mass fraction of $B_{10}H_{10}[(C_2H_5)_4N]_2$ (increase

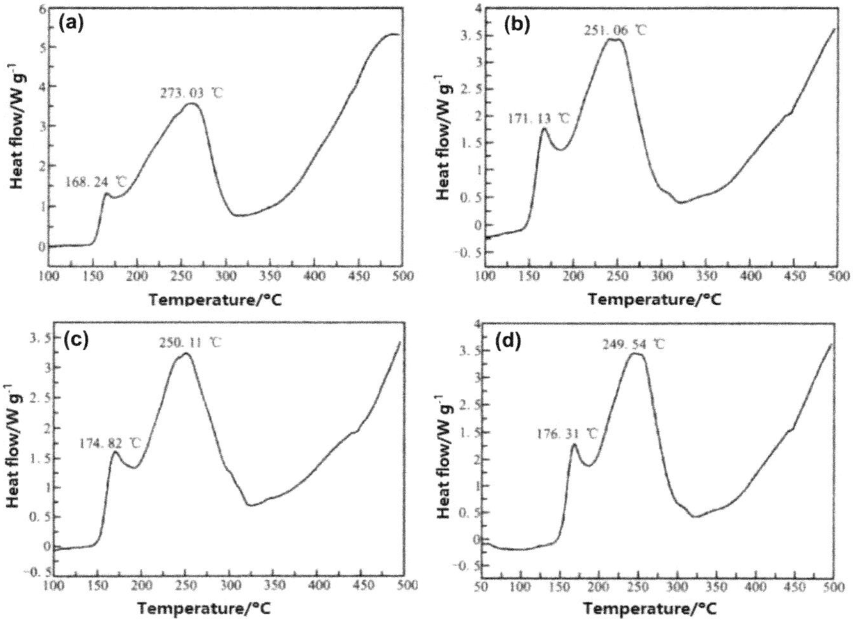

Figure 4.11 DSC curves of fuel-rich propellants containing different mass fraction of $B_{10}H_{10}[(C_2H_5)_4N]_2$. Reproduced from ref. 14.

from 268.5 °C at 0% of $B_{10}H_{10}[(C_2H_5)_4N]_2$ to 281.1 °C at 10% of $B_{10}H_{10}[(C_2H_5)_4N]_2$), making the high temperature gases heat feedback to the combustion surface of propellant reduce, thus the mass loss is reduced in the TG curves.

It can be seen from Figure 4.11 that there are two exothermic peak temperatures before 500 °C, which correspond to the first thermal decomposition stage in TG–DTG curves. There is no obvious exothermic or endothermic peak over the temperature range 322–462 °C, which may be the thermal decomposition temperature of II and III stages overlapping, causing only one large exothermic peak over this temperature range. It can also be found that the first thermal decomposition temperature increases with increasing mass fraction of $B_{10}H_{10}[(C_2H_5)_4N]_2$ in the formulation, causing needed heat in the initial thermal decomposition increase, while in the second exothermic thermal decomposition temperature decrease (from 273.0 to 249.5 °C), the temperature difference ΔT reduces. The reason may be that $B_{10}H_{10}[(C_2H_5)_4N]_2$ reduces the thermal decomposition reaction rate, and the heat of decomposition is reduced. The results indicate that the combustion agent $B_{10}H_{10}[(C_2H_5)_4N]_2$ plays a certain role in anti-catalytic deceleration of the main thermal decomposition reaction of fuel-rich propellant, and reduces the burning rate of the propellant.

4.3.3.5 *Effects of $B_{10}H_{10}[(C_2H_5)_4N]_2$ on the Combustion Properties of the Fuel-rich Propellants*

4.3.3.5.1 Burning Rate and Pressure Exponent. $B_{10}H_{10}[(C_2H_5)_4N]_2$, as one of the highest energy combustion agents, has a significant effect on the combustion properties of solid propellants.[33,34] The effects of different mass fractions of $B_{10}H_{10}[(C_2H_5)_4N]_2$ on the burning rate and pressure exponent of fuel-rich solid propellants are shown in Table 4.6. The data in Table 4.6 and Figure 4.12 show that the combustion behavior of the fuel-rich propellants changes significantly on addition of $B_{10}H_{10}[(C_2H_5)_4N]_2$ particles to the propellant formulation, compared with sample FRP-1 (reference formulation). The burning rate and pressure exponent of fuel-rich propellants decrease with an increase in the mass fraction of $B_{10}H_{10}[(C_2H_5)_4N]_2$ particles in the propellant formulation (from 0.44 to 0.41) over the pressure range 0.5–5 MPa. In particular, there is an obvious reduction for sample FRP-2 compared with the reference formulation. This may be attributed to the low oxygen content of the $B_{10}H_{10}[(C_2H_5)_4N]_2$ itself, meaning that its high energy cannot be released sufficiently when a large amount is added to the propellant formulation. Another interesting feature is that the burning rate and pressure exponent values for the fuel-rich solid propellants are dependent on the mass fraction of $B_{10}H_{10}[(C_2H_5)_4N]_2$ in the propellant composition, while the influence of the particle size of the $B_{10}H_{10}[(C_2H_5)_4N]_2$ should be investigated further.

At the same time, the effect of $B_{10}H_{10}[(C_2H_5)_4N]_2$ on the burning rate of HTPB composite solid propellant was investigated; the results are shown in Table 4.7. It was found that the burning rate of composite propellant decreases with increasing mass fraction of $B_{10}H_{10}[(C_2H_5)_4N]_2$ in the composition, while the pressure exponent changes little. The burning rate of

Table 4.6 Burning rate and pressure exponent of fuel-rich propellants with and without $B_{10}H_{10}[(C_2H_5)_4N]_2$. Reproduced from ref. 14.

Samples	Mass fraction of BHN/%	Burning rate/mm s^{-1}			
		0.5 MPa	1 MPa	3 MPa	5 MPa
FRP-1	0	14.11	17.81	32.68	37.04
FRP-2	3	11.63	15.87	25.30	30.49
FRP-3	6	11.07	14.45	23.73	28.85
FRP-4	10	11.04	14.43	22.73	28.70

Samples	Mass fraction of BHN/%	Pressure exponent (n)			
		0.5–1 MPa	1–3 MPa	3–5 MPa	0.5–5 MPa
FRP-1	0	0.34	0.55	0.25	0.44
FRP-2	3	0.45	0.42	0.37	0.42
FRP-3	6	0.38	0.45	0.38	0.42
FRP-4	10	0.38	0.42	0.46	0.41

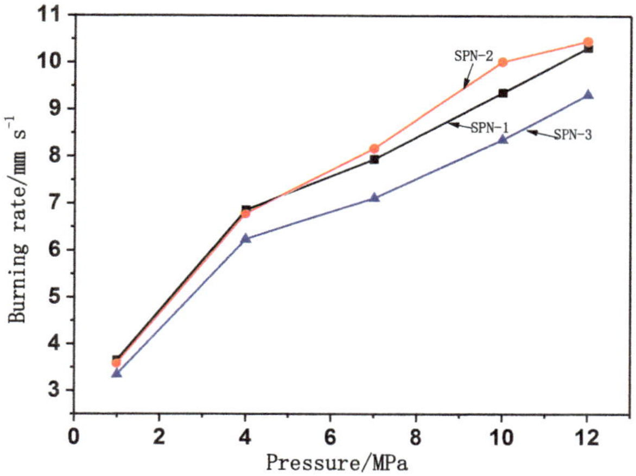

Figure 4.12 Effect of different mass fractions of $B_{10}H_{10}[(C_2H_5)_4N]_2$ on the burning rate of composite solid propellant over the pressure range from 1 MPa to 12 MPa.

Table 4.7 Burning rate of composite solid rocket propellant with and without $B_{10}H_{10}[(C_2H_5)_4N]_2$.

Samples	Mass fraction of BHN/%	Burning rate/mm s^{-1} 1 MPa	4 MPa	7 MPa	10 MPa	12 MPa
SPN-1	0	3.65	6.86	7.94	9.36	10.32
SPN-2	3	3.58	6.77	8.16	10.01	10.45
SPN-3	6	3.34	6.23	7.11	8.35	9.31
Samples	**Mass fraction of BHN/%**	**Pressure exponent (n) 1–4 MPa**	**4–7 MPa**	**7–10 MPa**	**10–12 MPa**	**1–12 MPa**
SPN-1	0	0.46	0.26	0.46	0.54	0.406
SPN-2	3	0.46	0.33	0.57	0.24	0.434
SPN-3	6	0.45	0.26	0.41	0.59	0.400

composite propellant without $B_{10}H_{10}[(C_2H_5)_4N]_2$ decreases by 10.45% compared to the composition with 6% of $B_{10}H_{10}[(C_2H_5)_4N]_2$ at 7 MPa.

4.3.3.5.2 Combustion Flame Structures. The combustion flame structures of the fuel-rich solid propellants containing different mass fractions of $B_{10}H_{10}[(C_2H_5)_4N]_2$ at 3 MPa and 5 MPa were recorded, and the images are shown in Figure 4.13.

It can be seen from Figure 4.13 that the combustion behavior of the fuel-rich propellant containing $B_{10}H_{10}[(C_2H_5)_4N]_2$ is similar to that of common composite propellants, which appears as a "multiple-flame structure". The bright flame is closer to the combustion surface and it becomes much brighter when the measured pressure is increased, in agreement with the combustion flame characteristics of common composite propellants.[27] Also, there are many

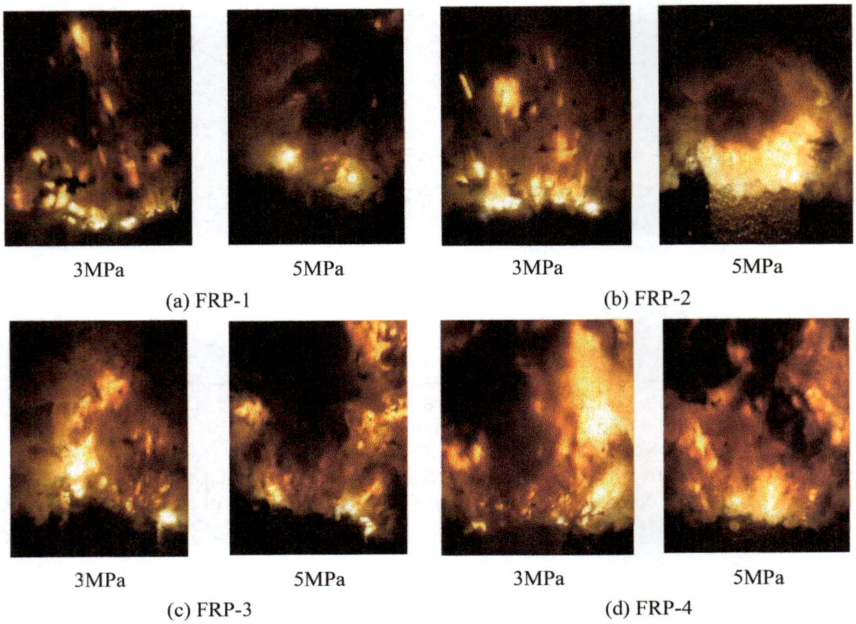

Figure 4.13 Combustion flame structures of the fuel-rich propellants with and without $B_{10}H_{10}[(C_2H_5)_4N]_2$. Reproduced from ref. 4.

sparks on the propellant surface during the combustion process, which can be attributed to the addition of metal particles to the propellant formulations.

4.3.3.5.3 Surface Morphologies. The high energy combustion agent, $B_{10}H_{10}[(C_2H_5)_4N]_2$, when added to a fuel-rich solid propellant, has a significant influence on the curing properties and the surface morphology of the propellant.[35] In order to analyze the effects of the $B_{10}H_{10}[(C_2H_5)_4N]_2$ particles on the physical structure of the fuel-rich solid propellant, the microstructure of the propellants, with and without $B_{10}H_{10}[(C_2H_5)_4N]_2$, is shown in Figure 4.14.

Figure 4.14 indicates that there are obviously many approximately spherical and granulated particles on the surface of the cured fuel-rich solid propellant, which may be attributed to the addition of spherical magnesium particles to the propellant formulation. The $B_{10}H_{10}[(C_2H_5)_4N]_2$ particles are compatible with the ingredients of the propellant systems, and additionally the granulated particles with smaller diameters can adequately fill the spaces between the larger grains. With the increased $B_{10}H_{10}[(C_2H_5)_4N]_2$ mass fraction in the propellant formulations, there is an increase in irregular particles on the cured surface of the propellant, caused by the approximately spherical magnesium particles being replaced by an equal mass fraction of irregular $B_{10}H_{10}[(C_2H_5)_4N]_2$ particles.

4.4 Effect of $B_{12}H_{12}[(C_2H_5)_4N]_2$ on the Properties of NEPE Propellants

4.4.1 Experimental

The basic formulation of nitrate ester plasticized polyether (NEPE) propellant is: AP 5–10%, Al 17–20%, RDX 45–50%, NG/DEGN 10–20%, the mass ratio of nitrate plasticizer to polymer is 2.8, and the curing parameter $R = [NCO]/[OH]$ is 1.0–1.2.

Preparation of samples: each component of the propellant was weighted according to its mass fraction, the mixture was stirred and vacuumized, then moved to the water oven at temperature 50 °C for 5 days.

4.4.2 Characterization Methods

DSC analysis: DSC-7 thermal analysis apparatus with N_2 atmosphere (PerkinElmer) was used, the heating rate is 10 °C min^{-1}.

Burning rate measurement: a constant pressure static burning rate meter is used to measure the burning rate of propellant over the pressure range 4–11 MPa.

4.4.3 Results and Discussion

4.4.3.1 *Compatibility of $B_{12}H_{12}[(C_2H_5)_4N]_2$ with Nitrate Esters*

NEPE propellant contains a large number of energetic plasticizers – mixed nitrate esters – which are sensitive to chemical and physical stimuli. The compatibility between the burning rate catalyst and nitrate ester is directly related to the curing quality, mechanical properties and combustion performance of the propellant. Therefore, studying the compatibility of the burning rate catalyst and nitrate ester is a key aspect in evaluating the application of burning rate catalysts. Table 4.8 shows the compatibility data of $B_{12}H_{12}[(C_2H_5)_4N]_2$ and nitrate esters obtained by means of the DSC technique.

It can be seen that addition of $B_{12}H_{12}[(C_2H_5)_4N]_2$ to the propellant formulation meant that the thermal decomposition peak temperature of NG was advanced by 18 °C and the decomposition activation energy reduced by 12 kJ mol^{-1}. At the same time, the thermal decomposition peak temperature of DEGDN was advanced by 3 °C and the decomposition activation energy reduced by 13 kJ mol^{-1}. The thermal decomposition peak temperature of NG/DEGDN was advanced by 7 °C and the decomposition activation energy reduced by 18 kJ mol^{-1}. According to the compatibility evaluation standard, the compatibility of $B_{12}H_{12}[(C_2H_5)_4N]_2$ with three types of nitrate esters is poor, indicating that $B_{12}H_{12}[(C_2H_5)_4N]_2$ cannot be applied in NEPE propellant directly. Furthermore, the solidified peak temperature and activation energy of propellant without $B_{12}H_{12}[(C_2H_5)_4N]_2$ is 83.28 °C and 41.4 kJ mol^{-1}, respectively, while the solidified peak temperature and activation energy of

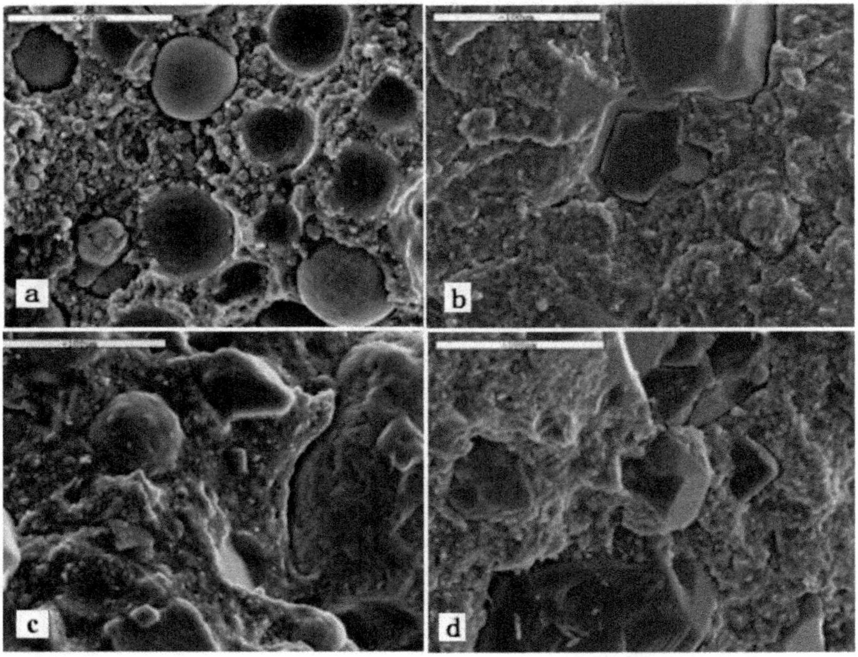

Figure 4.14 The surface morphologies of the fuel-rich propellants with and without $B_{10}H_{10}[(C_2H_5)_4N]_2$ (×500): (a) FRP-1; (b) FRP-2; (c) FRP-3; (d) FRP-4. Reproduced from ref. 4.

propellant containing $B_{12}H_{12}[(C_2H_5)_4N]_2$ is 83.37 °C and 42.3 kJ mol^{-1}, respectively.

4.4.3.2 *Effect of $B_{12}H_{12}[(C_2H_5)_4N]_2$ on Thermal Decomposition of Components*

The effect of $B_{12}H_{12}[(C_2H_5)_4N]_2$ on the thermal decomposition of AP were investigated by combining $B_{12}H_{12}[(C_2H_5)_4N]_2$ and AP (coarse/fine = 1 : 1) in a mass ratio of 1 : 6 mixture. The results indicate that addition of $B_{12}H_{12}[(C_2H_5)_4N]_2$ to AP meant that the melting peak temperature reduced by 3 °C and the thermal decomposition peak temperature in low temperature and high temperature changed little. At the same time, the effect of $B_{12}H_{12}[(C_2H_5)_4N]_2$ on the thermal decomposition of RDX and HMX were carried out by combining $B_{12}H_{12}[(C_2H_5)_4N]_2$ and RDX or HMX in a mass ratio of 1 : 10 mixture. The results indicate that addition of $B_{12}H_{12}[(C_2H_5)_4N]_2$ to RDX or HMX meant that the thermal decomposition peak temperature of RDX or HMX reduced by 19 °C and 54 °C, respectively, indicating that $B_{12}H_{12}[(C_2H_5)_4N]_2$ has a strong catalytic effect on the thermal decomposition of RDX and HMX.

Table 4.8 The compatibility of $B_{12}H_{12}[(C_2H_5)_4N]_2$ and nitrate esters. Reproduced from ref. 20.

Sample	T_p/°C	ΔT_p/°C	E/kJ mol^{-1}	$\lvert \Delta E/E \rvert \times$ 100%	Compatibility evaluation
NG	201.01	—	137.2	—	—
NG/$B_{12}H_{12}$-$[(C_2H_5)_4N]_2$	182.97	18.04	125.4	8.6	3
DEGDN	204.45	—	141.6	—	—
DEGDN/$B_{12}H_{12}$-$[(C_2H_5)_4N]_2$	201.69	2.76	128.6	10.9	3
NG/DEGDN	207.45	−6.44	154.1	0.6	1
NG/DEGDN/$B_{12}H_{12}$-$[(C_2H_5)_4N]_2$	200.44	7.01	136.4	13.0	3

Table 4.9 Burning rate of NEPE propellant with $B_{12}H_{12}[(C_2H_5)_4N]_2$. Reproduced from ref. 20.

	Burning rate/mm s^{-1}					n	
Catalyst	4 MPa	6 MPa	7 MPa	9 MPa	11 MPa	4–11 MPa	7–11 MPa
$B_{12}H_{12}[(C_2H_5)_4N]_2$	7.15	8.46	11.85	12.16	12.56	0.59	0.13
Lead citrate	6.42	8.24	8.75	10.07	11.55	0.57	0.61

4.4.3.3 *Effect of $B_{12}H_{12}[(C_2H_5)_4N]_2$ on the Combustion of NEPE Propellant*

The combustion performance of NEPE propellant was investigated, and the effect of $B_{12}H_{12}[(C_2H_5)_4N]_2$ on the burning rate of NEPE propellant was investigated. Table 4.9 shows the test results. It can be seen that the burning rate of NEPE propellant containing $B_{12}H_{12}[(C_2H_5)_4N]_2$ at 7 MPa increase by 35.63% compared to that of propellant with lead citrate. The reason for this may be that $B_{12}H_{12}[(C_2H_5)_4N]_2$ can catalyze the decomposition of nitramine directly and increase the burning rate of NEPE propellant by accelerating the decomposition rate of nitramine. It can also increase the burning rate of NEPE propellant by accelerating the thermal decomposition of nitrate ester. Moreover, the pressure exponent reduced from 0.61 to 0.13 over the pressure range 7–11 MPa, which may be due to the "platform" phenomenon in the high pressure range.

4.5 Stability and Infrared Spectrum of Boron Hydride (BH_n) ($n = 1$–5)

4.5.1 Theoretical Calculation Method

The possible geometrical structure of boron hydride (BH_n) ($n = 1$–5) has been calculated using optimized computation, based on the density functional theory B3P86 method and at the dunning related consistent base group cc-PVTZ level. The configuration geometric parameter, electronic structure, vibrational frequency and spectrum of the most stable structure were obtained, and given total energy (E_T), binding energy (E_{BT}), average binding energy (E_{av}), ionization potential (E_{ip}), energy crack (E_g), Fermi level (E_F) and so on. The second order difference on energy is $\Delta_2E(BH_n) = [E(BH_{n+1}) + E(BH_{n-1}) - E(BH_n)]$. The difference between the highest occupied orbital energy and the lowest altitude orbital energy is the energy gap (E_g). The highest occupied orbital energy is the Fermi level (E_F). The differential adsorption energy of hydrogen atom is $E_{diff} = -(E_{BH_n} - E_{BH_{n-1}} - 1/2E_{H_2})$.

4.5.2 Calculation Structure

4.5.2.1 The Balanced Geometry

The B3P86 method in DFT and dunning related consistent basis set cc-pVTZ are used to comprehensively optimize the bond length and bond angle of all possible geometric structures and multiplicities of ($B_{10}H_{10}[(C_2H_5)_4N]_2$) ($n = 1$–5). According to the principle of the minimum energy, the most stable ground state structure of ($B_{10}H_{10}[(C_2H_5)_4N]_2$) ($n = 1$–5) is obtained, as shown in Figure 4.15. The binding energy is in brackets. It can be seen that (BH_n) ($n = 1$–5) stable structures show all forms of hydrogen bridge bonds, and the structural symmetry of n is better when n is an odd number of hydrides, among which the symmetry of the BH_3 planar triangular structure is the best.[36]

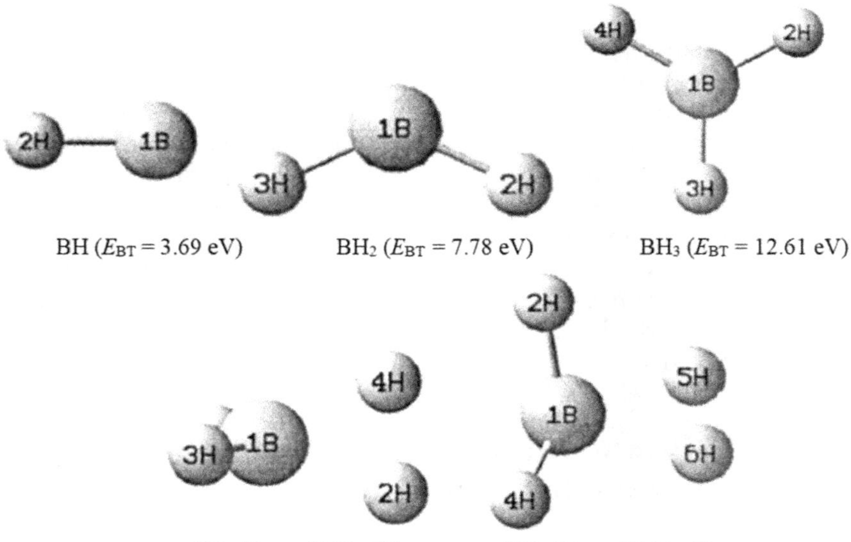

Figure 4.15 The ground state structure and binding energy of (BH_n) ($n = 1$–5). Reproduced from ref. 36.

Table 4.10 The geometric parameters of the most stable configuration (BH_n) ($n = 1$–5). Reproduced from ref. 36.

Molecule	Symmetry	Structure	Energy/ a.u.	Bond length/Å		Bond angle/°	
BH	$C_{\infty V}$	$^1\Sigma_g$	−25.394	R(1,2)	1.2325	—	—
	$C_{\infty V}$				1.2328[a]		
					1.2327[b]		
					1.2322[c]		
BH_2	C_{2V}	^2A	−26.063	R(1,2) = R(1,3)	1.1881	A(213)	129.31
			−26.060[d]		1.1899[d]		129.4[d]
			−26.061[c]		1.1860[e]		129.3[c]
BH_3	C_{3h}	^1A	−26.759	R(1,2) = R(1,3) = R(1,4)	1.1901	A(132)	120.08
			−26.506[f]				
BH_4	C_4	^2A	−27.321	R(1,2)	1.2858	A(214)	48.24
				R(1,3)	1.1826	A(315)	129.50
				R(1,4)	1.2848	D(1354)	153.71
				R(1,5)	1.1818	D(1352)	−153.58
BH_5	C_1	^1A	−27.991	R(1,2)	1.2036	A(213)	114.74
				R(1,3)	1.1921	A(314)	120.85
				R(1,4)	1.1922	A(214)	114.62
				R(1,5)	1.3680	D(3145)	130.35
				R(1,6)	1.3841	D(4136)	−98.16

[a] From ref. 38.
[b] From ref. 39.
[c] From ref. 37.
[d] From ref. 40.
[e] From ref. 41.
[f] From ref. 42.

Table 4.10 lists the equilibrium geometric parameters such as electronic state, point group, energy, bond length and bond angle of the most stable structure (BH_n) ($n = 1$–5). The experimental[37] and literature values[38–42] are in the brackets. It can be seen that the electronic state of n with odd hydride is $^1\Sigma_g$ and ^1A, n is even electronic states, ^2A, the bond length and the bond angle are closer to the experimental values than the literature values, and the energy is lower than the literature values. For hydrides BH_4 and BH_5 with a hydrogen bridge bond, the bond length of the latter is shorter than that of the former, indicating that the shorter the hydrogen bridge bond, the more stable the structure.

4.5.2.2 *Infrared Spectrum*

The most stable structured vibrational spectra of (BH_n) ($n = 1$–5) hydride were calculated using the B3P86 method at the cc-pVTZ level. The infrared IR spectra are shown in Figure 4.16. The calculated vibrational frequencies are positive when less than 6 and negative when greater than 5, which indicates

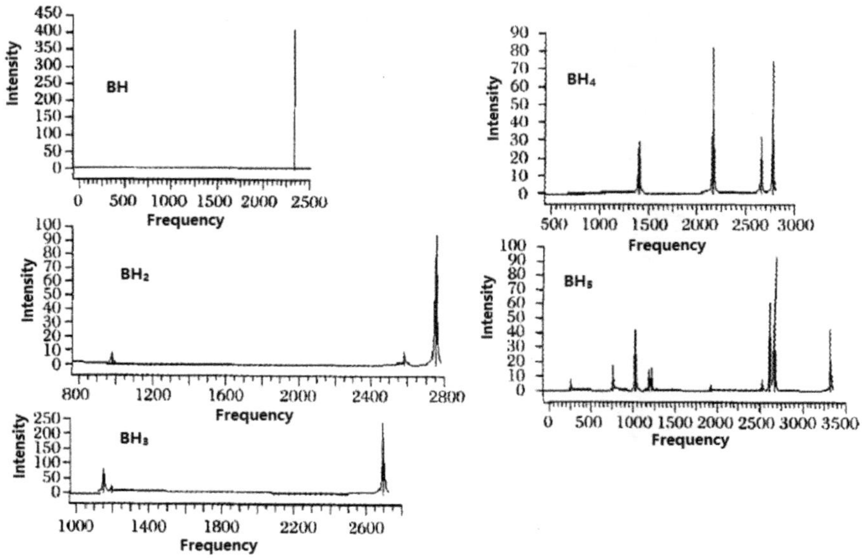

Figure 4.16 The infrared spectrum of BH–BH_5 (frequency: cm^{-1}; intensity: km mol^{-1}). Reproduced from ref. 36.

that the most stable structures of borohydride $n = 1$–5 are the minimum points on the potential energy surface, while n greater than 6 is an unstable structure.

The strongest vibration peak of IR spectrum of BH hydride is 2335.46 cm^{-1} (2366.73 cm^{-1}[37]), which is mainly reflected in the stretching vibration of the H atom, and its infrared intensity is 410.18 km mol^{-1}. At the same time, the strongest vibration peak of IR spectrum of BH_2 hydride is 2754.21 cm^{-1} (2759.30 cm^{-1}[38]), which is mainly reflected in the antisymmetric stretching vibration of the H atom, and its infrared intensity is 94.79 km mol^{-1}. The IR spectra of the BH_3 hydride are mainly concentrated at 1149.14 cm^{-1} and 2695.79 cm^{-1}, and the strongest vibration peak is at 2695.79 cm^{-1}, which is mainly realized as the antisymmetric stretching vibration of the end bond H atom, and its infrared intensity is 122.81 km mol^{-1}. The IR spectra of the BH_4 hydride are mainly concentrated at 1404.18 cm^{-1}, 2158.74 cm^{-1}, 2647.21 cm^{-1} and 2767.67 cm^{-1}, respectively. The strongest vibration peak is at 2158.74 cm^{-1}, which is mainly realized as the symmetric stretching vibration of the hydrogen bridge H atom, and its infrared intensity is 82.21 km mol^{-1}. The IR spectra of the BH_5 hydride are mainly concentrated at 1031.73 cm^{-1}, 2610.2l cm^{-1}, 2665.56 cm^{-1} and 3322.44 cm^{-1}, respectively. The strongest vibration peak is 2665.56 cm^{-1}, which is mainly realized as the antisymmetric stretching vibration of the end bond H atom. Its infrared intensity is 93.45 km mol^{-1}, indicating that the (BH_n) ($n = 1$–5) hydride is mainly characterized by the antisymmetric stretching vibration of the end bonded hydrogen atom.

4.5.2.3 Energy and Stability

At the B3P86/cc-PVTZ level, the energy of (BH_n) ($n = 1–5$) stable structures of each hydride molecule is optimally calculated. By defining $\Delta_2E(BH_n) = [E(BH_{n+1}) + E(BH_{n-1}) - E(BH_n)]$ from the second-order energy difference, the second-order energy difference value of (BH_n) ($n = 1–5$) can be calculated, and the results are shown in Figure 4.16. It can be seen that there is a significant "odd even" oscillation effect, the second-order difference value of energy with hydride containing an odd number of hydrogen atoms is larger than that of the adjacent even number hydrogenation, indicating that a hydride containing an odd number is more stable, especially for BH_n. At the same time, the mulliken atomic charge of each hydride was calculated and the charge of the B atom in BH–BH_5 is −0.008, −0.130, −0.139, −0.406 and −0.525, respectively. This indicates that the electrons migrate from the H_n atom to the B atom, and the electrons obtained by the B atom are negatively charged. Since the outer electrons of B atom are 3, which is odd, there are an odd number of H atoms to provide an even number of electrons to the B atom, which are easier to combine, showing that the stability with odd number hydrides is higher than that of even number ones.

4.5.2.4 Electronic Properties

The ionization potential E_{ip} and energy gap E_g, average binding energy E_{av}, Fermi energy level E_F and hydrogen differential adsorption energy E_{diff} of (BH_n) ($n = 1–5$) hydride molecules were calculated by B3P86 method at the cc-pVTZ level. Figure 4.17 shows the Δ_2E of (BH_n) ($n = 1–5$) along with the numbers of H. It can be seen that E_{ip}, E_g and E_{av} of the BH_3 molecule is the highest among the five hydrides, indicating that the hydride is relatively the most stable, and the Fermi level has a significant "odd even" oscillation effect. The ionization potential and energy gap also have obvious odd even oscillation effect, and it is considered that the ionization potential and energy gap of odd numbered hydrides are larger than those of adjacent even numbers, among which BH_3 is the largest ($E_{ip} = 12.52$ eV, $E_g = 7.59$ eV), indicating that BH_3 is the most stable among all hydride hydrogen storage materials (Figure 4.18).

The E_{diff} with odd n clusters is larger than that with even n clusters, and BH_3 has the largest E_{diff} among the adjacent hydride cluster molecules. In the analysis of the infrared spectrum, the average binding energy, the ionization potential, the energy gap and Fermi level characteristics, it is shown that BH_3 is the most stable molecule in (BH_n) ($n = 1–5$) for its energy gap, the ionization potential are maximal values, the H–B bridge bond key is longer than the terminal linkage, and the infrared intensity of strongest peak is at its maximal value.

4.6 Conclusions

(1) The combustion efficiency of small molecular weight combustible gas with high calorific value released from the decomposition process of

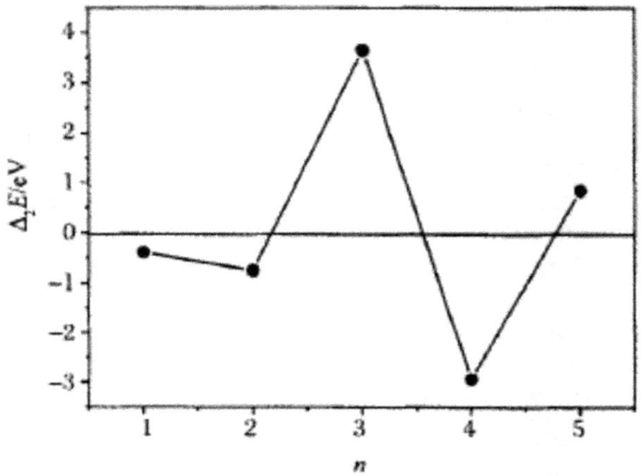

Figure 4.17 The Δ_2E of (BH_n) ($n = 1$–5) along with the numbers of H. Reproduced from ref. 36.

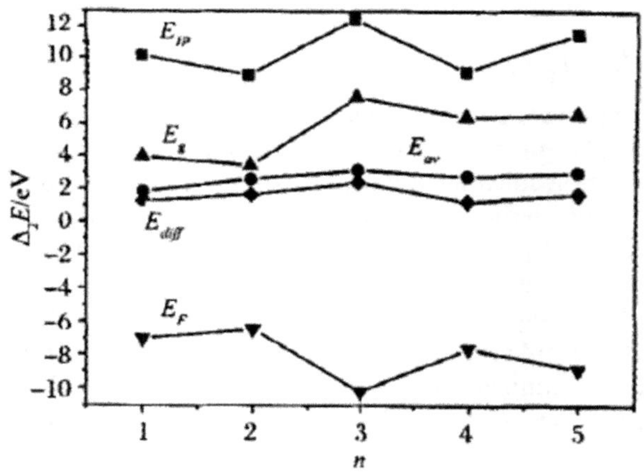

Figure 4.18 The E_{ip}, E_{av}, E_g, E_F, and E_{diff} of (BH_n) ($n = 1$–5) along with the numbers of H. Reproduced from ref. 36.

$B_{10}H_{10}[(C_2H_5)_4N]_2$/nAl composite is high, which can improve the gas work ability of the propellant. At the same time, the solid boron produced in the decomposition process can also release a lot of energy in the oxidation or fluorination processes, which can greatly improve the energy release ability of the propellant. The composite fuel has a potential application prospect in propellants and explosives.

(2) The heat of combustion of fuel-rich propellant increases with an increase in the mass fraction of BHN, but the density decreases, which has a certain negative influence on the propellant charge with limited volume of combustion chamber.

(3) The mechanical sensitivity of BHN is low, which indicates that BHN is safe, while BHN has a significant effect on the mechanical sensitivity of fuel-rich propellants, which can increase the mechanical sensitivity of the fuel-rich propellant obviously; in particular the characteristic drop height increases 7.9 times higher than the reference formulation.

(4) The burning rate and pressure exponent of fuel-rich solid propellants decrease with increasing mass fraction of BHN in the propellant formulation, the burning rate of fuel-rich propellant containing 10% mass fraction of BHN at 3.0 MPa decreases by 30.45%, and the pressure exponent decreases from 0.442 to 0.415.

References

1. W. Weiqiang, X. Yunna and Y. Jianming, *et al.*, Review of boron hydrides used in high burning rate propellant, *Chin. J. Energ. Mater.*, 2012, **20**(1), 132–136.
2. B. R. S. Hansen, M. Paskevicius and H. W. Li, *et al.*, Metal boranes: progress and applications, *Coord. Chem. Rev.*, 2016, **323**, 60–70.
3. S. Zixing, S. Lili and Y. Rongjie, Research progress icosahedronal polyhydroborate $B_{12}H_{12}^{2-}$ anion compounds and their application in propellants and explosives, *Chin. J. Explos. Propellants*, 2017, **40**(3), 1–16.
4. P. Weiqiang, Z. Fengqi and X. Yunna, *et al.*, Synthesis and characterization of a high energy combustion agent (BHN) and its effects on the combustion properties of fuel-rich solid rocket propellants, *Cent. Eur. J. Energ. Mater.*, 2015, **12**(3), 537–552.
5. C. Dillier, A. R. Demko, J. C. Thomas, *et al.*, Evaluation of composite propellants utilizing various nano-scale aluminum and boron incorporation methods, *AIAA Scitech 2019 Forum*, AIAA, Reston, 2019, p. 1238.
6. H. Belal, C. W. Han and I. E. Gunduz, *et al.*, Ignition and combustion behavior of mechanically activated Al-Mg particles in composite solid propellants, *Combust. Flame*, 2018, **194**, 410–418.
7. N. Zakiyyan, A. Wang and R. Thiruvengadathan, *et al.*, Combustion of aluminum nanoparticles and exfoliated 2D molybdenum trioxide composites, *Combust. Flame*, 2018, **187**, 1–10.
8. V. Bagalkote, D. Grinstein and B. Natan, Energetic nanocomposites as burn rate catalyst for composite solid propellants, *Propellants, Explos., Pyrotech.*, 2018, **43**(2), 136–143.
9. D. Li, Y. Ye and X. Liao, *et al.*, A novel method for preparing and characterizing graphene nanoplatelets/aluminum nanocomposites, *Nano Res.*, 2018, **11**(3), 1642–1650.
10. Y. Zhou, W. G. Jiang and X. Q. Feng, *et al.*, In-plane compressive behavior of graphene-coated aluminum nano-honeycombs, *Comput. Mater. Sci.*, 2019, **156**, 396–403.

11. A. Bhadauria, L. K. Singh and T. Laha, Effect of physio-chemically functionalized graphene nanoplatelet reinforcement on tensile properties of aluminum nanocomposite synthesized via spark plasma sintering, *J. Alloys Compd.*, 2018, **748**, 783–793.
12. Y. Jiang, S. Deng and S. Hong, *et al.*, Energetic performance of optically activated aluminum/graphene oxide composites, *ACS Nano*, 2018, **12**(11), 11366–11375.
13. J. Fan, N. Yulei and X. Xingle, *et al.*, Preparation and properties of tetraethylammonium dodecahydrodedecaborates/nano-aluminium composite, *Sci. Technol. Eng.*, 2019, **19**(36), 1671–1815.
14. P. Weiqiang, F. Xuezhong and Z. Fengqi, *et al.*, Effects of high energetic combustion agent (BHN) on the properties of fuel-rich solid propellant, *J. Solid Rocket Technol.*, 2013, **36**(5), 647–653.
15. M. Kaltchev, A. W. Toompson and W. T. Tysoe, Reflection-absorption infrared spectroscopy of ethylene on palladium (1 1 1) at high pressure, *Surf. Sci.*, 1997, **391**(1–3), 145–149.
16. F. E. Pinkerton, M. S. Meyer and G. P. Meisner, *et al.*, Phase boundaries and reversibility of $LiBH_4/MgH_2$ hydrogen storage material, *J. Phys. Chem. C*, 2007, **111**(35), 12881–12885.
17. S. Wengang, S. Tiegang and Z. Guodong, Effects of carborane derivatives on the combustion of solid propellants, *J. Solid. Rocket Technol.*, 1995, **18**(3), 24–26.
18. D. V. Peryshkov, A. A. Popov and S. H. Strauss, Direct perfluorination of $K_2B_{12}H_{12}$ in acetonitrile Occurs at the gas bubble - solution interface and is inhibited by HF. Experimental and DFT study of inhibition by protic acids and soft. polarizable anions, *J. Am. Chem. Soc.*, 2009, **131**, 18393–18403.
19. V. V. Drozdova, K. Y. Zhizhin and E. A. Malinina, *et al.*, Reaction of the closo - decaborate anion $B_{10}H_{10}{}^{2-}$ with dichloroethane in the presence of hydrogen halides, *Russ. J. Inorg. Chem.*, 2007, **52**(7), 996–1001.
20. X.-G. Wu, Q.-L. Yan and X. Guo, *et al.*, Combustion efficiency and pyrochemical properties of micron-sized metal particles as the components of modified double-base propellant, *Acta Astronautica*, 2011, **68**(7–8), 1098–1112.
21. C. Futai, T. Huimin and L. Yunjun, Effect of $B_{12}H_{12}[N(C_2H_5)_4]_2$ on combustion property of NEPE propellant, *Chin J. Explos. Propellants*, 2000, **27**(3), 19–21.
22. T. DeLuca Luigi, L. Galfetti and F. Maggi, *et al.*, Innovative metallized formulations for solid rocket propulsion, *Chin. J. Energ. Mater.*, 2012, **20**(4), 465–474.
23. O. G. Glotov, V. E. Zarko, V. V. Karasev, *et al.*, Effect of binder on the formation and evolution of condensed combustion product of metallized solid propellants, *28th International Annual Conference of ICT*, Karlsruhe, Federal Republic of Germany, Jun, 1997.
24. S. V. Chuiko and F. S. Sokolovskii, Effect of small additives on the combustion of composite solid propellants, *Russ. J. Phys. Chem. B*, 2009, **3**(6), 926–935.

25. G. Guoyiqibayi, Z. Rui and Y. Hong, Recent development on organometallic chemistry containing carborane ligand, *Chin. J. Inorg. Chem.*, 2010, **26**(5), 733–743.
26. N. Yong, C. Haiyan and M. Jinling, Advances in the chemistry and applications of the borane anion closo- $[B_{10}H_{10}]^{2-}$ and its derivatives., *Chin. J. Organic Chem.*, 2009, **29**(6), 822–834.
27. S. M. Vuga, Effects of liquid burn rate catalysts on rheological properties of high-energy composite propellants, *Propellants, Explos., Pyrotech.*, 1991, **16**, 293–298.
28. Development of Sensitivity Tests at the Explosive Research Laboratory, ed. D. H. Mallory, Bruceton Pennsylvania, NAVORD, Report No. 4236, 1960.
29. J. K. G. Peters, Production program of julius peter company for members of M. B. B., Course-81, Berlin, 1921, p. 14.
30. J. Šelešovský and J. Pachman, Probit analysis - a promising tool for evaluation of explosive's sensitivity, *Cent. Eur. J. Energ. Mater.*, 2010, **7**(3), 269–278.
31. F. Maggi, A. Bandera and L. T. DeLuca, *et al.*, Agglomeration in solid rocket propellants: novel experimental and modeling methods, *Prog. Prop. Phys.*, 2011, **2**, 81–98.
32. P. Weiqiang, F. Xuezhong and X. Yunna, *et al.*, Study on the compatibility of tetraethylammonium decahydrodecaborate (BHN) with some energetic components and inert materials, *Propellants, Explos., Pyrotech.*, 2013, **38**, 278–285.
33. I. B. Sivaev and V. I. Bregadze, Chemistry of nickel and iron bis(dicarbollides). A review, *J. Organomet. Chem.*, 2000, **614–615**, 27–36.
34. Z. Ren, L. Jiahua and W. Wujun, Development of ferrocene burning rate catalysts, *J. Propulsion Technol.*, 1994, **3**, 62–65.
35. L. Ruibin and L. Zhanxiong, Synthesis and thermal property of energetic caged - carborane derivatives, *Chin. J. Synth. Chem.*, 2010, **18**(2), 229–231.
36. Y. Ling, W. Donglan and Z. Xuefeng, The stability and infrared spectrum of boron hydride (BH_n) ($n = 1–5$), *J. Sichuan Univ.*, 2012, **49**(5), 1067–1072.
37. K. P. Huber and G. Herzberg, Molecular spectra and molecular structure. IV, *Constants of Diatomic Molecules*, Van Nostrand Reinhold Co., 1979.
38. S. Jinfeng, W. Jiemin and S. Deheng, *et al.*, Structure and analytic potential energy functions of the molecules BH_2 and AlH_2, *Acta Phys. Sin.*, 2006, **55**(9), 4490–4495.
39. W. Xinqiang, Y. Chuanlu and S. Tao, *et al.*, Analytical potential energy functions and spectroscopic properties of the ground and excited states of BH molecule, *Acta Phys. Sin.*, 2009, **58**(10), 6873–6878.
40. Y. Shiying, The molecular structure and potential energy function of the ground state of BH_2 molecule, *Acta Phys. Sin.*, 2006, **55**(7), 3408–3412.

41. B. S. Ursic, Density functionaI theory study of multiprotonated BH_5, *Theochem*, 2000, **505**, 67.
42. B. J. Duke, C. Liang and H. F. Schaefer III, Properties of small group IIIA hydrides including the cyclic and pertacoordinate structures of trialane ($A1_3H_9$) and trigallane (Ga_3H_9): can dialane be isolated?, *J. Am. Chem. Soc.*, 1991, **113**, 2884.

CHAPTER 5

Performance of Solid Propellants Containing Organic Borohydride Metal Additives

5.1 Introduction

Polyhedral borohydride has the characteristic of a high burning rate. It can be used as a burning rate regulator of solid propellant.[1,2] It also has a high combustion calorific value and combustion efficiency. Therefore, it is one of the high energy fuels with application potential. In recent years, the advantages of borohydride compounds, such as high calorific value and complete combustion, have gradually aroused the interest of researchers at home and abroad, especially for $B_{10}H_{10}^{2-}$ and $B_{12}H_{12}^{2-}$ salts.[3,4] Combined with the research results of the thermal decomposition characteristics of $B_{10}H_{10}^{2-}$ and $B_{12}H_{12}^{2-}$ salts,[5–7] it is considered that the stability of $B_{12}H_{12}^{2-}$ salts is higher than that of $B_{10}H_{10}^{2-}$ salts, but in terms of combustion, the $B_{12}H_{12}^{2-}$ salt decomposition temperature is higher than 700 °C and combustion needs to absorb a lot of energy, while the thermal decomposition temperature of $B_{10}H_{10}^{2-}$ salts is about 300 °C, which is an exothermic reaction. Therefore, $B_{10}H_{10}^{2-}$ salts are considered to be more valuable in propellant applications. So, two kinds of $B_{10}H_{10}^{2-}$ salts (BH-1-10 and BH-2-10) were prepared and the application performance aspects were investigated to improve application in solid propulsion. Moreover, there are few reports on the combustion properties of composite solid propellants containing an organic borohydride iron salt (BH–Fe), organic borohydride copper salt (BH–Cu), organic borohydride lead salt (BH–Pb) additives, *etc.*

High-energy Combustion Agents of Organic Borohydrides: Synthesis, Characterization and Applications
By XiaoHong Zhang, WeiQiang Pang, Luigi T. DeLuca and Yu Zhao

Published by the Royal Society of Chemistry, www.rsc.org

In this chapter, the characteristics of the BH–Fe, BH–Cu, and BH–Pb particles are analyzed by using the diagnostic techniques of scanning electron microscopy (SEM) and grain size distribution. Different mass fractions of BH–Fe, BH–Cu, BH–Pb, and $B_{10}H_{10}^{2-}$ salt (BH-1-10 and BH-2-10) particles were added to the formulations and three different propellant compositions with and without additives were produced. The focus of this chapter is on how organic borohydride metal additives affect the combustion properties (burning rate and pressure exponent) of composite solid propellants, placing the emphasis on the thermal decomposition, combustion flame structure, and combustion residue performance of the solid propellants, which could be used for solid rocket motor applications.

5.2 Experimental

5.2.1 Raw Materials

The binder system consists of hydroxyl-terminated polybutadiene (HTPB, $E_{OH} = 7.8 \times 10^{-4}$ mol g^{-1}) plasticized with di-2-etylhexyl sebacate (DOS) to increase the processibility, which is cured by 2,4-toluene diisocyanate (TDI). The additive isophthaloyl-bis-(2-methylaziridine) (HX-752) and bi-(2-methy-1-aziridinyl) phosphine oxide (MAPO) are used as a cross-linker and a bonding agent, respectively. Aluminum powders with an average particle size of 5 μm were used in the formulations (purchased from Nanjing Chemical Industry Co., Ltd), and parts of AP were replaced by organic borohydride iron (BH–Fe) or organic borohydride copper (BH–Cu) additives. The involved propellant formulations also contain catocene (GFP) and chromite copper (CC) as burning rate modifier. Ammonium perchlorate (AP), as oxidizer was always used, and kept tri-modal. The first and the second modal of AP consisted of pure research grade AP (>99% pure) with an average particle size of 0.178–0.250 mm and 0.105–0.150 mm, respectively. The third modal of AP was made by grinding AP (>99% pure) in a fluid energy mill to an average particle size of around 1–3 μm.

5.2.2 Tested Ingredients and Formulations

The composite solid propellant used as the baseline is an HTPB/AP/Al/catalyst formulation in the mass ratio 11.5/65/18/5.5 for all involved samples. Different mass fractions of BH–Fe and BH–Cu additives, respectively, were used as a catalyst additive by replacing AP particles with 0.178–0.250 mm diameter. Except where otherwise stated, all propellants were manufactured, processed, and tested at Xi'an Modern Chemistry Research Institute under identical conditions and using identical procedures. The samples SPF-1, SPF-2, SPF-3, SPF-4 are solid propellant containing 0%, 1%, 3%, 5% of BH–Fe additives, respectively. The samples HMA-1, HMA-2, HMA-3, HMA-4, HMA-5, HMA-6 are solid propellant containing 0% organic borohydride metal additive, 1% BH–Fe, 1% BH–Cu, 3% BH–Cu, 1% BH–Pb, 2% BH–Cu, respectively. The reference compositions of SPF-1 and HMA-1 are of different formulation with different particle size and size distribution of oxidizers.

5.2.3 Processing

Propellant formulations were mixed in 500 g batches using a vertical planetary mixer of 2 L capacity. All batches were mixed and cast under a vacuum by a slurry cast technique.[8] The propellant was cured at 50 °C for 3 days in a water jacketed oven to solidify the propellant samples.

5.2.4 Characterization Methods

5.2.4.1 BET and SEM Tests

The specific surface area measurement was computed from the nitrogen adsorption isotherm, obtained by static volumetric measurement at a liquid nitrogen boiling temperature (77 K). Samples were out-gassed at 100 °C for at least 4 h at an absolute pressure of less than 10^{-3} torr. All measurements were carried out on a completely automated instrument (ASAP 2010 from Micromeritics, USA), leading to the final value of the specific surface areas expressed in $m^2\,g^{-1}$. Electron microscopy was used to study the shape, size, morphology, and defects of the powders. The morphologies of the solid particles were examined using scanning electron microscope (SEM) technology.

5.2.4.2 Particle Size Distribution Experiments

Particle size and size distribution of samples were performed through laser scattering (Malvern Master sizer 2000) using a dry dispersion unit (Scirocco). Tests were performed with the same feed rate (45%) and varying feeding pressure in order to achieve sensitivity on this parameter. The quantity of material per test was about 0.07–0.09 g, fed through the "Micro Volume" tray. Obscuration filtering was switched on and set to values within the range 0.5–10%.

5.2.4.3 Density Test

The density (ρ, $g\,cm^{-3}$) measurement of solid propellants was carried out on a Model AG 104 METTLER TOLEDO balance with samples of 30 mm × 30 mm × 10 mm rectangular sticks in a steeping medium of liquid paraffin at the temperature of (20 ± 2) °C.

5.2.4.4 Hazardous Properties Test

The impact instrument made in Xi'an Modern Chemistry Research Institute was used to test the impact sensitivity of the propellant with different metric metal particle samples. Each sample (35 mg) was tested 25 times to obtain the H_{50} values (the H_{50} value represents the height from which dropping a 5 kg weight results in an explosive event in 50% of the trials). With five peering tests, an average value of H_{50} was calculated. Friction instrument (90°, 3.92 MPa), also made in Xi'an Modern Chemistry Research Institute, was employed to test the friction sensitivity of the samples. Each sample (20 mg) was tested 25 times and an explosive probability P (%) was obtained. An average value of P was estimated with five peering tests.[9,10]

5.2.4.5 TG–DTG Analysis

The thermal decomposition processes of the propellants were measured by differential thermal analysis (DTA) and thermogravimetry (TG). The equipment was operated under the condition of the flowing nitrogen at atmospheric pressure. TG and DTA tests were carried out with a heating rate of 10 K min^{-1}, while 3.0 mg samples were heated in Al_2O_3 pans without cover.

5.2.4.6 Burning Rate Test Method

The strand burning rate of the propellants were determined in the pressure range 1–12 MPa by means of a fuse-wire technique.[8] The methodology involved the combustion of strands (ignited by means of a nichrome wire) with the dimensions of 150 mm × 5 mm × 5 mm in a nitrogen pressurized steel bomb. For the trial conducted, the burning rates were computed from the time that was recorded at each pressure level for each sample.

5.2.4.7 Combustion Flame Structure Tests

The combustion flame structure tests were carried out in a transparent combustion bomb with four sides. The combustion behavior of the composite propellant, containing different aluminum samples with the dimensions of 1.5 mm × 5 mm × 15 mm, were performed at the scheduled pressure under N_2 atmosphere, and ignited by chromium–nickel wires from top to bottom.

5.3 Results and Discussion

5.3.1 SEM and Grain Size Distribution Analysis

The BH–Fe particles with and without modification with different particle sizes were tested "as received". The visual inspections of the collected SEM morphologies and particle size distribution of the tested BH–Fe powder were investigated and the results are shown in Figure 5.1. Table 5.1 summarizes the typical characteristic data of the BH–Fe additive obtained using a Malvern laser granulometer and reports the calculated average diameters.

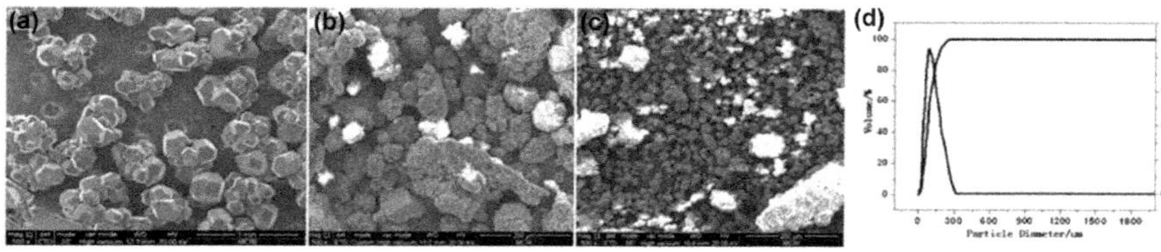

Figure 5.1 SEM images and grain size distribution of organic borohydride metal additives. (a) BHN; (b) BH–Cu; (c) BH–Fe; (d) particle size distribution of BH–Fe particles. Reproduced from ref. 11, https://doi.org/10.1088/1742-6596/1721/1/012006, under the terms of the CC BY 3.0 license https://creativecommons.org/licenses/by/3.0/.

Table 5.1 The characteristics of typical BH–Fe powder. Reproduced from ref. 11, https://doi.org/10.1088/1742-6596/1721/1/012006, under the terms of the CC BY 3.0 license https://creativecommons.org/licenses/by/3.0/.

Sample	$D[3,2]$/μm	$D[4,3]$/μm	d_{10}/μm	d_{50}/μm	d_{90}/μm
BH–Fe	72.12	101.54	40.77	93.51	186.88
Sample	Span	SSA[a]/m^2 g^{-1}	SSA[b]/m^2 g^{-1}	Density/g cm^{-3}	
BH–Fe	1.562	0.083	0.088	0.981 ± 0.002	

[a] Determined by the Malvern Master sizer.
[b] Measured by the BET method; d_{10}, particle diameter corresponding to 10% of the cumulative undersize distribution, μm; d_{50}, median particle diameter, μm; d_{90}, particle diameter corresponding to 90% of the cumulative undersize distribution, μm; span = $[D(0.9)-D(0.1)]/D(0.5)$.

It can be seen from Figure 5.1 and Table 5.1 that the microstructure of the tested BH–Fe particles presents an irregular shape, and tend to disperse or not stick together into cold clots. The d_{50} of BH–Fe particles is 93.51 μm, and the particle size distribution curve is a little bit coarse. Corresponding to the high values of d_{50}, the width of BH–Fe particles is 1.562, and the specific surface area of BH–Fe is 0.083 m^2 g^{-1} (Malvern size measurement) and 0.088 m^2 g^{-1} (BET method). For the micro-sized sample, particles show no cold cohesion, a phenomenon occurring at room temperature during storage, handling, and manufacture, which is responsible for micro-sized clusters reducing the specific surface.

At the same time, SEM photos of the two borohydride salts are shown in Figure 5.2. It can be seen that most of the particles of BH-1-10 have a particle size of 10 μm, and the surface of the particles is irregular, which is made up of many small particles overlapping, resulting in more defects and micropores on the surface of particles. The particle size distribution of BH-2-10 is wide, about 10–200 μm; the particle size of most particles is 100 μm. The surface is regular and smooth, and only a few particles have irregular defects.

One point we must emphasize is that dry laser granulometry may be affected by errors if the material creates clots. The accuracy of these

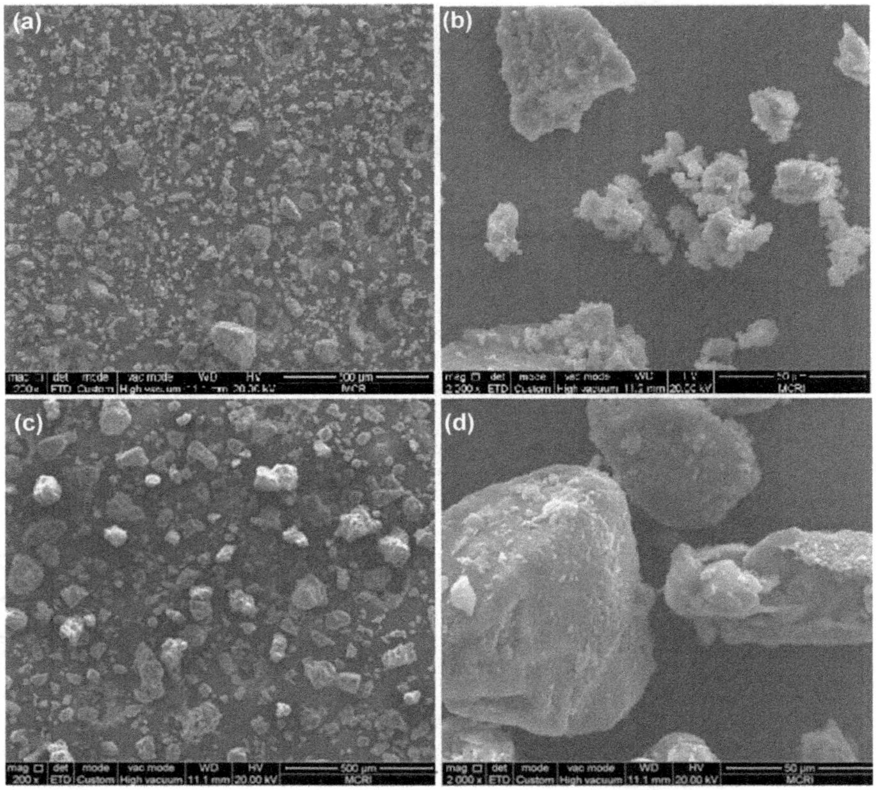

Figure 5.2 SEM photos of boron hydride salts. (a) BH-1-10-10, × 200; (b) BH-1-10-10, × 2000; (c) BH-2-10, × 200; (d) BH-2-10, × 2000. Reproduced from ref. 8.

measurements is also affected by the fact that well fluid particles are not found freely but sometimes are available as an agglomerate/aggregation powder, thus making it difficult to measure the definition of a particle mean diameter.

5.3.2 Analysis of Surface and Interface Properties of Organic Borohydrides

In order to explain the interface interaction mechanism between two organic borohydride salts and fuel-rich propellant components from the microscopic point of view, the contact angles of organic borohydride, Mg, AP, HTPB, and agglomerated boron particles (ABP) were measured using a surface/interfacial tension meter. The results are shown in Table 5.2. The surface free energy γ, nonpolar component γ^d and polar component γ^p were fitted, and the surface free energy and its components of the above components were obtained (Table 5.3). The surface free energy of two kinds of

Table 5.2 Contact angle of boron hydride salts, AP, Mg, agglomerated boron, and HTPB. Reproduced from ref. 8.

Sample	Dispersants	$\Theta/°$
BH-1-10	Ethyl acetate	53.79
	Ethyl alcohol	41.59
BH-2-10	Ethyl acetate	60.33
	Ethyl alcohol	67.05
Mg	Glycol	25.5
	Glycerin	63.4
AP	Glycerin	86.95
	H_2O	89.98
ABP	Glycerin	32.11
	H_2O	20.45
HTPB	Glycerin	84.62
	H_2O	87.92
	Glycol	70.62

Table 5.3 Surface free energies of boron hydride salt, Mg, AP, agglomerated boron, and HTPB. Reproduced from ref. 8.

Sample	γ/mN m^{-1}	γ^d/mN m^{-1}	γ^p/mN m^{-1}
BH-1-10	38.2	2.40	35.80
BH-2-10	57.5	36.35	21.15
Mg	38.77	25.57	13.20
AP	182.4	1.07	181.33
ABP	59.33	48.43	10.90
HTPB	21.50	12.23	9.27

borohydride, Mg, AP, can be seen, and also that of agglomerated boron is higher than that of the HTPB binder, and the HTPB binder can be wetted with the above solid filler surface.

The calculated bonding work *W* and spreading coefficient *S* between two borohydride, AP and HTPB, adhesive interfaces were also obtained, and the results are shown in Table 5.4. From the data in Table 5.4 it can be seen that the bonding work and spreading coefficient of the interface between Mg, AP, agglomerated boron, and the HTPB binder are larger, which is consistent with the good adhesion and bonding properties of these materials in a HTPB propellant system. The spreading coefficient of BH-2-10 is close to that of other solid fillers, which indicates that BH-2-10 has good bonding properties with a HTPB adhesive system, but the spreading coefficient of BH-1-10 and HTPB is small, which indicates that the interaction between BH-1-10 and HTPB is weak and it is difficult for HTPB to wet the surface of BH-1-10. The spread coefficient and adhesion work are in the order of: $[W_{AP/HTPB}] > [W_{BH\text{-}2\text{-}10/HTPB}] > [W_{AB/HTPB}] > [W_{Mg/HTPB}] > [W_{BH\text{-}1\text{-}10/HTPB}]$, indicating that the order of interaction between these fillers and the HTPB binder is [AP/HTPB] > [BH-2-10/HTPB] > [AB/HTPB] > [Mg/HTPB] > [BH-1-10/HTPB].

Table 5.4 Results of adhesions and spread coefficient of filler/HTPB systems. Reproduced from ref. 8.

Sample	W/N m^{-1}	S/N m^{-1}
BH-1-10/HTPB	47.27	4.27
BH-2-10/HTPB	70.21	27.21
Mg/HTPB	57.43	14.44
AP/HTPB	89.22	46.22
ABP/HTPB	68.78	25.78

5.3.3 Effect of BH on the Energetic Features of Propellant

For the compositions of propellant, the mass fraction of each formulation containing 25% of HTPB, 35% AP, 40% of agglomerated boron and boron hydride salt, the details are as follows: B-1 40% of agglomerated boron; B-2, 10% of agglomerated boron and 30% BH-1-10; B-3, 10% of agglomerated boron and 30% BH-2-10; B-4, 30% of agglomerated boron and 10% BH-2-10.

The heat of combustion of propellants with and without boron hydride salts were conducted and the results are shown in Table 5.5. The results show that the combustion temperature of boron containing propellant decreases and the mole number of products increases, but the amount of C in the products also increases significantly. The analysis shows that in the primary combustion process, although the borohydride decomposes and releases heat at low temperature, the heat release is small, and the thermal decomposition at higher temperature belongs to an endothermic process. Similar to the decomposition of $B_{10}H_{10}[(C_2H_5)_4N]_2$ in argon, the thermal decomposition of $B_{10}H_{10}{}^{2-}$ ions also produces a large amount of boron ions and H^+ in the structure of borohydride, In the competitive oxidation reaction of H^+ and carbon ions, the ability to generate H_2O is stronger, and boron ions also consume part of the oxygen, which makes it difficult to fully oxidize C in primary combustion. Therefore, the amount of C in the product increases greatly.

Table 5.6 summarizes the tested density of composite solid propellants containing different mass fractions of BH–Fe powder. From the results in Table 5.6, one can see that the measured density of composite solid propellant formulations, with common aluminum particles, is larger than those of propellants with BH–Fe powder, which can be attributes to the modest porosity of BH–Fe powders during the manufacture of the propellant formulation. It can be concluded that increasing the BH–Fe powder percentages in the composition leads to a decrease in propellant density, which may be because the density of BH–Fe (1.09 g cm^{-3}) is lower than that of Al (2.70 g cm^{-3}), and the poor shape of BH–Fe and powder packing during the manufacturing process is worse than that of Al-based composition.

Table 5.5 Effect of boron hydride salt mass fraction on the combustion efficiency of propellants. Reproduced from ref. 8.

Sample	ABP/BH-1-10 (BH-2-10)	Burning temperature/°C	Mole of balance product/mol	C/mol kg^{-1}	Heat of combustion/MJ kg^{-1}	
					Theoretical value	Measured value
B-1	40/0	1924	48.7	6.95	32.5	24.5
B-2	10/30 (BH-2-10)	1843	79.3	24.30	30.9	28.9
B-3	10/30 (BH-2-10)	1847	77.9	23.08	30.7	28.6
B-4	30/10 (BH-2-10)	1896	57.9	11.96	31.9	25.6

Table 5.6 Comparing densities of the tested composite solid rocket propellants with and without BH–Fe. Reproduced from ref. 11, https://doi.org/10.1088/1742-6596/1721/1/012006, under the terms of the CC BY 3.0 licence https://creativecommons.org/licenses/by/3.0/.

Samples	Measured density/g cm^{-3}	Porosity/%
SPF-1	1.754 ± 0.004	Negligible to modest
SPF-2	1.750 ± 0.006	Negligible
SPF-3	1.742 ± 0.002	Modest
SPF-4	1.740 ± 0.002	Negligible

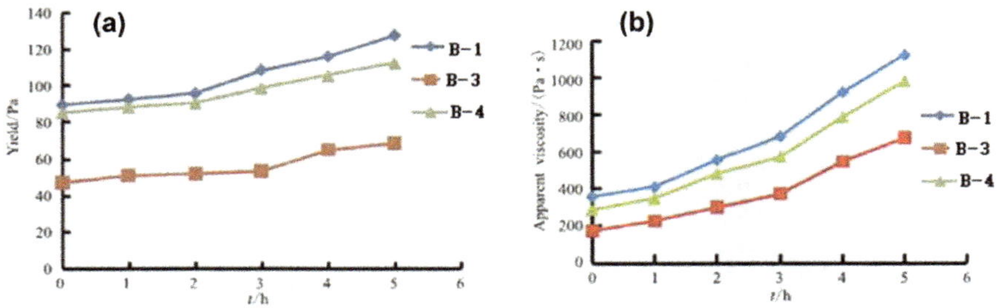

Figure 5.3 Effect of BH-2-10 mass fraction on yield stress and apparent viscosity of propellants. (a) Yield stress *vs.* time; (b) apparent viscosity *vs.* time. Reproduced from ref. 8.

5.3.4 Rheological Properties

The rheological properties of fuel-rich propellant formulations B-2 and B-3 (in Table 5.5) are studied, and the results are shown in Figure 5.3. It can be seen that with the increase in borohydride metal salt mass fraction and the decrease in agglomerated boron particles content, the yield stress and apparent viscosity of the propellant slurry decrease obviously. For the B-2 formulation with higher BH-2-10 content, the yield value and apparent viscosity of propellant slurry are 44.5 Pa and 188 Pa s, respectively. The reasons may be attributed to the fact that the particle size distribution of BH-2-10 is smaller than that of the agglomerated boron particles in the fuel-rich propellant, and the surface of the agglomerated boron particles is more regular than that of the agglomerated boron particles. The interface effect with HTPB adhesive is stronger than that of the agglomerated boron powder. Therefore, the interface bonding with HTPB adhesive system is good, and there is no interface reaction. Replacing the agglomerated boron powder will significantly improve the process performance of fuel-rich solid propellant.

5.3.5 Thermal Decomposition Analysis

Additives such as BH–Fe have important effects on the combustion properties of HTPB-based composite solid propellant. For example, the influence mechanism can be related with the thermal decomposition of propellants.[12] Figure 5.4 shows the TG–DTG curves of composite solid propellants with different mass fractions of BH–Fe particles at the heating rate of 10 K min^{-1}. It can be seen that the HTPB-based composite solid propellants with and without BH–Fe particles show three main mass loss stages (stages I–III). The first stage begins at about 150.05 °C and ends at 199.8 °C, accompanied by about 4.5% mass loss, corresponding to the mass of burning rate catalyst GFP that evaporates and decomposes. The second stage (SPF-1 sample) begins at 224.7 °C and ends at 297.2 °C, accompanied by 16.8% mass loss. This is mainly attributed to the partial decomposition of the polymer binder HTPB and AP, which is in agreement with the mass fraction amounts of AP in the formulation. The third stage begins at 301.5 °C and ends at 347.8 °C, and is accompanied by a 35.3% mass loss, which is in agreement with the decomposition of part of the AP binder.[13,14] In the temperature range of 348.1–450.0 °C, the mass loss is more than 40.0%, which indicates that there are a few remains at the end of the decomposition. In essence there is no visible difference in the excess stage for the three types of solid propellant. Also, the thermal decomposition process of the propellant containing BH–Fe particles indicates that the gaseous products formed during thermal decomposition exert higher feedback on the deflagrating propellant surface, which can increase the burning rate. Consequently, it is in agreement with the burning rate results.

The TG–DTG curves of solid propellant containing different organic borohydride metal additives are shown in Figure 5.5. It can be seen that there are three thermal decomposition processes for propellants containing BH–Pb organic borohydride metal additive, and the third one is the main thermal decomposition stage of propellant, which is the same as the propellant with BH–Fe. The main thermal decomposition peak temperature of propellant containing BH–Pb is the highest (343.3 °C), while that of propellant containing BH–Fe is the lowest (328.8 °C). The propellant containing BH–Cu organic borohydride metal additive has two thermal decomposition processes, and the main thermal decomposition peak temperature is 333.2 °C. The results indicate that the thermal decomposition peak temperature of the propellant containing BH–Fe is advanced and the catalytic effect of BH–Fe on propellants is obvious, which is consistent with the burning rate of propellants.

5.3.6 Combustion Characterization

5.3.6.1 Burning Rate and Pressure Exponent

Propellant burning rates can influence the rate of gas generation, which determines the pressure inside the motor and the overall thrust. Also, it can be affected by changes in composition (ingredients, mass fraction, and particle size distributions) and conditions (pressure and initial temperature).[15] The focus of this section is on the different types of organic

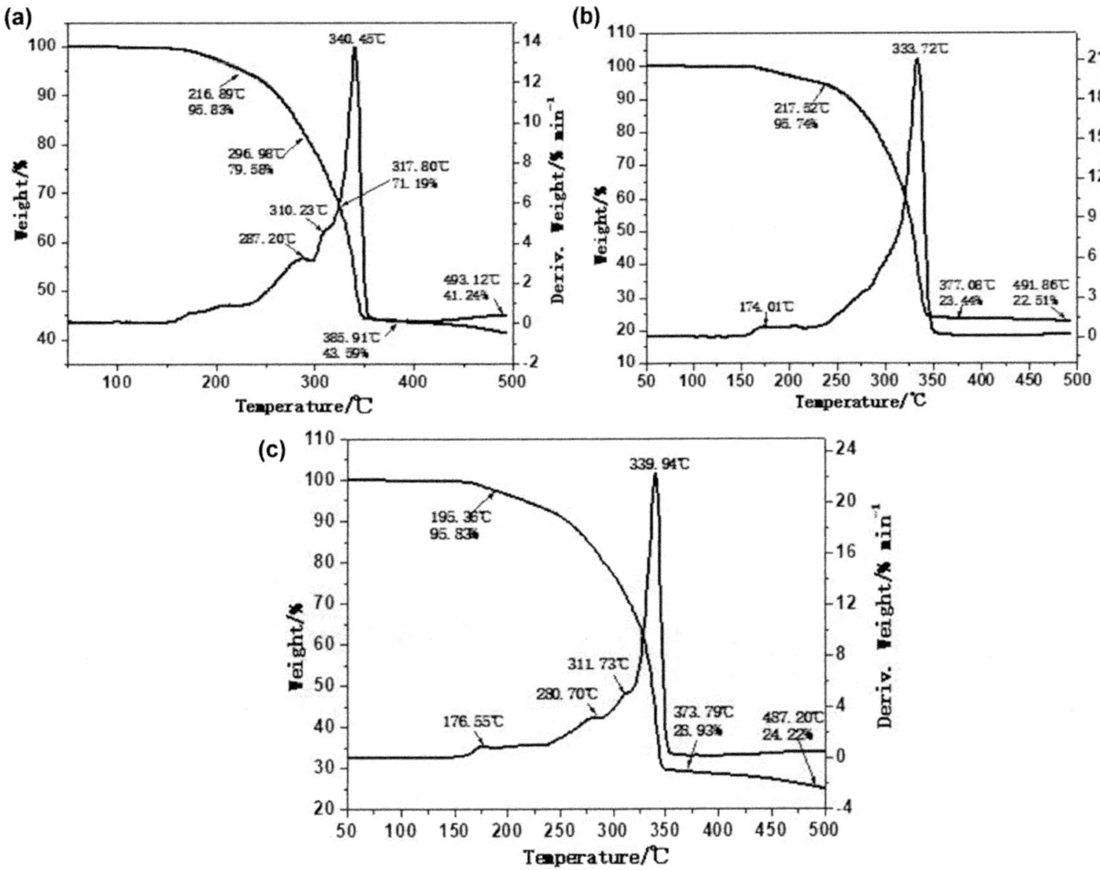

Figure 5.4 The thermal decomposition curves of composite propellant with and without BH–Fe particles. (a) SPF-1; (b) SPF-2; (c) SPF-3. Reproduced from ref. 11, https://doi.org/10.1088/1742-6596/1721/1/012006, under the terms of the CC BY 3.0 license https://creativecommons.org/licenses/by/3.0/.

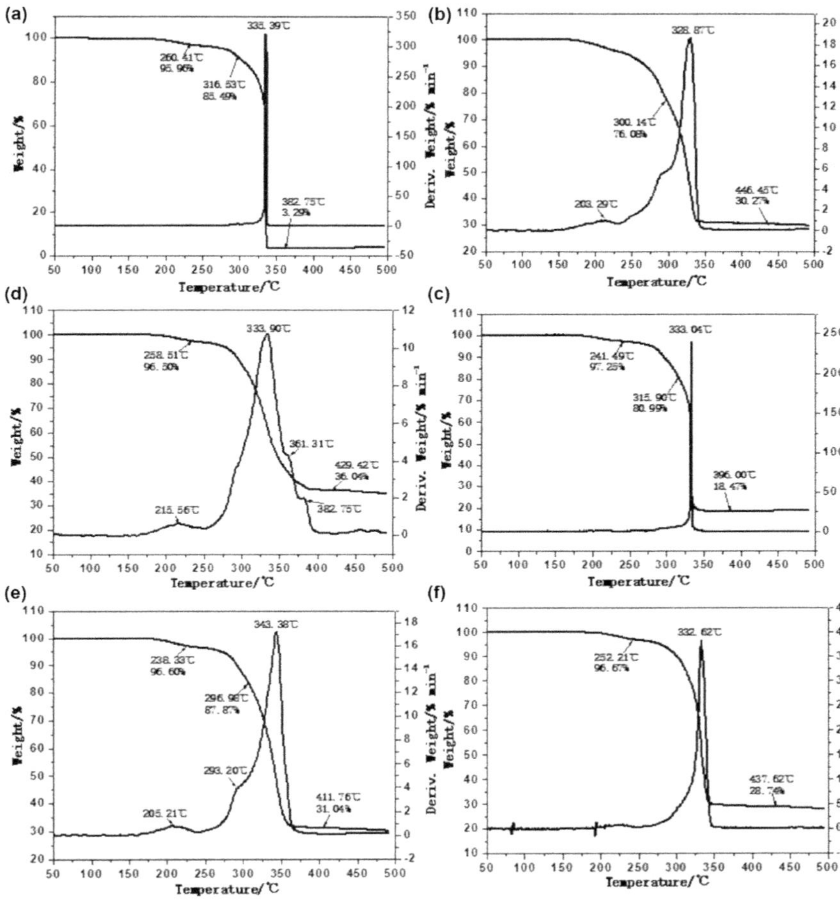

Figure 5.5 The thermal decomposition curves of composite propellant containing different organic borohydride metal additives. (a) HMA-1; (b) HMA-2; (c) HMA-3; (d) HMA-4; (e) HMA-5; (f) HMA-6.

borohydride metal additives, and how they interact with the RDX and binder under high pressure and temperature, thus affecting the burning rates; the data are shown in Table 5.7 and Figure 5.6. It can be seen that when the oxidizers and catalysts are the same, the BH–Fe additives can affect the homogeneous combustion behavior and change the burning rate effectively. A change to the PAT mass fraction of BH–Fe additives may boost the burning rate of the composite propellant even though the referenced HTPB/AP/Al composite formulation exhibits a high burning rate here. The burning rate can be increased more than 12.1% for 3% mass fraction of BH–Fe replaced micro-sized Al powder in the formulations; however, the pressure exponent cannot be changed much from 0.30–0.31 (1–12 MPa). Thus, it can be concluded that the increasing mass fraction of BH–Fe powder can not only

Table 5.7 Burning rate of the composite solid rocket propellant with and without BH–Fe additives. Reproduced from ref. 11, https://doi.org/10.1088/1742-6596/1721/1/012006, under the terms of the CC BY 3.0 license https://creativecommons.org/licenses/by/3.0/.

Sample	Mass fraction of BH–Fe/%	Burning rate/mm s^{-1} 1 MPa	4 MPa	7 MPa	10 MPa	12 MPa
SPF-1	0	19.63	27.29	33.07	37.64	42.74
SPF-2	1	20.22	27.55	33.52	37.85	43.18
SPF-3	3	20.68	30.49	37.08	39.34	44.08
SPF-4	5	21.02	33.24	39.58	41.01	46.22
		Pressure exponent (n)				
Sample	Mass fraction of BH–Fe/%	1–4 MPa	4–7 MPa	7–10 MPa	10–12 MPa	1–12 MPa
SPF-1	0	0.24	0.34	0.36	0.70	0.30
SPF-2	1	0.22	0.35	0.24	0.61	0.35
SPF-3	3	0.32	0.35	0.17	0.37	0.31
SPF-4	5	0.33	0.31	0.10	0.65	0.31

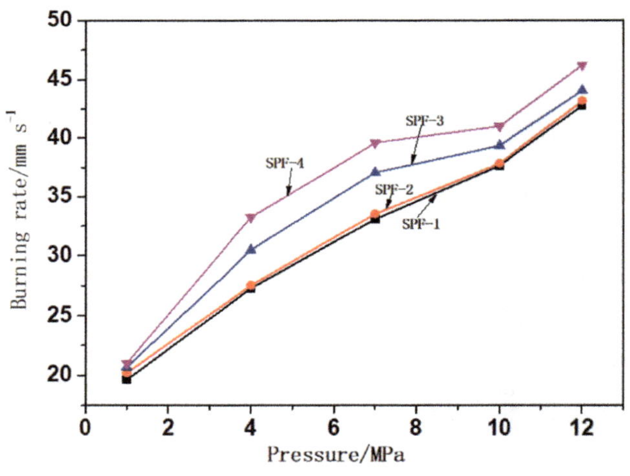

Figure 5.6 Effect of different mass fractions of BH–Fe additives on the burning rate of composite solid rocket propellant at the different pressures (pressure range: 1–12 MPa; initial temperature: $T_0 = 293$ K). Reproduced from ref. 11, https://doi.org/10.1088/1742-6596/1721/1/012006, under the terms of the CC BY 3.0 license https://creativecommons.org/licenses/by/3.0/.

increase the specific impulse but can also increase the burning rate for a composite propellant. An interesting feature is that the burning rate values depend on the mass fraction of BH–Fe powder, however the pressure exponent did not change much. Moreover, the influence of the particle size of BH–Fe powder for a composite propellant could be investigated further.

The addition of BH–Fe additives to the formulation can increase the burning rate of propellant, which may be attributed to the heat release and heat feedback to the combustion surface. From the energetic viewpoint,[16–18] the heat release and heat feedback to the combustion surface for BH–Fe particles are higher than those of Al, and its promoting effect on the propellant combustion process is the main function in the experimental pressure range. From the heat transfer viewpoint, the addition of BH–Fe powder to the propellant formulation can increase the heat adsorption effectively during the combustion process. The enhanced heat transfer to the propellant surface, due to large energy releases and reduced flame standoff distance in the gas phase at elevated pressure, overrides the influence of the preconditioned temperature in determining the energy balance at the surface, and consequently decreases the temperature sensitivity of the burning rate. As shown in the results, the basic assumptions in the following description of the burning rate model are as follows: (1) one-dimensional burning; (2) steady-state burning at a fixed pressure; (3) radiative energy from the gas phase is absorbed at the burning surface.[19,20]

Furthermore, the combustion performance of BH–Pb and BH–Cu in HTPB composite propellants was investigated, the results are shown in Table 5.8 and Figure 5.7. It can be seen that when the mass fraction of organic borohydride metal additives in propellant are the same, the burning rate of propellants are obviously different. The catalytic effect and combustion effect of BH–Fe on the propellant is improved obviously. When the composition contains 1% BH–Fe, the burning rate (9.67 mm s^{-1}) of the propellant is higher than that of the one without a catalyst (12.72 mm s^{-1}) at 7 MPa, and the burning rate of the propellant with BH–Cu is a bit increased

Table 5.8 Burning rate of composite solid rocket propellant with different organic borohydride metal additives.

Sample	Mass fraction of additives/%	Burning rate/mm s^{-1} 1 MPa	4 MPa	7 MPa	10 MPa	12 MPa
HMA-1	0	6.67	8.23	9.67	12.12	13.39
HMA-2	1%, BH–Fe	7.77	10.96	12.72	13.87	15.40
HMA-3	1%, BH–Cu	5.48	8.06	9.40	10.71	11.70
HMA-4	3%, BH–Cu	6.62	9.54	10.62	12.01	14.09
HMA-5	1%, BH–Pb	4.56	7.17	8.47	9.70	10.52
HMA-6	2%, BH–Cu	6.45	8.82	9.99	11.45	12.42
		Pressure exponent (n)				
Sample	Mass fraction of additives/%	1–4 MPa	4–7 MPa	7–10 MPa	10–12 MPa	1–12 MPa
HMA-1	0	0.23	0.26	0.36	0.54	0.27
HMA-2	1%, BH–Fe	0.25	0.27	0.24	0.57	0.27
HMA-3	1%, BH–Cu	0.28	0.27	0.37	0.49	0.30
HMA-4	3%, BH–Cu	0.26	0.19	0.34	0.88	0.31
HMA-5	1%, BH–Pb	0.33	0.30	0.38	0.45	0.36
HMA-6	2%, BH–Cu	0.23	0.22	0.38	0.45	0.30

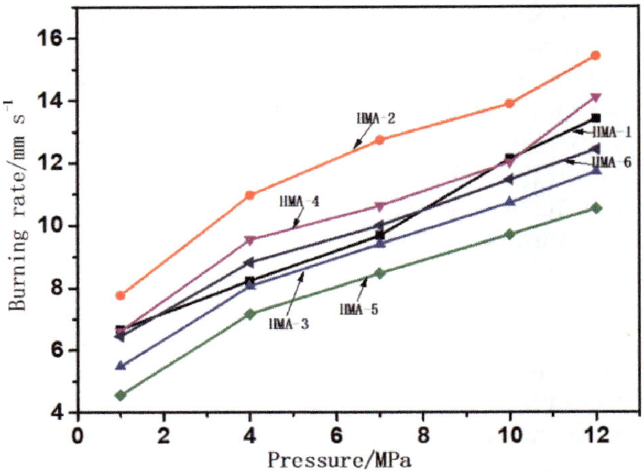

Figure 5.7 Effect of different mass fractions of different organic borohydride metal additives (HMA) on the burning rate of composite solid propellant over the pressure range from 1 MPa to 12 MPa.

Table 5.9 Effect of boron hydride salts mass fraction on the burning rate of propellants. Reproduced from ref. 8.

	Burning rate/mm s^{-1}		
Sample	0.5 MPa	1 MPa	3 MPa
B-1	2.41	3.00	7.20
B-3	1.23	1.73	2.64
B-4	1.88	2.53	3.83
	Pressure exponent n		
Sample	0.5–1 MPa	1–3 MPa	0.5–3 MPa
B-1	0.31	0.70	0.55
B-3	0.49	0.39	0.43
B-4	0.43	0.38	0.40

compared to the reference formulation, while the burning rate of the propellant with BH–Pb decreased. Moreover, the burning rate of the propellant increases with increase in the mass fraction of BH–Cu. When the mass fraction of BH–Cu is 1%, the burning rate could be increased from 9.40 mm s^{-1} to 10.62 mm s^{-1} at 7 MPa.

Furthermore, the burning rates of propellant with and without BH-2-10 were measured and the results are shown in Table 5.9 and Figure 5.8. It can be seen that without the addition of a burning rate catalyst, the burning rate of the basic formula is higher than that of BH-2-10 containing borohydride, and the pressure exponent is also higher. With the increase of borohydride metal additive mass fraction, the burning rate of the propellant decreases.

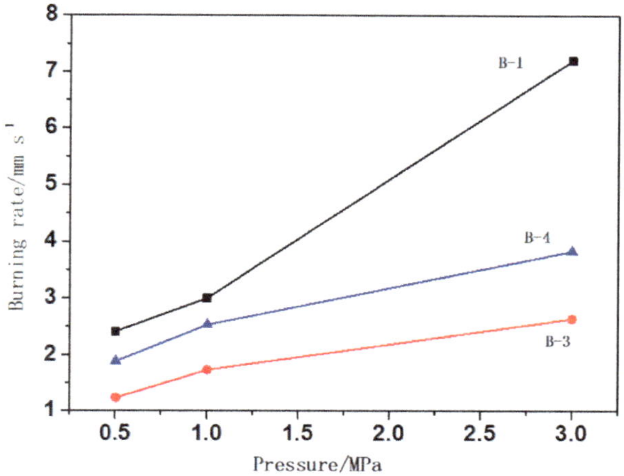

Figure 5.8 Burning rate and pressure exponent of propellants with and without BH-2-10. Reproduced from ref. 8.

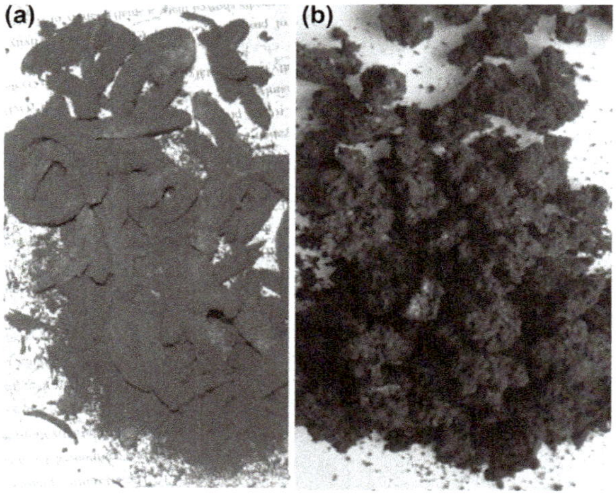

Figure 5.9 Photos of the combustion residue of samples at 1 MPa. (a) B-3; (b) B-4. Reproduced from ref. 8.

Furthermore, the combustion residues of samples B-2 and B-3 at 1 MPa were collected and the photos are shown in Figure 5.9. It can be seen that there is much residue after combustion; most of B-3 residue is fast, not dispersed into small particles, and keep the strip shape before test, while in B-4, the mass fraction of borohydride metal additives is reduced to 10%, and the condensed-phase combustion products are dispersed into large particles. Combined with the theoretical calculation results of combustion product components of propellants, it is considered that borohydride metal additives can greatly increase the content of C in the product and reduce the combustion temperature, which leads to an increase in condensed-phase products content and the decrease of gaseous content in the product. It is difficult for condensed phase product to be blown away from the combustion surface by gaseous products, forming massive residue. With the decrease in the borohydride metal additives mass fraction, the spherical morphology of the condensed-phase combustion products is significantly improved.

Lastly, one point we must emphasize is that the thermal decomposition at a lower temperature indicates that the gaseous products formed during thermal decomposition exert higher feedback to the deflagrating propellant surface. Hence, more energy is released in the combustion process (at the surface), which supports the augment of the burning rate.[21,22]

In summary, this kind of massive residue will greatly reduce the injection efficiency of propellant in the gas generator, which indicates that at higher borohydride metal additives content it is difficult to maintain the continuous thermal decomposition and injection of propellant. In the future, we should continue to optimize the fuel-rich propellant formula and explore its appropriate content, and according to the characteristics of this kind of compound in the oxygen environment, the heat release is greatly increased.[23,24] For example, the exothermic peak temperatures of $B_{10}H_{10}[(C_2H_5)_4N]_2$ decomposed at 300 °C in argon and oxygen are 112.8 J g^{-1} and 4967 J g^{-1}, respectively, and the exothermic heat of oxygen is 44 times that of inert atmosphere. That is to say, the incoming air directly contacts with the burning surface of the propellant, and the thermal decomposition products of the propellant react with oxygen directly to release heat.

5.3.6.2 Combustion Flame Structure

Typical combustion flame structures of composite solid rocket propellants containing different mass fractions of BH–Fe particles at 1 MPa and 3 MPa are shown in Figure 5.10. It can be seen that the combustion flame structures of composite solid propellants, with and without BH–Fe particles at various pressure ranges, present multi-flame structures, and the brightness of the flame structures is increased with an increase in pressure. During the propellant combustion, there are many sparks on/near the propellant burning surface, which can be attributed to the aluminum particles addition to the composite propellant formulations. Although the metal oxidation process follows a common set of events, aggregation/agglomeration phenomena near the burning surface may be noticeably different depending on the enforced operating conditions and details of the solid propellant formulation.

5.3.6.3 Combustion Residue Analysis

(1) Combustion Residues Rate. The combustion residue analysis is one of the most important parameters for evaluating the combustion

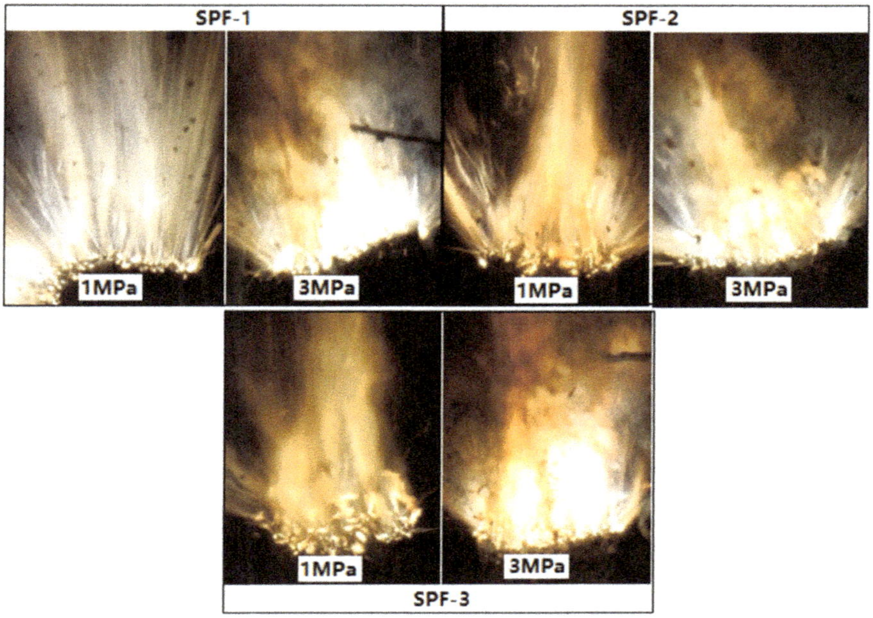

Figure 5.10 Combustion flame structures of composite solid rocket propellant with and without BH–Fe particles (×500). Reproduced from ref. 11, https://doi.org/10.1088/1742-6596/1721/1/012006, under the terms of the CC BY 3.0 license https://creativecommons.org/licenses/by/3.0/.

Table 5.10 The combustion residue rate of propellant containing different organic borohydride metal additives at different pressures.

Sample	Mass fraction of additives/%	Combustion residue rate/%		
		1 MPa	3 MPa	5 MPa
SPF-1	0	3.5	2.6	3.5
SPF-2	1%, BH–Fe	5.0	3.3	2.5
SPF-3	3%, BH–Fe	3.2	4.2	4.0
SPF-4	5%, BH–Fe	3.1	4.3	3.8
HMA-1	0	9.2	8.3	6.0
HMA-2	1%, BH–Fe	9.9	8.3	6.3
HMA-3	1%, BH–Cu	10.6	7.5	8.3
HMA-4	3%, BH–Cu	13.4	16.0	12.4
HMA-5	1%, BH–Pb	9.3	10.1	7.6
HMA-6	2%, BH–Cu	13.7	9.8	9.2

efficiency of propellants. The combustion residue rate of propellant containing different organic borohydride metal additives at 1 MPa, 3 MPa, 5 MPa were measured and calculated, and the results are shown in Table 5.10. It can be seen that the combustion residue rate of the propellant with BH–Fe increases with increasing mass fraction of

BH–Fe at 3 MPa, while the combustion residue rate decreases with pressure for the sample containing 3% mass fraction of BH–Fe.

(2) Morphology Analysis of Combustion Residues. The morphologies and the typical elemental compositions of propellant combustion residue at different pressures were studied by means of SEM and elemental analysis techniques, and the results are shown in Figure 5.11.

(3) Combustion Residue Elements. The combustion residue elements of propellant samples at different pressures were analyzed and several typical results are shown in Figure 5.12. It is shown that the main elements are Al and O in HMA-3 and HMA-4, while there is Cu, Pb, Cr, *etc.* in the other samples, which may be attributed to the different additional ingredients in the preparation of propellant samples.

5.3.7 Mechanical Properties Tests

In order to analyze the mechanical behavior of different BH–Fe particles in composite solid propellants, the mechanical properties and the surface microstructures of composite propellants with and without BH–Fe particles are shown in Table 5.11 and Figure 5.13, respectively. It can be seen that the elastic moduli (E) at +20 °C of fuel-rich propellants prepared above are high (>7.80 MPa), and the elastic moduli (E) and maximum tensile strengths (σ_m) of propellant sample SPF-1 are larger than those of the other propellants. Contrary to the referenced propellant samples SPF-1, the elongation at maximum tensile strength (ε_m) of the other propellant samples did not change very much at 20 °C (Table 5.11). It can be also seen that there are obviously many spherical particles on the surface of cured composite solid rocket propellants, which may be attributed to the addition of AP particles to the formulation, and the particles with smaller diameters can fill the spaces between the bigger grains effectively (Figure 5.13).

5.3.8 Hazardous Properties

The impact sensitivity of an energetic material to unintended initiation is an important safety factor in their use. There are many uses for knowing the hazardous properties of a material's impact and friction. These can help in the evaluation of material sensitivity over time as the determination of one material replaces another one in a particular process for comparing the sensitivity of two materials. The sensitivity of the different aluminum particles and propellants with aluminum powder can be confirmed this way. Table 5.12 shows the hazardous properties of BH–Fe powder and composite propellants with different mass fraction of organic borohydride metal additives. From the results in Table 5.12, it can be concluded that the tested BH–Fe powders are insensitive for friction and impact sensitivity (0%, and >125 cm), the designed and prepared propellants containing different mass fractions of BH–Fe particles were sensitive to impact and friction, and the impact sensitivity and friction sensitivity are increase with increasing in the

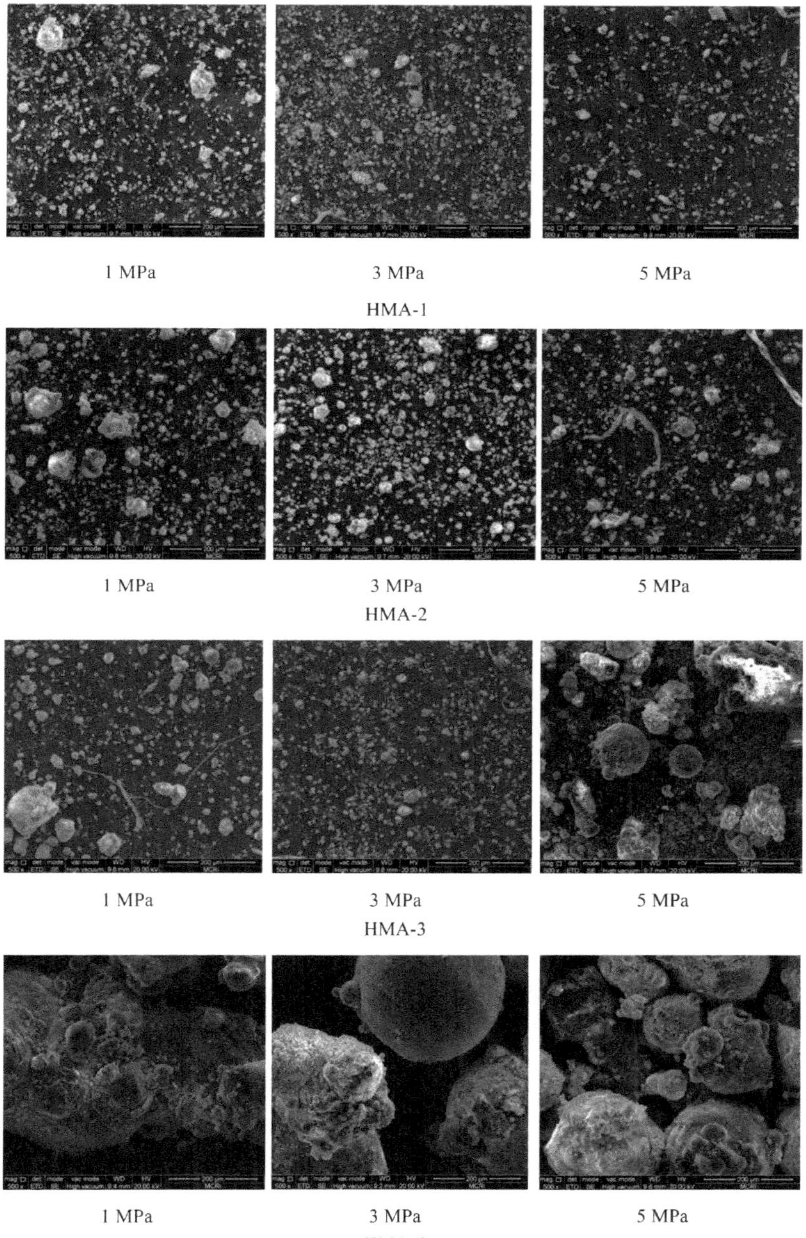

Figure 5.11 SEMs of combustion residues for solid propellants containing different organic borohydride metal additives.

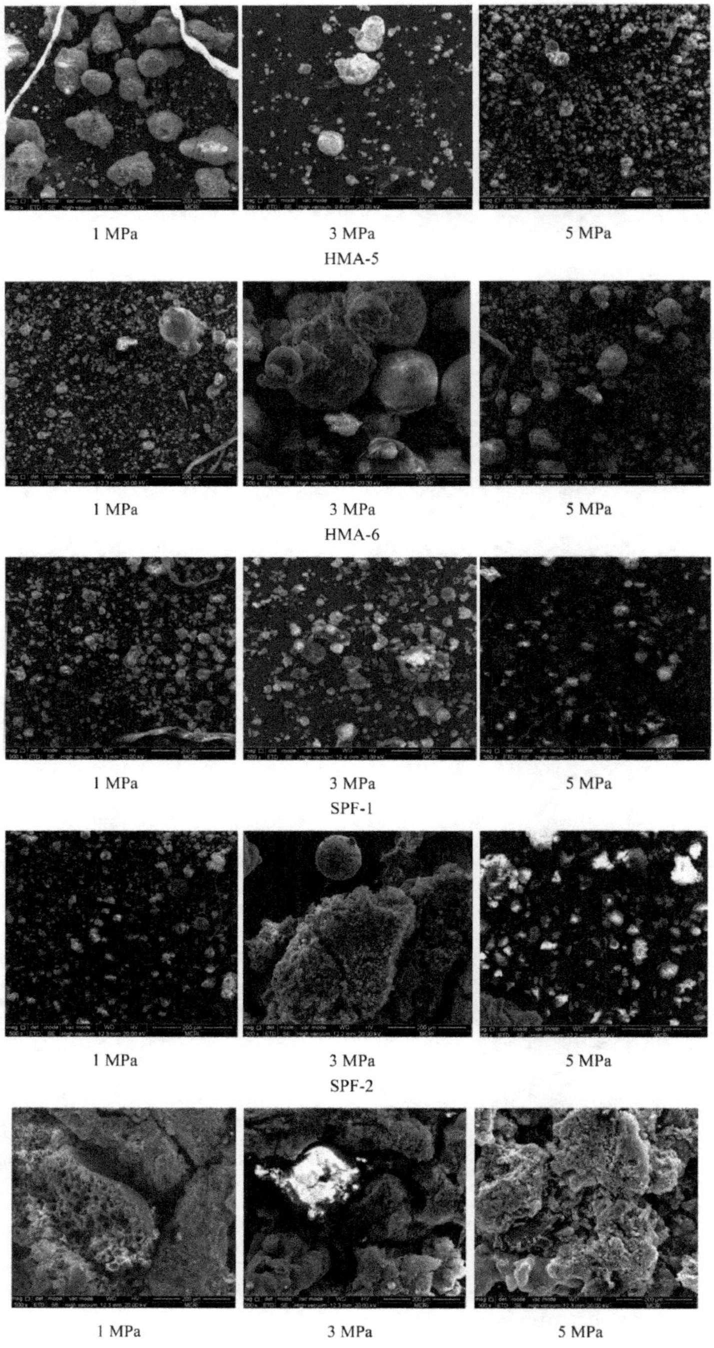

Figure 5.11 (Continued).

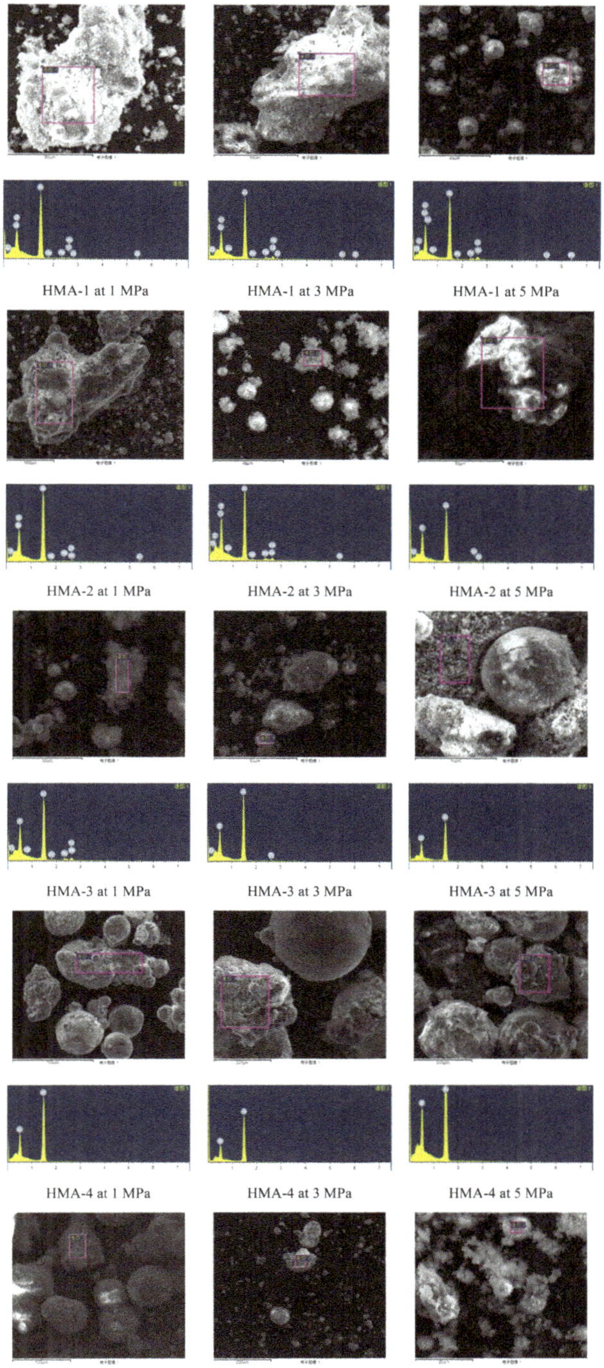

Figure 5.12 Combustion residues and elements analysis of solid propellant containing different organic borohydride metal additives.

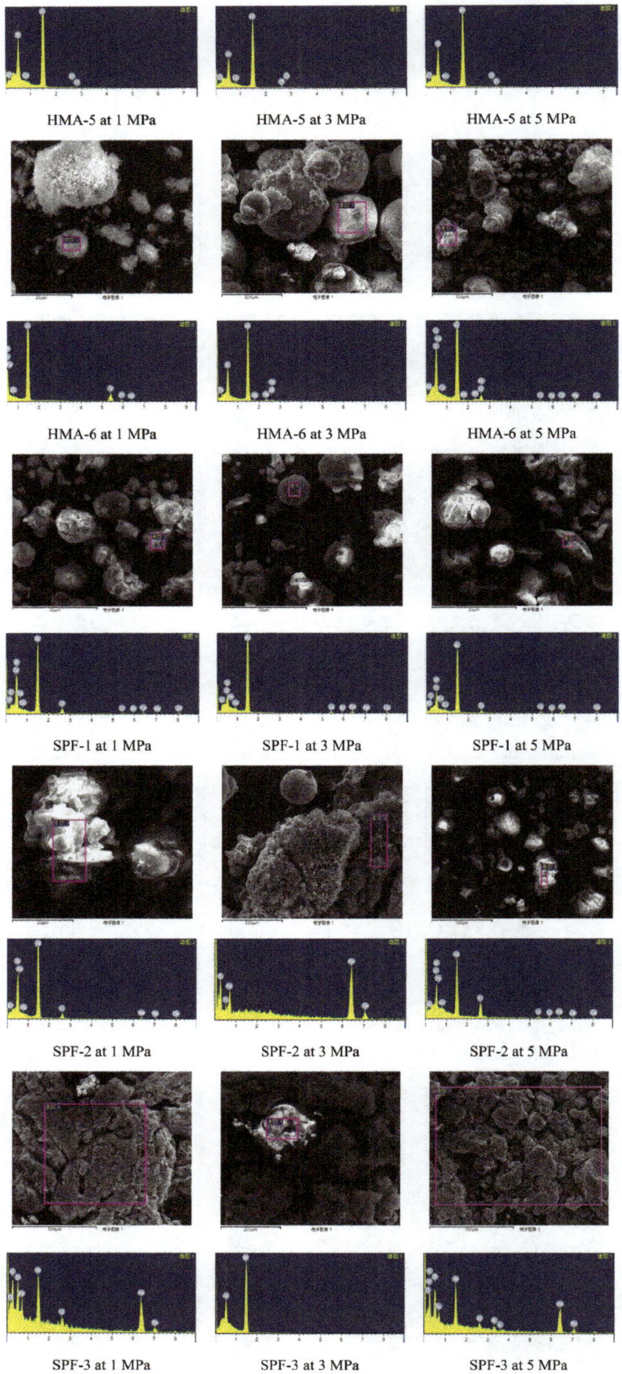

Figure 5.12 (Continued).

Table 5.11 Mechanical properties of fuel-rich propellants with different mass fractions of BH–Fe particles. Reproduced from ref. 11, https://doi.org/10.1088/1742-6596/1721/1/012006, under the terms of the CC BY 3.0 license https://creativecommons.org/licenses/by/3.0/.

	Mechanical properties[a]/(+20 °C)		
Samples	σ_m/MPa	ε_m/%	E/MPa
SPF-1	1.78	56.6	8.18
SPF-2	1.69	60.1	7.98
SPF-3	1.65	55.8	7.89

[a] σ_m, the maximum tensile strength, MPa; ε_m, the elongation at maximum tensile strength, %; E, the elastic modulus, MPa.

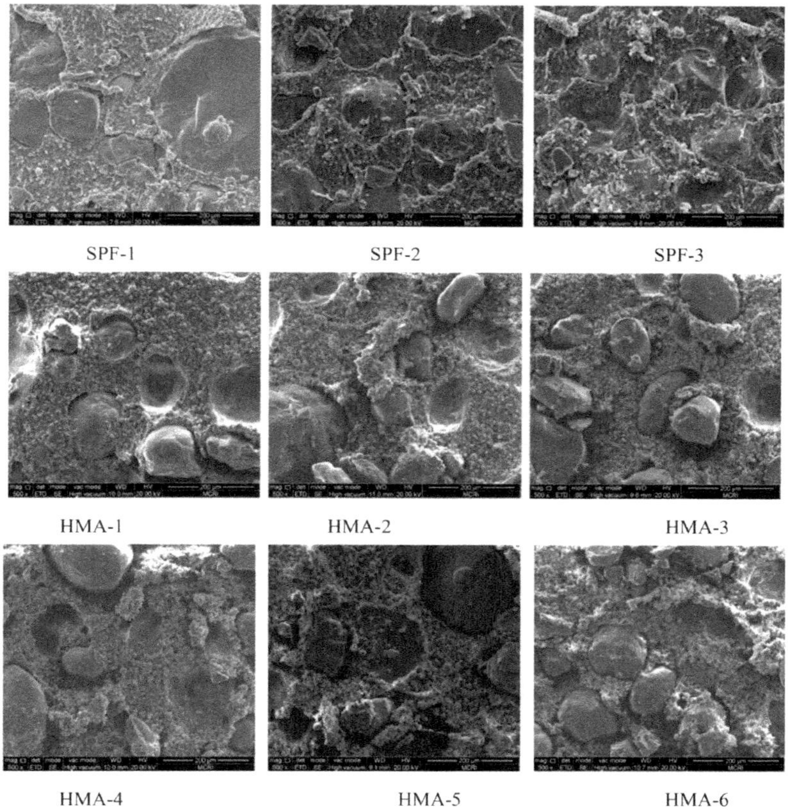

Figure 5.13 SEM surface images of solidified propellant with different organic borohydride metal additives.

mass fraction of BH–Fe powder in the formulation. The friction and impact sensitivity of the reference propellant without BH–Fe powder are 92% and 30.2 cm, respectively. The effects of BH–Fe powder on the impact and

Table 5.12 The hazardous properties of BH–Fe powder and composite propellants with different mass fraction of organic borohydride metal additives.

	Impact sensitivity		Friction sensitivity	
Sample	H_{50}/cm	Standard deviation S (logarithmic value)	P/%	Confidence level of 95% belief level
BH–Fe	>125	—	0	(0%, 14%)
HMA-1	77.6	0.10	56	(35%, 76%)
HMA-2	17.4	0.10	68	(47%, 85%)
HMA-3	27.5	0.10	72	(51%, 88%)
HMA-4	37.6	0.06	92	(74%, 99%)
HMA-5	15.8	0.05	96	(80%, 100%)
HMA-6	17.8	0.09	92	(74%, 99%)
SPF-1	30.2	0.09	92	(74%, 99%)
SPF-2	17.0	0.08	92	(74%, 99%)
SPF-3	10.0	0.07	96	(80%, 100%)
SPF-4	9.4	0.06	96	(80%, 100%)

friction sensitivity of solid propellants is not significant when comparing formulations without BH–Fe powder. The sensitivity depends on a lot of factors (such as molecular structure, active group, interaction, superfine particles, the aggregation state, *etc.*). The interaction may be possible for CC and BH–Fe additives, which increases the interfacial contact resulting in higher activity; the activity is responsible for hot spot generation. In addition, the large specific surface area of the CC powder with high surface activity may be one the reasons for increasing the mechanical sensitivity.[25,26] As for particle size of BH–Fe influencing the mechanical sensitivity, this should be investigated in detail in the future. Thus, research must be done to decrease the mechanical sensitivity of the designed composite solid rocket propellant with organic borohydride iron additives.

5.4 Conclusions

(1) Micro-sized BH–Fe particles are insensitive to friction and impact (0%, >125 cm), while the propellant formulations containing different mass fractions of BH–Fe particles are sensitive to friction and impact (96%, 10.0 cm), which is close to the reference propellant formulation without BH–Fe powder (92%, 30.2 cm). The density of composite propellant with BH–Fe powder decreases with increasing mass fraction of BH–Fe powder.

(2) The burning rate and pressure exponent of composite solid propellant with BH–Fe additive samples increase with increasing mass fraction of BH–Fe in the formulation. The burning rate of solid propellant containing 3% BH–Fe additive was increased by 12.1% at 7 MPa compared with that of the reference propellant, and the pressure exponent was increased little from 0.30 to 0.31 over the experimental pressure range from 1 to 12 MPa.

(3) The burning rate of propellant with BH–Cu increased a bit compared to the reference formulation, while the burning rate of propellant with BH–Pb decreased. The combustion residue rate of propellant with BH–Fe increased with increasing mass fraction of BH–Fe at 3 MPa, while the combustion residue rate decreased with pressure for samples containing 3% BH–Fe.

References

1. W. Weiqiang, X. Yunna and Y. Jianming, *et al.*, Progress of boron hydrides used in high burning rate propellant, *Chin. J. Energ. Mater.*, 2012, **20**(1), 132–136.
2. V. A. Babuk, V. A. Vasilyev and V. V. Sviridov, Formation of Condensed Combustion Products at the Burning Surface of Solid Rocket Propellant, in *Solid Propellant Chemistry, Combustion, and Motor Interior Ballistics, Progress in Aeronautics and Astronautics*, ed. V. Yang, T. B. Brill and W. Z. Ren, AIAA, Reston, 2000, pp. 749–776.
3. S. Wengang, S. Tiegang and Z. Guodong, Effects of carborane derivatives on the combustion of solid propellants, *J. Solid Rocket Technol.*, 1995, **18**(3), 24–26.
4. D. V. Peryshkov, A. A. Popov and S. H. Strauss, Direct perfluorination of $K_2B_{12}H_{12}$ in acetonitrile Occurs at the gas bubble - solution interface and is inhibited by HF. Experimental and DFT study of inhibition by protic acids and soft. polarizable anions, *J. Am. Chem. Soc.*, 2009, **131**, 18393–18403.
5. V. V. Drozdova, K. Y. Zhizhin and E. A. Malinina, *et al.*, Reaction of thecloso-decaborate anion $B_{10}H_{10}{}^{2-}$ with dichloroethane in the presence of hydrogen halides, *Russ. J. Inorg. Chem.*, 2007, **52**(7), 996–1001.
6. P. Weiqiang, F. Xuezhong and Z. Fengqi, *et al.*, Effects of high energetic combustion agent (BHN) on the properties of fuel-rich solid propellant, *J. Solid Rocket Technol.*, 2013, **36**(5), 647–653.
7. C. Futai, T. Huimin and L. Yunjun, Effect of $B_{12}H_{12}[N(C_2H_5)_4]_2$ on combustion property of NEPE propellant, *Chin. J. Explos. Propellants*, 2000, **27**(3), 19–21.
8. X. Huixiang, H. Yonggang and L. Yonghong, *et al.*, Surface characteristic of two dodecahydrododecaborate metal salts and its application in fuel-rich propellants, *J. Solid Rocket Technol.*, 2020, **43**(3), 310–316.
9. P. Weiqiang, X. Yunna and F. Xuezhong, *et al.*, Thermal behavior of tetraethylammonium dodecahydrododecaborates (BHN) and its compatibility with main components of propellant, *Chin. J. Energ. Mater.*, 2012, **20**(3), 280–285.
10. X. Yunna, L. Jiaoyi and H. Zhijun, *et al.*, Synthesis, thermal decomposition mechanism and kinetic equation of polyhedral boron hydride compound $[(C_2H_5)_4N]_2B_{10}H_{10}$ with high enthalpy of combustion, *Chem. J. Chin. Univ.*, 2015, **36**(2), 375–380.

11. P. Weiqiang, L. T. DeLuca and W. Ke, *et al.*, Effect of hydroborate iron additives (BH–Fe) on the properties of composite solid rocket propellants, *J. Phys.: Conf. Ser.*, 2021, **1721**, 012006.
12. P. J. Kaste, Studies of the Effect of hivelite and other boron compounds on nitramine decomposition by pyrolysis GC-FTIR, AD-A206802, 1988.
13. Y. L. Zhu, Q. J. Jiao, H. Huang and H. Ren, Effect of aluminum particle size on thermal decomposition of AP, *Chem. J. Chinese Univ.*, 2013, **34**(3), 662–667.
14. P. Weiqiang and L. T. DeLuca, *Nano and Micro-scale Energetic Materials: Propellants and Explosives*, Wiley, 2023.
15. P. Weiqiang, L. T. DeLuca, and L. Daolun, *et al.* Effect of hydroborate metal additives on the properties of composite solid rocket propellants. To be published at Cambridge Press as a chapter in 2023.
16. X.-G. Wu, Q.-L. Yan and X. Guo, *et al.*, Combustion efficiency and pyrotechnical properties of micron-sized metal particles as the components of modified double-base propellant, *Acta Astronaut.*, 2011, **68**(7–8), 1098–1112.
17. F. Maggi, A. Bandera and L. T. DeLuca, *et al.*, Agglomeration in Solid Rocket Propellants: Novel Experimental and Modeling Methods, *Prog. Prop. Phys.*, 2011, **2**, 81–98.
18. A. S. Michael, Possible chemical mechanism for boron hydrides acceleration of nitramine decomposition: literature review, *J. Propul. Power*, 2012, **14**(6), 981–990.
19. T. Song-qing and D. Hong-xun, The latest review of boron hydrides as high burning rate modifier of solid propellant, *J. Propul. Technol.*, 1983, **2**, 35–51.
20. Z. Yuanyang and W. Minwen, High burning rate modifier polyhedral boron hydride, *Chin. J. Explo. Propellants*, 1989, **3**, 32–37.
21. B. R. S. Hansen, M. Paskevicius and H. W. Li, *et al.*, Metal boranes: progress and applications, *Coord. Chem. Rev.*, 2015, **323**(4), 60–70.
22. C. E. Housecroft, R. Snaith and K. Wade, Bond enthalpies of borane anions $BnHn^{2-}$, *Inorg. Nucl. Chem. Lett.*, 1979, **15**(9–10), 343–347.
23. D. W. VanKrevelen, *Properties of polymers, estimation of the properties of a polymer and its relationship to chemical structure*, ed. X. Yuanze, Z. Delu and W. Dacheng, Science Press, Beijing, 1981.
24. S. Zixing, S. Lili and Y. Rongjie, Research progress of icosahedronal polyhydroborate $B_{12}H_{12}{}^{2-}$ anion compounds and their application in propellants and explosives, *Chin. J. Explos. Propellants*, 2017, **40**(3), 1–16.
25. N. Yong, C. Hai-yan and M. Jin-ling, Advances in the chemistry and applications of the borane anion closo-$[B_{10}H_{10}]^{2-}$ and its derivatives, *Chin. J. Org. Chem.*, 2009, **29**(6), 822–834.
26. L. Yanhua, S. Zixing and S. Lili, *et al.*, Compatibility of dodecahydrododecaborate with propellants ingredients, *Chin. J. Explos. Propellants*, 2017, **40**(6), 95–100.

CHAPTER 6

Performance of Energetic Composites Containing Carborane Derivatives

6.1 Introduction

Since the 1970s, carborane derivatives have been used as propellant burning rate catalysts in the United States, Japan and the former Soviet Union, as they can make the burning rate reach a high level that is possible for solid propulsion. For example, *n*-hexylcarborane can not only increase the burning rate of the solid propellant, but also improve the technological and mechanical properties of the propellant, and the reduction in the contrast impulse is small.[1] When carborane acrylic acid is used in AP/NG propellant, the burning rate can reach 170 mm s^{-1} (14 MPa),[2] and when 12% 1-isopropenyl-2-propyl-(2-oxygen) carborane (IPCA) is added in AP/TEGDN propellant, the burning rate can reach 101.6 mm s^{-1} (14 MPa).[3] Carborane derivatives are similar to boron cages in structure, but do not contain unstable top hydrogen atoms. During their combustion, they are attacked by oxidizing groups and decomposed, releasing a large amount of heat, which improves the burning rate of solid propellants.

Due to the poor atomization performance and insufficient combustion of aluminum powder of high-energy fuel composites containing aluminum,[4,5] their ignition and combustion performance are not ideal, and the advantages of the fuel are difficult to utilize, which restricts the engineering application of this new fuel in ramjets. To improve the ignition and combustion performance of high-energy fuel containing aluminum by adding additives with specific functions is an important strategy for the development of high-energy fuels containing aluminum, and also an important development direction for the application research of high-energy fuel

High-energy Combustion Agents of Organic Borohydrides: Synthesis, Characterization and Applications
By XiaoHong Zhang, WeiQiang Pang, Luigi T. DeLuca and Yu Zhao

Published by the Royal Society of Chemistry, www.rsc.org

composites containing aluminum in the future. Boron cage compounds have a closed polyhedral compact configuration, no active hydrogen bridge bonds in the structure, high thermal stability, a high burning rate, high density, high calorific value and low toxicity. They have been used as burning rate regulators and combustion promoters in solid propellants.[6,7] Although boron cage compounds have excellent burning rate regulation and combustion promotion effects, the application of boron cage compounds to high-energy fuels containing aluminum has not been reported.

Furthermore, of all of the elements of the periodic table, only neighboring carbon and boron share the properties of self-bonding (catenation) and the support of electron-delocalized structures based upon these catenated frameworks. Carbon catenation, of course, leads to the immense field of organic chemistry. Boron catenation provides the *nido*-, *arachno*-, and *hypho*-boranes, which may be considered the borane equivalents of aliphatic hydrocarbons, and the discrete families of *closo*-borane derivatives which bear a formal resemblance to the aromatic hydrocarbons, heterocycles, and metallocenes. Aside from these analogies, boron and carbon chemistries are also important to each other through their extravagant ability to mix in ways not available to other element pairs. Thus, the conflux of boron and carbon chemistries effectively provides a new elementpair for exploitation in a variety of novel ways.

6.2 Experimental

6.2.1 Raw Materials

Carborane methyl acrylate (CMP), purity >98%, Liming Research Institute of Chemical Industry; liquid high density hydrocarbon fuel (LHDF), imitating JP-10, purity >99%. The average particle size of nano-sized aluminum powder is 80 nm. The surface is passivated and the activity is >85%. Auxiliary dispersant gelling agent, composed of four elements C, H, O and N, with purity >99%.

6.2.2 Tested Ingredients and Formulations

AP/NG formulation: AP 27.3%, nitrocellulose (NC) 20.4%, nitroglycerin (NG) 37.5%, Al 7.0%, triacetin (TA) 4.5%, additives 2.3%; RDX/NG formulation: RDX 36%, NC 36%, NG 34.3%, additives 3.7%. CMP was added to the formulations additionally, and propellants were prepared by using a slurry casting process in vacuum.

Preparation of fuels containing aluminum (Al): according to the composition of high-energy fuel containing Al, weigh the gelling agent and LHDF of corresponding quality into a 250 mL three necked flask, and stir with a high shear machine for 1 h at 25 °C. Then, add the surface modified nano-sized aluminum powder, continue stirring for 3 h, stop stirring, and stand for 10 min to form stable and dispersed aluminum containing high-energy fuels.

The boron cage modified aluminum containing high-energy fuels can be obtained by adding boron cage compound and nano-sized aluminum powder together.

6.2.3 Characterization Methods

Burning rate of propellant testing: a nichrome fine wire with 0.1 mm in diameter was threaded through the top of the propellant strand, with an alternating voltage of 100 V to ignite the strand (diameter = 5–6 mm, length = 140 mm) at an initial temperature of 20 °C. The samples were placed vertically on the combustion rack in the sealed chamber which was filled with a nitrogen atmosphere. The burning rates of the propellant strands were determined in the pressure range 2.0 MPa $< P <$ 12.0 MPa by means of the fuse-wire technique.[8–10] The burning rates were computed from the combustion time that was recorded for the tests conducted at each pressure, for each pair of fuse wires, separation distance 10 mm, and each pressure, and were tested five times.

The density of the aluminum containing high-energy fuel prepared above was measured by the density bottle method, and the measurement accuracy was $\leq \pm 1\%$. The measurement method of volumetric calorific value is as follows: firstly, the combustion heat of high-energy fuel containing aluminum is measured by the oxygen bomb method on a E-5200 calorimeter. Then, the mass net calorific value (GB/T 384-1988) of high-energy fuel containing aluminum is calculated according to LHDF and hydrogen content in boron cage compound. Finally, the volume net calorific value of high-energy fuel containing aluminum is obtained by mathematical conversion from the density and mass net calorific value, and the measurement accuracy is $\leq \pm 1.5\%$.

Static shock tube ignition and combustion performance test device: Figures 6.1 and 6.2 respectively show the structure and principle diagram of the static shock tube ignition and combustion performance test device. The device can be divided into: (1) shock tube ignition area; (2) ignition and combustion zone; (3) CCD high-speed photography area; (4) data acquisition

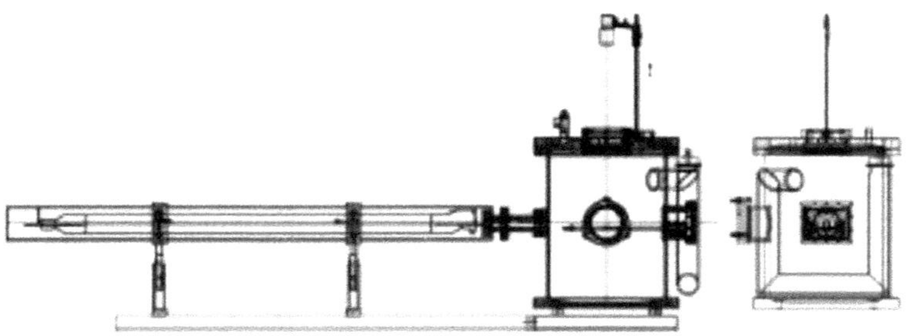

Figure 6.1 Structure diagram of shock-induced ignition and combustion performance evaluation apparatuses. Reproduced from ref. 11.

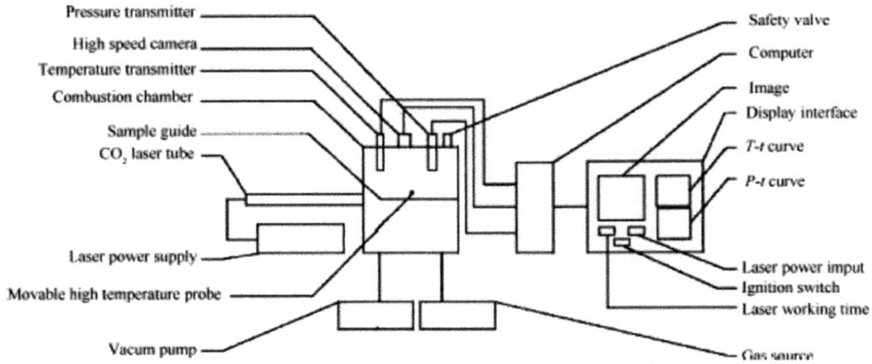

Figure 6.2 Schematic diagram of shock-induced ignition and combustion performance evaluation apparatuses. Reproduced from ref. 11.

and processing area. The performance parameters of the device are as follows: test pressure: 0.1–0.8 MPa, pressure control error: $\leq \pm 1\%$; test temperature: 20–2000 °C, temperature control error: $\leq \pm 1$ °C; test mass: 0.1–20.0 g, weighing error: ± 0.01 g; combustion chamber volume: 35.01 ± 0.01 L; ignition delay measurement error: $\leq \pm 3$ ms.

6.3 Results and Discussion

6.3.1 Combustion of NG-based Propellants with Carborane Methyl Acrylate

Table 6.1 shows the experimental results of the effect of CMP mass fraction on the burning rate of AP/NG propellant under different pressures. It can be seen that CMP can significantly increase the burning rate of AP/NG propellant, the burning rate increase with an increase in the range 83–155%, and the catalytic efficiency of CMP increases with an increase in the mass fraction. Compared with the propellant without CMP, the pressure exponent of the propellant with CMP increases slightly over the pressure range 2.0–12 MPa, but when the CMP mass fraction increases, the pressure exponent of the propellant decreases. When the pressure increases to 4.0–10.0 MPa, the catalytic effect of CMP on the combustion is less affected by the pressure.

Table 6.2 shows the experimental results of the effect of CMP mass fraction on the burning rate of RDX/NG propellant under different pressures. It can be seen that CMP can greatly improve the burning rate of RDX/NG propellant and the burning rate increases with an increase in the range 8–50%. With the increase of CMP mass fraction, the burning rate of propellant and the catalytic efficiency of CMP to the propellant increase, except for abnormality within 6–8 MPa. The catalytic efficiency of CMP reaches the maximum value within 8–10 MPa, that is CMP has the strongest effect on improving the burning rate of RDX/NG propellant within this pressure range. Compared with the

Table 6.1 Effect of CMP on the burning rate of AP/NG propellant. Reproduced from ref. 12.

Mass fraction of CMP	Burning rate/mm s^{-1} 2 MPa	4 MPa	6 MPa	8 MPa	10 MPa	12 MPa	Pressure exponent, *n*
0	16.31	26.95	36.10	43.48	51.02	55.25	—
8	29.94	53.19	69.44	83.33	97.09	107.53	0.71
10	30.21	53.76	69.70	84.46	94.04	109.95	0.70
12	31.55	56.50	72.99	106.38	106.38	119.05	0.69

Table 6.2 Effect of CMP on the burning rate of RDX/NG propellant. Reproduced from ref. 12.

Mass fraction of CMP	Burning rate/mm s^{-1} 2 MPa	4 MPa	6 MPa	8 MPa	10 MPa	12 MPa	Pressure exponent, *n*
0	3.28	6.01	8.70	11.24	14.08	16.42	0.90
2	3.52	6.51	9.47	12.47	16.03	18.12	0.93
4	4.07	8.20	12.11	16.72	20.24	22.42	0.97
6	4.44	8.42	12.02	16.26	21.14	24.27	0.96

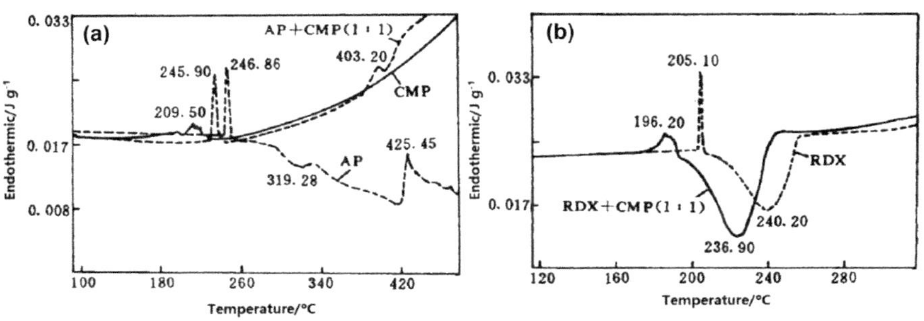

Figure 6.3 DSC curves of AP/CMP and RDX/CMP compositions. (a) AP/CMP; (b) RDX/CMP. Reproduced from ref. 12.

propellant without CMP, the pressure exponent of the propellant containing CMP increases, and the pressure exponent of the propellant increases with an increase in the mass fraction of CMP.

In order to explain the reason for the burning rate increase by addition of CMP to the propellant formulation, the effect of CMP on the thermal decomposition of AP, RDX was investigated. Figure 6.3 shows the DSC curves of AP/CMP and RDX/CMP compositions. It can be seen from Figure 6.3a that the addition of CMP increased the phase transition temperature of AP from 249.86 °C (orthorhombic) to 251.46 °C (cubic), which is the temperature at which produced lattice defects increased, while the low-temperature autocatalytic thermal decomposition exothermic at 300 °C was covered up by the endothermic process of CMP over this temperature range. In the high-temperature thermal decomposition stage, CMP reduces the temperature of the high-temperature thermal decomposition peak by 20 °C, and CMP catalyzes the high-temperature thermal decomposition of AP, resulting in an increase in the burning rate of AP/NG propellant.

It can be seen from Figure 6.3b that the addition of CMP not only reduced the melting peak temperature of RDX from 205.10 °C to 196.20 °C, but also reduced the thermal decomposition peak temperature of RDX from 240.20 °C to 236.90 °C, indicating that CMP promotes the thermal decomposition of RDX. The melting of RDX increases the reaction contact area and increases the burning rate of RDX/NG propellant. The increasing thermal decomposition of RDX is also beneficial to the increasing burning rate of propellant.[13]

6.3.2 Boron Cage Compounds on Combustion of Liquid–Solid High-energy Fuels

6.3.2.1 Effect of Composition on the Dispersion Stability of High-energy Fuel Containing Al

Table 6.3 and Figure 6.4 show the dispersion stability liquid–solid gel-type high-energy fuels (LSGHEFs) with different compositions results and photos of LSGHEFs stored for 6 months. LHDF is similar to JP-10 with purity greater than 99%, the average of nano-sized Al is 80 nm, and the gelatinizer is composed of C, H, O, N elements with purity greater than 99%. It can be seen that after GNRL-1 and GNRL-7 have been standing for 6 months, a small amount of thin layer liquid is precipitated from the upper layer of the fuel, but this phenomenon is not found for other fuels. Comparing the fuel composition in Table 6.3, it can be seen that the gelling agent and Al/LHDF mass ratio have obvious effects on the dispersion stability of aluminum containing high-energy fuel. Increasing the Al/LHDF mass ratio or the mass fraction of gelling agent can improve the dispersion stability of fuel, and the effect of the gelling agent is more obvious.

Table 6.3 Dispersion stability of LSGHEFs with different compositions.[a] Reproduced from ref. 11.

Sample	LHDF/wt%	Nano-Al/wt%	Gelatinizer/wt%	Liquid precipitation
GNRL-1	87	10	3	Yes
GNRL-2	77	20	3	No
GNRL-3	67	30	3	No
GNRL-4	57	40	3	No
GNRL-5	68	30	2	No
GNRL-6	76	20	4	No
GNRL-7	79	20	1	Yes

[a] Stored for 6 months.

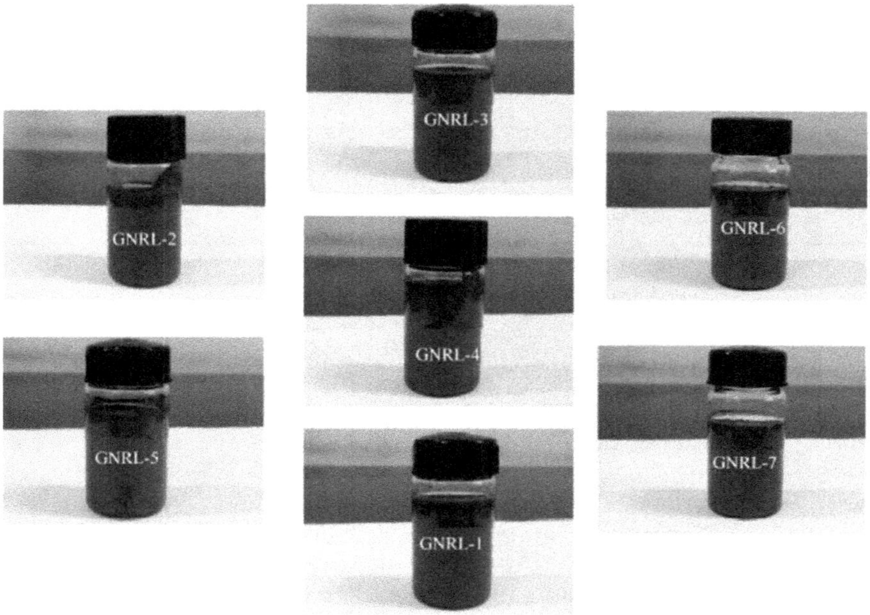

Figure 6.4 LSGHEFs with different compositions stored for 6 months. Reproduced from ref. 11.

Generally speaking, the factors affecting the gel effect mainly include temperature, pH value, solid content, gelling agent content, *etc.* Increasing the solid mass fraction or gelling agent mass fraction is conducive to forming a good gel effect.[14] From the perspective of the gel process and principle, the gelling agent is added to the liquid hydrocarbon fuel LHDF and nano-sized aluminum powder system. At a certain temperature, the liquid hydrocarbon fuel molecules and nano-sized aluminum powder particles form a three-dimensional network integrated unit structure through gelation[15] so as to form a stable and dispersed high-energy fuel containing aluminum. The low mass fraction of the gelling agent weakens the binding and anchoring effect in the high-energy fuel system containing aluminum, which is not conducive to the formation of a three-dimensional network structure. If the mass ratio of Al to LHDF is low, more liquid hydrocarbon fuels are filled around the gelling agent molecules, which is not conducive to forming a three-dimensional network structure.

6.3.2.2 *Effect of Boron Cage Compounds on the Physico-chemical Properties of High-energy Fuel Containing Al*

Table 6.4 gives the density, calorific value and other key physico-chemical performance parameters of 1# to 4# aluminum containing high-energy fuels with different chemical compositions. Both boron cage S and boron cage T

Table 6.4 Key physicochemical performance parameters of LSGHEFs. Reproduced from ref. 11.

Sample	Compositions	Density/ g cm^{-3}	Calorific value/MJ kg^{-1}	Calorific value/MJ L^{-1}
1#	GNRL-4	1.33	36.91	49.09
2#	GNRL-4 + 16.7% BCC-S	1.29	34.64	44.69
3#	GNRL-4 + 16.7% BCC-T	1.07	41.18	44.06
4#	GNRL-4 + 16.7% BCC-A	1.07	41.48	44.38

are boron cage compounds containing metal elements, and the molecular structure is $M_xB_yH_z$. Boron cage A does not contain metal elements and its molecular formula is $C_aB_bH_cN_d$. It can be seen that the density of high-energy fuel containing aluminum decreases to different degrees after the addition of the boron cage compound. Among them, after adding boron cage S, the density of aluminum containing high-energy fuel decreases slightly, about 3.0%, while after adding boron cage T and boron cage A, the density of aluminum containing high-energy fuel decreases greatly, about 19.5%. The reason for analysis is that the density of the three solid powder boron cage compounds is quite different. The density of the boron cage S is 2.36 g cm^{-3}, while the density of the boron cage A and the boron cage T is only 0.93 g cm^{-3}. After adding the high-energy fuel containing aluminum, this density difference causes the density of the high-energy fuel containing aluminum to decrease in varying degrees.

Different from the density change trend, the mass calorific value of aluminum containing high-energy fuel decreases slightly (6.1%) after adding boron cage S, while the mass calorific value of aluminum containing high-energy fuel is significantly increased by boron cage T and boron cage A, and the increase is similar, 12.0%. The reason maybe that the mass calorific value of boron cage compounds has great influence on the mass calorific value of high-energy fuel containing aluminum. The mass calorific value of boron cage S is low, while the mass calorific value of boron cage T and boron cage A are high. In addition, it can also be seen that the addition of boron cage compounds has different effects on the mass calorific value and volume calorific value of fuel. The volumetric calorific value of fuel decreased after the boron cage compound was added.

Theoretically, the volumetric calorific value of high-energy fuel containing aluminum is affected by both density and mass calorific value, that is, under other conditions with the same composition, it ultimately depends on the density and molecular structure of the boron cage compound added.

It is worth noting that during the combustion heat test of 1# to 4# aluminum containing high-energy fuel oxygen bomb method, it was found that 0.65 g of the 4# fuel sample directly melted the bottom of the crucible (as shown in Figure 6.5). After reducing the mass of the 4# fuel sample to 0.30 g, the crucible bottom was still partially melted. Finally, the mass of

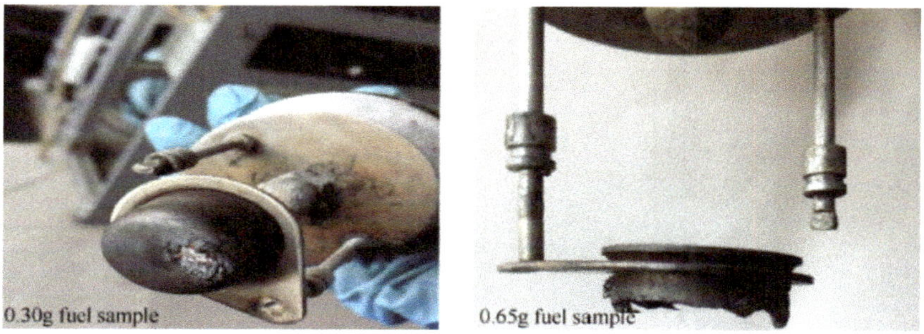

Figure 6.5 Special phenomenon during the calorific value test of the 4# fuel sample. Reproduced from ref. 11.

the 4# fuel sample was reduced to 0.1 g, and the crucible remained intact after the test. However, 0.65 g of the 1# and 2# fuel samples did not show this special phenomenon during the heat of combustion test. Since the random test sequence of the fuel sample is 1#, 2#, 4#, and 3#, and several crucibles were burned during the repeated test of the 4# fuel sample, in order to ensure that the calorimeter was not damaged and keep the only crucibles, the sample quality of the 3# fuel sample was set at 0.1 g during the combustion heat test, which ensured the smooth completion of the test.

The oxygen bomb method tests the samples according to the mass. The mass calorific value of the 4# fuel sample is 41.48 MJ kg^{-1}, which belongs to the calorific value range of conventional hydrocarbon fuel. However, it melts the bottom of the crucible, indicating that the fuel produces a very high combustion temperature during the test, and the heat release rate is fast, resulting in the crucible material unable to withstand huge thermal stress.

6.3.2.3 Effect of the Boron Cage Compound on the Combustion Temperature of High-energy Fuel Containing Al

Figure 6.6 shows the change curve of the combustion temperature of different high-energy fuels containing aluminum with time. As can be seen from Figure 6.6, there is a peak in the combustion temperature with the extension of the combustion time. Among them, the peak combustion temperature of blank fuel sample 1# is 1011 °C. After adding boron cage compounds, the peak combustion temperature increases significantly. Among them, boron cage S, boron cage T and boron cage A increase the combustion peak temperature of the high-energy fuel containing Al by 214, 420, and 535 °C, respectively, with an increase in 21.1–52.9%. The combustion temperature time curves of boron cage S and boron cage A are smooth without obvious inflection points, indicating that the combustion process is relatively stable. Moreover, the combustion temperature of fuel

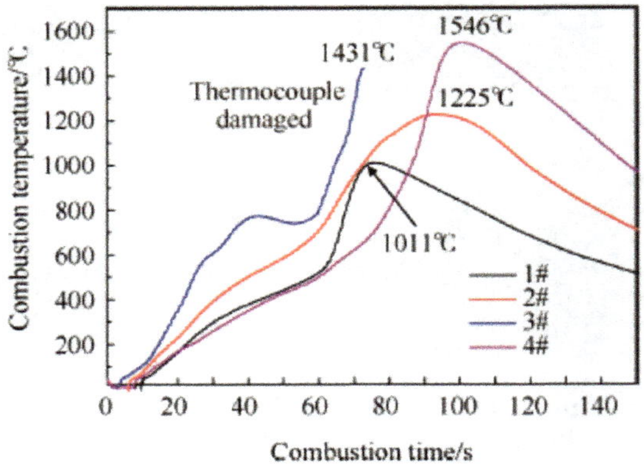

Figure 6.6 Temperature–time curve during the combustion of LSGHEFs. Reproduced from ref. 11.

sample 4# is as high as 1546 °C, which is 535 °C and 321 °C higher than that of fuel samples 1# and 2#, respectively, inferring that boron cage A can greatly promote the combustion of high-energy fuel containing aluminum from the perspective of combustion temperature compared with blank fuel sample 1#.

From the calorific value of mass measured by the oxygen bomb method, [4#] > [3#] > [1#] > [2#], but it is found from Figure 6.6 that the combustion temperature of 2# is still higher than that of 1#. It can be seen that under the working conditions of static shock tube ignition and the combustion performance test, the combustion efficiency of 2# is higher than that of 1#. The reason for this is that the oxygen pressure measured by the oxygen bomb method is 3 MPa. In a confined volume, the high-energy fuel containing aluminum can burn completely under oxygen rich conditions. Therefore, the combustion temperature of the high-energy fuel containing aluminum is mainly determined by its mass calorific value. In the static shock tube ignition and combustion performance test, the oxygen pressure for the ignition and combustion performance test of the high-energy fuel containing aluminum is 0.6 MPa (equivalent oxygen pressure), and the high-energy fuel containing aluminum is not completely combusted. Therefore, the combustion temperature of the high-energy fuel containing aluminum is determined by its calorific value of mass and its combustion efficiency. The mass calorific value of the 1# fuel sample is higher than that of the 2# fuel sample, but the combustion temperature is lower, indicating that the boron cage S added to the 2# fuel sample can improve the combustion efficiency of high-energy fuel containing aluminum. In general, boron cage A has the best effect, which not only greatly increases the peak combustion temperature of high-energy fuel containing aluminum and promotes its combustion, but also improves the stability of the combustion process.

6.3.2.4 Effect of the Boron Cage Compound on the Combustion Pressure of High Energy Fuel Containing Al

Figure 6.7 shows the change curve of combustion pressure with time for different aluminum-containing high-energy fuels. As can be seen from the figure, with the extension of the combustion time there is a peak (P_{max}) for the combustion pressure of high-energy fuel containing Al, and the order of magnitude is: [4#] ≈ [1#] > [3#] ≈ [2#]. It can be seen from the pressure change rate in the pressurization section (500 kPa to P_{max} kPa) $\Delta P/\Delta T$ that [4#] > [3#] ≈ [1#] > [2#].

Figure 6.8 shows the change curve of the intensity of each molecular ion peak in the pyrolysis product of boron cage A with temperature. It can be seen from the figure that boron cage A is rapid nears pyrolysis close to 310 °C, and the MS ion charge mass ratio m/z of the gas-phase product

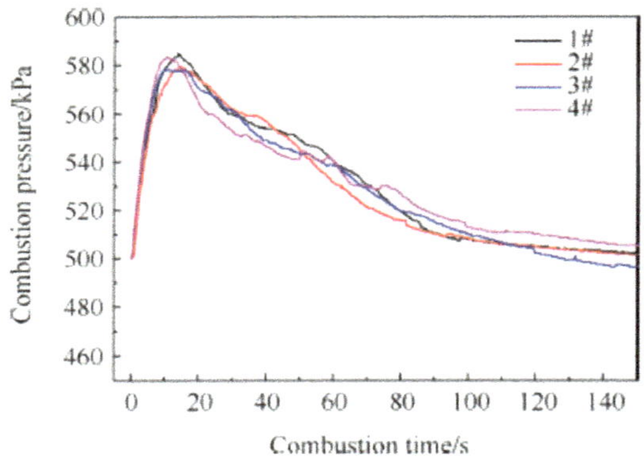

Figure 6.7 Pressure–time curve during the combustion of LSGHEFs. Reproduced from ref. 11.

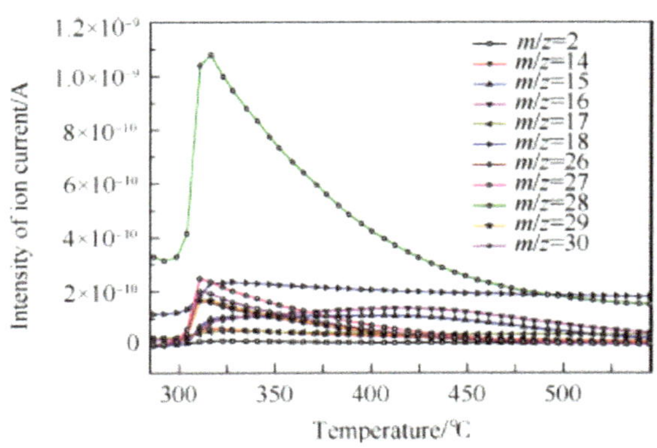

Figure 6.8 Intensity of ion current evolution from thermal cracking of BCC-A. Reproduced from ref. 11.

mainly includes molecular ion peak information such as 2, 14, 15, 16, 17, 18, 26, 27, 28, 29, 30, which can be classified as H_2, CH_4, C_2H_6, C_2H_4 and other small molecule gases. From the ionic current intensity, it can be seen that the peak corresponding to C_2H_4 is higher in the pyrolysis product, while the peak corresponds to H_2 and other molecules.

Based on the above molecular composition of boron cage compounds and their pyrolysis products, the pyrolysis products of boron cage A mainly generate CO_2, H_2O (gaseous), B_2O_3 (solid) and N_2 after combustion. The pyrolysis products of boron cage S and boron cage T mainly generate H_2O (gaseous), B_2O_3 (solid) and M_aO_b (solid) after combustion. Moreover, due to the large atomic weight of metal M, its hydrogen content is low, and less H_2O (gaseous) is generated.

Blank fuel 1# is composed of Al powder, LHDF and a small amount of gelling agent (3 wt%). There is no pyrolysis in the combustion process, and the combustion products are mainly Al_2O_3, CO_2 and H_2O (gaseous). Compared with the gas-phase products of blank fuel 1# combustion, the gas production of boron cage S and boron cage T is less, so the pressure is lower. Boron cage A does not contain metal elements, and the gas production is large, which is equivalent to the gas production of blank fuel 1# combustion, so the pressure peak is similar. For the pressure rising section (500 kPa to P_{max} kPa), the order of the pressure change rate is different from the pressure peak value, which may be caused by the synergistic effect of combustion products and temperature in the early stage of combustion.

In general, after boron cage A is added, in addition to the high combustion temperature of 4# fuel, relatively more flammable small molecule gas products are produced due to pyrolysis, which is conducive to improving the ignition and combustion performance of high-energy fuel containing aluminum. In addition, when extrapolated to the application condition of a high Mach ramjet, the high-energy fuel containing aluminum burns in millisecond contact in air, and the pressure of the 4# fuel increases relatively fastest ($\Delta P/\Delta T_{max}$), that is, the expansion work ability of high-energy fuel containing aluminum is enhanced after the addition of boron cage A.

It should be pointed out that because the combustion chamber is large and non-adiabatic, and the sample volume is small, there is a certain heat dissipation phenomenon in the system, so that the pressure drops after 20 s of combustion, which is not synchronized with the temperature change time.[16]

6.3.2.5 *Effect of the Boron Cage Compound on the Ignition Delay Time of the Liquid–Solid Gel Type High-energy Fuel Containing Al*

Figure 6.9 shows the charge coupled device (CCD) image captured by the ignition delay of LSGHEFs. Figure 6.9a is the first electric arc light of the shock tube captured by the high-speed CCD camera, and Figure 6.9b is the first bright spot captured by the CCD camera. The time difference between the two is the ignition delay of the LSGHEFs.

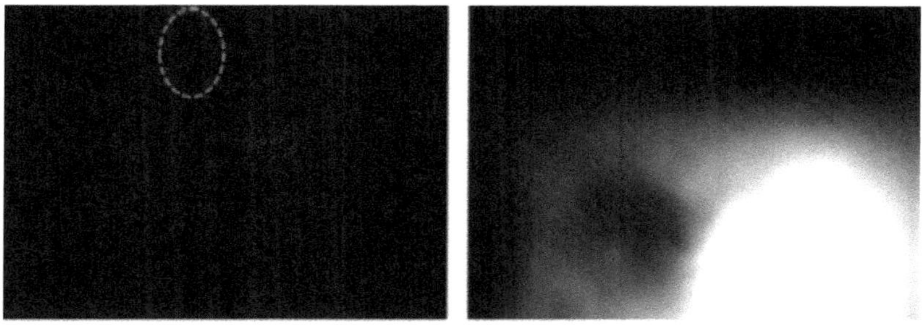

(a) Image of captured arclight from CCD; (b) image of the first bright spot.

Figure 6.9 Captured CCD images used to determine the ignition delay time of LSGHEFs. Reproduced from ref. 11.

Table 6.5 Ignition delay times of LSGHEFs. Reproduced from ref. 11.

Sample	Ignition delay time/ms	Degree of reduction/%
1#	155	Baseline
2#	86	44.5
3#	60	61.3
4#	54	65.2

Table 6.5 shows the ignition delay data of different LSGHEFs. It can be seen from the table that, after boron cage compounds are added to the LSGHEFs containing aluminum, the ignition delay time of the fuel is significantly shortened, in the order of [4#] > [3#] > [2#] > [1#]. The ignition delay time of 4# is shortened from the original 155 ms to 54 ms, which is 65.2%, indicating that the addition of boron cage compounds to the liquid–solid gel propellant system containing aluminum can significantly improve the ignition performance of the LSGHEFs containing aluminum. The reason can be attributed to the special boron cage structure of the boron cage compound itself, that is, the boron cage framework collapses and disintegrates rapidly when heated. Compared with hydrocarbon fuel and aluminum powder, its decomposition rate is faster the energy released is higher, and the thermal decomposition products contain rich H_2, CH_4, C_2H_6, C_2H_4 and other small molecule gases that are easily ignited. After boron cage A is added, the ignition delay time of LSGHEFs is the lowest, which may be mainly related to the production of more gas products.

6.3.3 Recent Advances in Camouflaged Carboranes and Polyhedral Boranes

The three isomers of the icosahedral $C_2B_{10}H_{12}$ (1,2- or *ortho*-, 1,7- or *meta*- and 1,12- or *para*-) carboranes and the isostructural polyhedral borane dianion $B_{12}H_{12}^{2-}$ are uniquely suited to play the role of building blocks in the construction of molecular scaffolding and stereochemically rigid platforms. These versatile components are often used for functionalization of the macrostructure or to add other interesting chemical properties. This versatility has led to the formation of several different types of structural motifs.[17–20] This section will focus on just one of these structural types, the camouflaged carboranes and polyhedral borane anions. The term camouflaged carborane or polyhedral borane describes a molecule in which all or nearly all, the B–H vertices have been substituted by a typical organic substituent. The resulting polyhedral borane derivative, which has a borane core stabilized by multicenter electron delocalization, is shrouded by a periphery of organic substituents. Hence, camouflaged carboranes demonstrate chemical properties indicative of their hybrid nature between that of the thermally and chemically stable carboranes and boranes, and the versatile aliphatic hydrocarbons.

Electrophilic halogenation or alkylation of hydrogen at the B–H vertices of the carboranes is a well-known reaction.[21–23] The methylation of *para*-carborane with methyl triflate and a catalytic amount of triflic acid at reflux temperature for 20 h cleanly affords the per-B-methylated carborane in 89% yield.[24] Under these conditions, only a small portion, less than 5%, of undecamethylcarborane is observed in the gas chromatography/mass spectrometry (GC/MS) of the crude reaction mixture. This demonstrates the remarkable chemoselectivity of methylation towards the B–H vertices. The camouflaged per-B-methylated carborane leaves the C–H vertices so inaccessible that, even though they can be lithiated by phenyl lithium, they cannot be methylated by methyl iodide. The complete methylation of all the vertices of *para*-carborane can be accomplished if 1,12-dimethyl-1,12-dicarba-*closo*-dodecaborane[25] is methylated using methyl triflate and triflic acid, producing dodecamethyl-1,12-dicarba-*closo*-dodecaborane (12) in excellent yield, 91%.[24] Figure 6.10a presents a view of a space-filling model of this

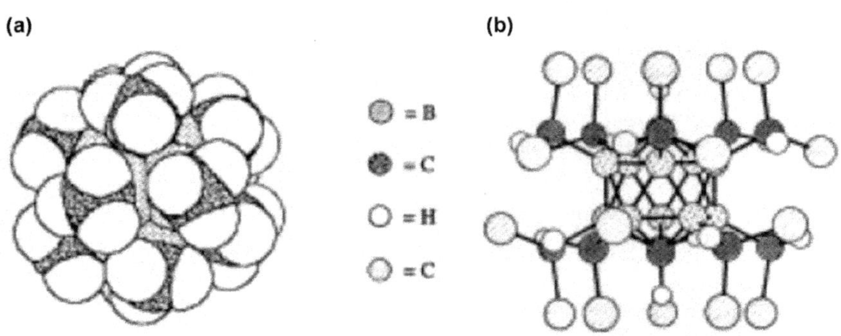

Figure 6.10 (a) Space-filling representation of dodecamethyl-*para*-carborane; (b) ball and stick representation of deca(dichloromethyl)-*para*-carborane. Reproduced from ref. 26 with permission from John Wiley & Sons, Copyright © 2000 Wiley-VCH Verlag GmbH & Co. KGaA, Weinheim.

novel icosahedral hydrocarbon-like molecule. Radical photochlorination of one of these apparent hydrocarbons, decamethyl-*para*-carborane, by chlorine in carbon tetrachloride under ultraviolet (UV) irradiation at ambient temperature for eight hours forms decakis(dichloromethyl)-1, 12-dicarba-*closo*-dodecaborane (12) in high yield, 88%. This further camouflaged carborane is depicted in Figure 6.10b as a ball and stick structure showing the rigorous stereochemical orientation of the $-CHCl_2$ substituents.

The extensive steric crowding observed in the spacefilling diagram of deca(dichloromethyl)-*para*-carborane accounts for the selective chlorination of only two of the three methyl hydrogen atoms. Systematic chlorination using stoichiometric amounts of *N*-chlorosuccinimide showed that initial chlorination of the methyl group activates the second C–H bond toward chlorination. Likely due to the extensive steric crowding around decamethyl-*para*-carborane camouflaged carborane as well as unfavorable bond strengths, the bromination of decamethyl-*para*-carborane was unsuccessful.

An application which is of particular interest is the possible adoption of these camouflaged icosahedral carboranes as pharmacophore groups possessing hydrophobic properties and potential functionalization for further reactions.[27] A comparison of the calculated van der Waals diameters of dodecamethyl-*para*-carborane (990 pm), deca(dichloromethyl)-*para*-carborane (1440 pm) and a well-known hydrophobe, C_{60} (1070 pm) shows similar volumes of these three molecules. In order to permit either of these camouflaged carboranes to be used as pharmacophore groups, further functionalization of the carboranes needs to be developed. The photochlorination results demonstrate that reactions characteristic of aliphatic hydrocarbons can be utilized to functionalize the organic shroud surrounding the delocalized polyhedral cage bonding of the carborane scaffolding. This hydrocarbon-like reactivity was further exploited through adaptation of the Barton photochemical oximation reaction.[28]

By focusing on functionalization at one of the ten available methyl groups of per-B-methyl-*para*-carborane, isomerically pure products can be prepared. While the radical chlorination resulted in only one isomer due to steric constraints, radical-mediated reactions are generally nonselective. However, the Barton reaction is an example of an intramolecular regiospecific radical reaction capable of activating methyl C–H bonds. The rigid geometry of an appropriately carbon-substituted deca-methyl-*para*-carborane is ideally suited for the Barton reaction which, through the photochemical decomposition of a nitrite ester function, produces NO and an alkoxy radical in close proximity to a C–H bond. It has been shown that the Barton reaction occurs through a cyclic six-membered transition state containing the alkoxy radical and the activated C–H bond.[29] The oxime product is then formed following a 1,5-hydrogen shift, collapse of the NO–alkyl radical pair, and isomerization of the nitroso derivative.

The appropriate functionality on the carbon atom of the carborane, a nitrite, was prepared from the reaction of deca-B-methyl-1-hydroxymethyl-1,12-dicarba-*closo*-dodecaborane (12)[28] and nitrosyl chloride in pyridine in excellent yield, 95%. The ensuing photolysis of this product in benzene produces the boron-substituted hydroximinoalcohol in 42–25% yields with varying amounts of starting material and the N–N linked nitroso dimer intermediate isolated from the reaction mixture.

Having successfully developed a synthetic approach to decamethyl- and dodecamethyl-*para*-carboranes by triflate mediated electrophilic methylation of all available B–H vertices, it became feasible to attempt the similar methylation of 1,2- and 1,7-$C_2B_{10}H_{12}$ and at the same time perhaps increase the yields of decamethyl- and dodecamethyl-*para*-carborane, a new procedure employing $AlCl_3$ in neat CH_3I solvent easily methylated *para*-carborane in high yield by substitution of all available B–H vertices. This procedure allowed for the preparation of 4,5,7,8,9,10,11,12-octamethyl-1,2-dicarba-*closo*-dodecaborane (12) and 4,5,6,8,9,10,11,12-octamethyl-1,7-dicarba-*closo*-dodecaborane (12) derivatives of 1,2- and 1,7-$C_2B_{10}H_{12}$ after 50 h using either 1 or 10 equivalents of $AlCl_3$.[30] In both cases the two equivalent B–H vertices nearest the 1,2- or 1,7-pairs of C–H vertices escaped reaction. This is not an unexpected result due to the reduced ground-state electron density at these positions. In fact, the "exhaustive" iodination of 1,2-, 1,7- and 1,12-$C_2B_{10}H_{12}$ resulted in the exact same selectivity and extent of substitution as the methylation reactions described above.[31]

This same synthetic scheme was employed to methylate 1-[1′-1′,2′-dicarba-*closo*-dodecaborane (12)]-1,2-dicarba-*closo*-dodecaborane (12),[32,33] bis-*ortho*-carborane, resulting in quantitative formation of 8,9,10,12,8′,9′,10′,12′-octamethyl-bis-*ortho*-carborane after 5 days. Again, methylation occurred at the eight B–H vertices of the dimer with the highest electron densities.

The class of camouflaged carboranes was extended to include a *nido*-carborane with the preparation of 1,2,4,5,6,9,10,11-octamethyl-7,8-dicarba-undecabo rate (−1) through the base-mediated deboration[34,35] of *closo*-1,2-$C_2B_{10}Me_8H_4$. Not surprisingly, the octa-B-methylated, *closo*-carborane was quite resistant to degradation. However, complete conversion to the corresponding *nido*-carborane was accomplished by treatment of the *closo*-carborane with an excess of potassium ethoxide in dimethoxyethane solution in a high-pressure vessel at 200 °C for 14 h. The crude potassium salt was converted to the bis(triphenylphosphoranylidene) ammonium (PPN^+) salt by treatment of the crude reaction mixture with PPNCl in absolute ethanol followed by crystallization from a saturated solution. The acid form of this camouflaged *nido*-carborane was prepared by treatment of the crude reaction mixture with aqueous HCl and sublimation from the resulting oily solid.

Again the camouflage carborane demonstrates enhanced stability toward hydrolysis in organic solvents over that of the similar, non-camouflaged, *nido*-7,8-$C_2B_9H_{13}$.[36] While simple heating of *nido*-7,8-$C_2B_9H_{13}$ results in dehydrogenation to form *closo*-1,8-$C_2B_9H_{11}$, the camouflaged *nido*-carborane must be passed over a bed of silica in order to induce dehydrogenation. Interestingly, *nido*-7,8-$C_2B_9H_{13}$ is inert in the presence of silica. The resulting crystalline product is a mixture of two isomers of *closo*-1,8-$C_2B_9Me_8H_3$, resulting in differing positions of the unique B–H vertex. From this isomeric mixture, a single pure compound was prepared by methylation of the single B–H vertex to produce 2,3,4,5,6,7,9,10,11-nonamethyl-1,8-dicarbaundecaborane (11), *closo*-$C_2B_9Me_9H_2$, by reaction with MeI/$AlCl_3$ at reflux temperature for 8 h, in 95% yield.

With the permethylation results summarized above and the report of the permethylation of the $[CB_{11}H_{12}]^-$ by Michl and coworkers[37] using an alkylation method based upon methyl triflate that followed, the series of neutral, monoanionic, and dianionic camouflaged carboranes was complete with the recent report of the methylation of $[B_{12}H_{12}]^{2-}$.[38] Permethylation occurred when $Cs_2B_{12}H_{12}$ was first treated with a mixture of methyl iodide and trimethylaluminum (6 days, 45 °C) giving $Cs_2B_{12}(CH_3)_{11}I$ followed by the reaction of this intermediate with neat trimethylaluminum. Figure 6.11a presents an ORTEP representation of this new ion. The hypothesis that methylation occurred by successive iodination followed by methylation was confirmed when [*closo*-$B_{12}H_{11}I]^{2-}$ was selectively converted to [*closo*-$B_{12}H_{11}CH_3]^{2-}$ when treated with trimethylaluminum and heated at reflux temperature. Note that conversion of B–I bonds to B–alkyl bonds in this manner offers an alternative route to the palladium-catalyzed alkylation of [*closo*-$B_{12}H_{12}]^{2-}$ with Grignard reagents.[39] As in the cases of the permethylated carborane and monocarbon anion analogs, the icosahedral surface of $[B_{12}(CH_3)_{12}]^{2-}$ presents a virtual forest of hydrophobic methyl groups even though this species' salts display solubility in water. This amphiphilic ion may serve as a weakly coordinating dianion in a variety of applications. In addition, functionalized modifications of this species could serve as pharmacophores. Of particular interest was the fact that cyclic voltammetry in acetonitrile produced a reversible one-electron oxidation at 0.41 V. Chemical oxidation with a Ce^{4+} ion in the same solvent led to the isolation of the bright blue, air-stable anion radical $[B_{12}(CH_3)_{12}]^-$

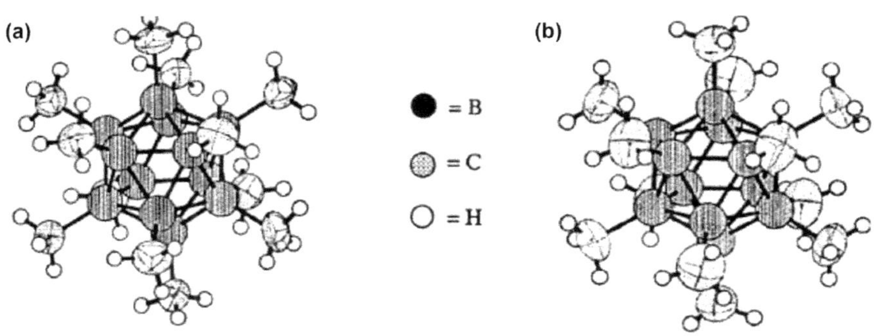

Figure 6.11 ORTEP representations of (a) $[B_{12}(CH_3)_{12}]^{2-}$ and (b) the $[B_{12}(CH_3)_{12}]^-$ radical. Reproduced from ref. 26 with permission from John Wiley & Sons, Copyright © 2000 Wiley-VCH Verlag GmbH & Co. KGaA, Weinheim.

in good yield (66%).[40] Figure 6.11b is an ORTEP representation of this anion radical. Bonding features of the diamagnetic dianion and the paramagnetic anion radical are essentially identical. The analogous neutral $[CB_{11}(CH_3)_{12}]$ radical reported by King *et al.*[41] is also bright blue and structurally very similar to the isoelectronic anion radical.

The methyl camouflaged species described above demonstrate the amplification of aromatic polyhedral borane chemistry by the prudent application of organic chemistry and the emergence of a novel family of organoboranes, the camouflaged carboranes and polyhedral boranes.

The development of successful routes for an electrophilic polymethylation of icosahedral carboranes and the $[B_{12}H_{12}]^{2-}$ ion provides species equipped with hydrophobic hydrocarbon surfaces, spawning the field of hydrophobicly camouflaged carboranes and polyhedral boranes. As a consequence, the possible existence of similar structures presenting hydrophilicly camouflaged surfaces offered a synthetic challenge. While the oxidation of boranes is highly exothermic and reminiscent of high-performance aircraft fuels, rocket propellants, and explosives, moderation of this reaction using relatively weak oxidants with stabilized icosahedral borane derivatives offered a challenging approach to polyhydroxylated products. This approach proved to be successful[42] using 30% hydrogen peroxide as the solvent and oxidant at reflux temperatures with $[B_{12}H_{12}]^{2-}$, $[CB_{11}H_{12}]^-$ and 12-$(CH_2OH)_2$-1,12-$C_2B_{10}H_{10}$ as substrates. Figure 6.3 presents the structure of $[B_{12}(OH)_{12}]^{2-}$. In each case all B–H vertices were converted to their B–OH counterparts in variable yield, Cs_2[*closo*-$B_{12}(OH)_{12}$] 65%, Cs[*closo*-1-H-1-$CB_{11}(OH)_{10}$] 31%, and *closo*-1,12-H_2-1,12-$C_2B_{10}(OH)_{10}$ 80%, respectively. The higher yield of the hydroxy-substituted carborane demonstrates the greater kinetic stability of the carborane toward degradation to the thermodynamic product, boric acid. The lower kinetic stability of [*closo*-1-H-1-$CB_{11}(OH)_{10}$]$^-$ is likely due to its overall molecular electronic polarization compared to the two centrosymmetric molecules.

The –CH_2OH substituents present on the *para*-carborane derivative provide some solubility of the carborane derivative in 30% H_2O_2 and are removed *in situ* by conversion to –COOH groups and subsequent decarboxylation. The properties of the $[B_{12}(OH)_{12}]^{2-}$ ion are of particular interest since this structure offers a unique geometric platform for formation of extended structures such as dendrimers. An unexpected property is the facile aggregation of the disodium salt which has very limited water solubility. Sodium ions appear to coordinate with the –OH surface and provide intercage bonding. The dicesium salt is freely soluble in water since cesium is too large to effectively coordinate with the $[B_{12}(OH)_{12}]^{2-}$ ion. While $[B_{12}(OH)_{12}]^{2-}$ may be viewed as an intermediate in the hydrolysis of $[B_{12}H_{12}]^{2-}$ to boric acid, it still retains aromatic character and 26 cage-bonding electrons available for chemical reactions. The chemistry of these novel polyhydroxylated carborane and borane derivatives is being actively pursued.

6.4 Conclusions

(1) The burning rate of propellants is greatly enhanced by the addition of CMP. With the increase of added CMP wt%, the burning rate of AP/NG propellant increases and the pressure exponent decreases. The best catalysis of CMP on RDX/NG propellants appears under 8–10 MPa. The pressure exponent of RDX/NG propellants increases with an increase in CMP wt%. Both the temperature of AP phase transition and RDX melting and the decomposition of AP and RDX are affected by the addition of CMP.

(2) The density and volumetric calorific values of high-energy fuel containing aluminum slightly decreased with the addition of boron cage compounds, but the mass calorific value increased by 11.6% and 12.4%, respectively, after the addition of boron cage T and boron cage A. Boron cage compounds can increase the peak combustion temperature of high-energy fuel containing aluminum by 21.1–52.9%, shorten the ignition delay by 44.5–65.2%, and significantly improve the ignition and combustion performance of high-energy fuel containing aluminum.

(3) The chemistry described above exemplifies the great similarity between aromatic polyhedral borane chemistry and the aromatic branch of organic chemistry. Modular syntheses with carboranes and the discovery of a variety of camouflaged derivatives clearly reveal the tentacles of organic chemistry reaching into the polyhedral borane field. Thus, the conflux of boron and carbon chemistries is broadened.

References

1. L. YiWei, Development trend of foreign Solid State Statistics Bureau in 1980s, The 210th Research Institute of ordnance industry, 1989.
2. D. C. Sayles, *Non-nitroglycerin containing composite modified double base propellant*, USP4707199, 1983.
3. W. E. Hill and L. R. Beason, Composite modified double base propellants with very high burning rates, USP3800966, 1974.
4. L. Galfetti, L. T. Deluca and F. Severini, *et al.*, Pre and Post-Burning Analysis of Nano-Aluminized Solid Rocket Propellants, *Aerosp. Sci. Technol.*, 2007, **11**(1), 26–32.
5. H. Tyagi, P. E. Phelan and R. Prasher, *et al.*, Increased Hot-Plate Ignition Probability for Nanoparticle-Laden Diesel Fue, *Nano Lett.*, 2008, **8**(5), 1410–1416.
6. A. S. Michael, Possible Chenmical Mechanisms for Boron Hydride Acceleration of Nitramine Decomposition: Literature Review, *J. Propul. Power*, 2012, **14**(6), 981–990.
7. X. Yunna, L. Jiaoyi and H. ZhiJun, *et al.*, Synthesis, Thermal Decomposition Mechanism and Kinetic Equation of Polyhedral Boron Hydride Compound $[(C_2H_5)_4N]_2B_{10}H_{10}$ with High Enthalpy of Combustion, *Chem. J. Chinese Universities*, 2015, **3**(2), 375–380.

8. W. Zhang, W. X. Xie and X. Z. Fan, *et al.*, Energetic Characteristics and Comprehensive Properties of Propellants Containing ADN, *Chin. J. Explos. Propellants*, 2015, **38**(2), 81–85.
9. H. Bergman, H. Edvinsson and H. Ostmark, Trends and Development of New Energetic Materials, *Militarteknisk tidskrift*, 2001, **70**(2), 26–28.
10. D. E. G. Jones, Characterization of ADN and ADN-based Propellants, *Propellants, Explos., Pyrotech.*, 2005, **30**(2), 526–528.
11. S. Daoan, L. Chunying and D. Yongmei, *et al.*, Effects of boron cage compounds on ignition and combustion performance of liquid-solid high energy fuels containing nano-Al, *J. Propuls. Technol.*, 2019, **40**(7), 1644–1650.
12. S. Wengang, S. Tiegang and Z. Guodong, Effects of carborane derivatives on the combustion of solid propellants, *J. Solid Rocket Technol.*, 1995, **18**(3), 24–26.
13. M. A. Schroeder, Critical analysis of nitramine decomposition data: Activation energies and frequency factors for HMX and RDX decomposition, AD-A 160543, 1985.
14. X. Min, M. Qing and F. Yu, Synthesis and gelation behavior of a new cholesterol based small molecule gelling agent, *J. Phys. Chem.*, 2013, **29**(9), 2005–2012.
15. H. Jian-feng, *Principle and technology of sol–gel*, Chemical Industy Press, Beijing, 2005.
16. X. G. Yang, W. Jiang, C. B. Knobler and M. F. Hawthorne, *J. Am. Chem. Soc.*, 1992, **114**, 9719.
17. P. Weiqiang, L. T. DeLuca and A. Gromov, *et al.*, *Innovative Energetic Materials: Properties, Combustion and Application*, Spinger, 2020.
18. W. Jiang, I. T. Chizhevsky and M. D. Mortimer, *et al.*, *Inorg. Chem.*, 1996, **35**, 5417.
19. M. F. Hawthorne and Z. P. Zheng, *Acc. Chem. Res.*, 1997, **30**, 267.
20. Z. Zheng, W. Jiang, A. A. Zinn, C. B. Knobler and M. F. Hawthorne, *Inorg. Chem.*, 1995, **34**, 2095.
21. V. I. Bregadze, *Chem. Rev.*, 1992, **92**, 209.
22. V. V. Grushin, V. I. Bregadze and V. N. Kalinin, *J. Organomet. Chem. Libr.*, 1988, **20**, 1.
23. R. N. Grimes, *Carboranes*, Academic Press, New York, 1970.
24. W. Jiang, C. B. Knobler, M. D. Mortimer and M. F. Hawthorne, *Angew. Chem., Int. Ed. Engl.*, 1995, **34**, 1332.
25. V. I. Stanko, A. I. Klimova and A. N. Kashin, *Zh. Obshch. Khim.*, 1969, **39**, 1895.
26. J. J. Rockwell, A. Herzog and T. Peymann, *et al.*, Recent advances in the field of camouflaged carboranes and polyhedral boranes, *Curr. Sci.*, 2000, **78**(4), 405–409.
27. Y. Endo, K. Yaguchi, M. Tsuji, K. Yamaguchi and K. Shudo, *Chem. Pharm. Bull.*, 1999, **47**, 699.
28. A. Herzog, C. B. Knobler and M. F. Hawthorne, *Angew. Chem. Int. Ed.*, 1998, **37**, 1552.

29. M. Akhtar and M. M. Pechet, *J. Am. Chem. Soc.*, 1964, **86**, 265.
30. A. Herzog, A. Maderna, G. N. Harakas, C. B. Knobler and M. F. Hawthorne, *Chem. – Eur. J.*, 1999, **5**, 1212.
31. R. R. Srivastava, D. K. Hamlin and D. S. Wilbur, *J. Org. Chem.*, 1996, **61**, 9041.
32. J. A. Dupont and M. F. Hawthorne, *J. Am. Chem. Soc.*, 1964, **86**, 1643.
33. T. E. Paxson, K. P. Callahan and M. F. Hawthorne, *Inorg. Chem.*, 1973, **12**, 708.
34. R. A. Wiesboeck and M. F. Hawthorne, *J. Am. Chem. Soc.*, 1964, **86**, 1642.
35. M. F. Hawthorne, D. C. Young, P. M. Garrett, D. A. Owen, S. G. Schwerin, F. N. Tebbe and P. A. Wegner, *J. Am. Chem. Soc.*, 1968, **90**, 862.
36. D. A. T. Young, R. J. Wiersema and M. Hawthorne, *J. Am. Chem. Soc.*, 1971, **93**, 5687.
37. B. T. King, Z. Janousek, B. Gruner, M. Trammell, B. C. Noll and J. Michl, *J. Am. Chem. Soc.*, 1996, **118**, 3313.
38. T. Peymann, C. B. Knobler and M. F. Hawthorne, *J. Am. Chem. Soc.*, 1999, **121**, 5601.
39. T. Peymann, C. B. Knobler and M. F. Hawthorne, *Inorg. Chem.*, 1998, **37**, 1544.
40. T. Peymann, C. B. Knobler and M. F. Hawthorne, *Inorg. Chem.*, 2000, **39**, 1163–1170.
41. B. T. King, B. C. Noll, A. J. McKinley and J. Michl, *J. Am. Chem. Soc.*, 1996, **118**, 10902.
42. T. Peymann, A. Herzog, C. B. Knobler and M. F. Hawthorne, *Angew. Chem., Int. Ed.*, 1999, **38**, 1062.

CHAPTER 7

Development Prospective of High-energy Combustion Agents in Propellants and Explosives

7.1 Introduction

Borohydride is an effective and practical burning rate regulator for solid propellants with high burning rate of 50–250 mm s^{-1}. The high burning rate propellants with borohydride compounds can be charged with various types by using existing charging technology and equipment. The propellants have higher energy and better technological and mechanical properties. *N*-Hexylcarborane is a kind of high burning rate regulator with high boiling point liquid and excellent comprehensive performance. It has been used in the propellant of the ''Agkistrodon'' anti-tank rocket launcher, and is being considered for use in future systems such as the "guard'' anti-ballistic missile.[1]

With the development of borohydride chemistry, the toxicity of new borohydride compounds has changed fundamentally. Toxicity tests show that *N*-hexylcarborane and other compounds have a new closed cage structure and stable performance, and their toxicity has been greatly reduced compared with the open structure of simple borohydride compounds (such as $B_{10}H_{14}$).[2] Under the existing industrial health protection conditions, they can be safely produced and used. The toxicity of ionic hydroborates is lower, and some are even non-toxic. These closed cage boron hydrogen derivatives are being studied as antitumor drugs.[3] Of course, in the synthesis of these substances, some boron hydrogen intermediates are still toxic, so we should

High-energy Combustion Agents of Organic Borohydrides: Synthesis, Characterization and Applications
By XiaoHong Zhang, WeiQiang Pang, Luigi T. DeLuca and Yu Zhao

Published by the Royal Society of Chemistry, www.rsc.org

pay attention to taking necessary protective measures and to minimize or avoid the generation of toxic substances in the synthesis route.

7.2 Innovative Combustion Agents

A combustion agent containing metal is one of the important components of modern solid propellants. It can not only improve the heat of the explosion and the density of propellants, but also inhibit the combustion oscillatory of propellants. In recent years, hydrogen storage alloys and ternary metal alloys have been of great interest to researchers worldwide.[4] This is because hydrogen storage alloy is a functional material that can store a large number of hydrogen atoms in the gap of crystal, and it has reversible hydrogen absorption and desorption. It can store hydrogen thousands of times the volume of the alloy itself. Its hydrogen absorption density exceeds the density of liquid hydrogen and solid hydrogen.[5] Moreover, the melting point and boiling point of the alloy are lower than that of the corresponding metal. Therefore, when a metal–metal solid solution is used as a combustion agent, its performance is expected to exceed that of pure metal.[6]

7.2.1 Hydrogen Storage Composites

Hydrogen storage composites were developed in the 1960s. They can absorb, store and release hydrogen many times.[7] When the hydrogen storage composite is heated to the hydrogen release temperature (generally expected to be controlled above 200 °C) in the oxidation environment, the hydrogen in the hydrogen storage composite is released and burned to produce heat and water vapor, and then it ignites the composite powder to produce a violent metal combustion reaction, release more heat, and play a role similar to metal hydride.[8,9] Therefore, a hydrogen storage composite can become a special metal combustion agent – hydrogen storage composites can be used as an important component of solid propellant to improve the combustion performance and increase the thrust of the propellant.

7.2.2 Metal Hydrides

7.2.2.1 Single Metal Hydrides

In the single metal hydride, the hydrogen content of light metal hydride is higher. Light metal hydrides include lithium hydride (LiH), sodium hydride (NaH), magnesium hydride (MgH_2), calcium hydride (CaH_2) and aluminum hydride (AlH_3).[10] The hydrogen storage capacity is more than 4.2% and the hydrogen storage capacity of LiH is up to 12.7%, which is much higher than that of hydrogen storage alloy hydrides. Obviously, a light metal hydride has obvious advantages over hydrogen storage alloy hydride in introducing H_2 into propellant combustion process.

AlH_3 is a covalent hydride with a hydrogen storage capacity of 10.1%. It has six crystal forms, and the decomposition and hydrogen release temperature of the different crystal forms are different. Among them, α-AlH_3 is a crystal form which is being deeply studied at present. Its hydrogen evolution temperature is lower than 150 °C and is related to particle size. In air, it is easy to form a dense Al_2O_3 oxide film on the surface of AlH_3, which prevents further oxidation reaction. Therefore, AlH_3 can be treated in air.[11] At present, AlH_3 is being successfully used in propellants. Compared with traditional propellants, the energy performance of propellant containing AlH_3 is greatly improved. For example, the measured specific impulse of a AlH_3/ADN system high-energy propellant developed in Russia has exceeded 2940 N s kg^{-1}, which is a propellant with the highest specific impulse.[12] DeLuca *et al.*[13] studied the physicochemical properties and ballistic properties of propellant containing AlH_3. It was found that replacing Al with AlH_3 would reduce the combustion temperature of the propellant and the content of CO_2, OH and H_2O in combustion products, so as to weaken the throat ablation and improve the specific impulse of the propellant. Flynn *et al.*[14] used AlH_3 in double base propellant, raised the theoretical specific impulse of propellant to 2874.34 N s kg^{-1}, and the measured specific impulse to 2675.4 N s kg^{-1}, showing excellent performance.

MgH_2 is a light metal hydride that is of interest to researchers. The hydrogen storage capacity of MgH_2 is 7.7%, and its thermal stability is higher than AlH_3. Ordinary MgH_2 can decompose and release H_2 at more than 300 °C.[10] Liu *et al.*[15] found that the catalytic effect of MgH_2 on the thermal decomposition of AP is significantly stronger than that of Mg, which can significantly reduce the low and high temperature decomposition temperature of AP and increase the apparent decomposition heat. In AP/Al/HTPB composite propellant, MgH_2 can reduce the thermal decomposition temperature of the propellant and increase its reaction heat by catalyzing the decomposition of AP. MgH_2 with mass fraction of 1.3% can increase the burning rate of propellant by 13.9%.[16] Yao *et al.*[17] studied the effect of MgH_2 on the thermal decomposition characteristics of RDX and found that MgH_2 reduced the apparent activation energy of RDX decomposition from 159.22 kJ mol^{-1} to 133.69 kJ mol^{-1}, and also reduced the stability of RDX. Researchers have studied the application of MgH_2 in emulsion explosives and found that both MgH_2 and TiH_2 can significantly improve the explosive properties of emulsion explosives,[18,19] which further proves that metal hydrides have great potential in improving the energy properties of energetic compounds.

In addition, the effects of various metal hydrides on the energy properties of propellants were investigated through theoretical calculations. Li *et al.*[12] calculated the chemical equilibrium of HTPB ternary and quaternary composite propellants containing metal hydrides based on the standard specific impulse of Al containing propellants by using the principle of minimum free energy. The single metal hydrides involved included AlH_3, MgH_2, TiH_2, CaH_2, ZrH_2, SrH_2, BaH_2 and CSH. It was found that AlH_3 and MgH_2

contribute more to the standard theoretical specific impulse than Al, while TiH_2, CaH_2, ZrH_2, SrH_2, BaH_2 and CSH contribute less than Al. The effects of AlH_3, MgH_2, TiH_2, ZrH_2, BaH_2 and CSH on the energetic characteristics of 3,3-diazomethoxyheterocyclobutane (BAMO) and 3-methyl-3-aziomethoxy-heterocyclobutane (AMMO) block copolymer P(BAMO–AMMO)-based propellants were studied.[20] Among the above hydrides, AlH_3 contributes the most to the propellant energy, the enhancement and promotion effect of MgH_2 is weaker than AlH_3, while TiH_2, ZrH_2, BaH_2 and CSH will reduce the propellant energy.

At present, the academic research on the application of light metal hydride in propellants mainly focuses on AlH_3 and MgH_2, and a series of achievements have been made. AlH_3 is considered to be an important development direction of combustors for high-energy propellants. It should be noted that AlH_3 is a dynamically stable hydride, and its decomposition and dehydrogenation is a weak exothermic reaction,[10] which means that it will decompose slowly during storage at room temperature, which is unfavorable to the long-term storage stability of propellant. Therefore, this issue needs more attention.

The hydrogen content of hydride is closely related to its energy contribution to propellant. High hydrogen content has an obvious effect on further improving the performance of propellant. In addition, from the point of view of production technology, the thermal stability of hydride is also an important criterion for its application in propellant. If its thermal stability is too low, it may decompose during production and processing, resulting in potential safety hazards. Compared with AlH_3 and MgH_2, alkali metal hydrides represented by LiH (hydrogen content 12.5%) generally have higher thermal stability and higher hydrogen storage capacity, and have great application potential in the field of propellant.

7.2.2.2 Bimetallic Hydrides

Bimetallic hydrides are mainly composed of transition metal elements, and hydrogen reacts with them to form interstitial hydrides. The metal elements that make up bimetallic hydrides can be divided into two categories according to the characteristics of reaction with hydrogen. The first type is mainly I A–V B group elements, and their reaction with hydrogen is exothermic reaction ($\Delta H < 0$), which can form stable hydride with hydrogen, which is called exothermic metal (represented by the symbol A), including Ti, Zr, La, MM (mixed rare earth), Mg, Ca, *etc.* The other is mainly transition metals of the VI B–VIII B group (except Pd), which have low affinity with hydrogen and usually do not form hydride, but hydrogen is easily migrated in its lattice. The reaction of these metals with hydrogen is endothermic ($\Delta H > 0$), which is called endothermic metal (represented by the symbol B), including Ni, Fe, Co, Mn, Cu and Al. According to the different proportion of elements A and B, bimetallic hydrides can be divided into AB_5 type (rare earth system), AB_2 type (zirconium system), AB type (titanium system) and

A_2B type (magnesium system). Among them, rare earth bimetallic hydrides represented by $LaNi_5$ have good hydrogen storage thermodynamic and kinetic properties, but their hydrogen storage capacity is generally less than 2% (mass fraction, the same below). The hydrogen storage capacity of magnesium alloys represented by Mg_2Ni is relatively high, up to 3.6%.

At present, researchers have applied magnesium-based alloy hydrides to propellants and achieved some good research results. It was found that Mg_2NiH_4 has a good catalytic effect on the thermal decomposition of ammonium perchlorate (AP), a common oxidant for propellants. It can significantly promote the low-temperature thermal decomposition of AP, reduce its high-temperature thermal decomposition temperature and increase the apparent decomposition heat.[21] The mechanism study shows that H_2, Mg and Ni produced by thermal decomposition of Mg_2NiH_4 can react with the decomposition products of AP, so as to promote its decomposition.[22] Liu *et al.*[23] studied the effect of magnesium copper alloy hydrogen storage material Mg_2Cu–H on the thermal decomposition process of AP, and found that it can significantly promote the thermal decomposition of AP, and the effect is better than Mg_2Cu. After introducing Mg_2Cu–H, the decomposition rate of AP increased, the high temperature thermal decomposition temperature decreased, and the apparent decomposition heat increased significantly. It was further found that Mg_2NiH_4 and Mg_2Cu–H can reduce the thermal decomposition temperature of AP/Al/HTPB composite propellant and increase its decomposition heat by catalyzing the decomposition of AP.[16] Mg_2NiH_4 and Mg_2Cu–H with mass fraction of 1.3% can increase the burning rate of AP/Al/HTPB composite propellant by 3.5% and 14.4%, respectively. In addition, the effect of magnesium-based alloy hydride on the thermal decomposition of AP/Al/HTPB composite propellant is positively correlated with its hydrogen content. The effect of hydrogen storage alloy hydride on the combustion of AP/HTPB propellant has been studied.[24,25] It was found that hydrogen storage alloy hydride can improve the combustion efficiency of organic components of propellant at the combustion surface, and its decomposition and hydrogen release, H_2 combustion, good ignition characteristics, high combustion heat and high combustion efficiency can significantly improve the combustion performance of AP/HTPB propellant.

In order to realize the application of hydrogen storage materials in propellant, in addition to considering its influence on propellant energy and combustion performance, it is also necessary to investigate the compatibility between hydrogen storage materials and propellant components and its influence on propellant density, characteristic signal, mechanical properties, storage life and other performance indexes. The basic parameters such as stability, density, oxygen consumption and combustion heat of alloy hydride containing Mg, Al, Ni, B and H were investigated.[26] It was found that it has good storage stability in dry air, the density and oxygen consumption are equivalent to Al, and the combustion calorific value is higher than Al, so it can be used as a high-energy combustion agent of propellant.

The compatibility between a hydrogen storage alloy hydride combustor and the common energetic components of propellant by DSC were studied.[27] It was found that it was compatible with ammonium perchlorate (AP), RDX, CL-20, 1/1-NG/DEGDN and nitrocellulose (NC), and it can be used in propellant. In addition, the researchers also found that hydrogen storage alloy hydrides significantly improved the performance of the new energetic binder GAP, reducing the curing activation energy of GAP, enhancing the gel fraction, promoting the formation of three-dimensional network and improving the mechanical properties.[28]

At present, research on the application of hydrogen storage alloy hydride in propellant mainly focuses on the role of hydrogen storage alloy hydride in promoting propellant component reaction and the role of hydrogen in its decomposition in propellant combustion. Relevant results have confirmed that hydrogen storage alloy hydrides can significantly enhance the combustion performance of composite propellants. In fact, compared with other chemical hydrogen storage materials, the hydrogen content of hydrogen storage alloy hydride is low, which is not the best choice from the perspective of introducing hydrogen into the combustion process of propellant. However, hydrogen storage alloy hydrides contain a wide range of elements, in which transition metal elements such as zirconium (Zr), iron (Fe) and copper (Cu) can regulate the combustion of propellants, and can be used as combustion catalysts of propellants.

7.2.3 Metal Borohydride Compounds

7.2.3.1 Aluminum Hydrogen Compounds

Common metal aluminum hydrogen compounds are mainly light metal aluminum hydrogen compounds, including lithium aluminum hydride ($LiAlH_4$, hydrogen storage capacity of 10.5%), sodium aluminum hydride ($NaAlH_4$, hydrogen storage capacity of 7.4%), potassium aluminum hydride ($KAlH_4$, hydrogen storage capacity of 5.7%), magnesium aluminum hydride ($Mg(AlH_4)_2$, hydrogen storage capacity 9.3%) and calcium aluminum hydride ($Ca(AlH_4)_2$, hydrogen storage capacity 7.8%), *etc.*[29]

The thermal stability of light metal aluminum hydrogen compounds is generally lower than that of corresponding boron hydrogen compounds, and can decompose and release hydrogen at lower temperature. $LiAlH_4$ melts at about 110 °C, then decomposes to produce Li_3AlH_6 and LiH and releases H_2. The dehydrogenation process is an exothermic reaction, and its products can continue to decompose and release hydrogen at a higher temperature. The whole dehydrogenation process ends at about 400 °C.[30] The dehydrogenation temperature of $NaAlH_4$ is slightly higher than that of $LiAlH_4$. The dehydrogenation starts at about 160 °C, which is an endothermic reaction.[31] The thermal stability of $KAlH_4$ is significantly higher than that of $LiAlH_4$ and $NaAlH_4$, and its decomposition and dehydrogenation reaction temperature is above 300 °C, which is also accompanied by an endothermic effect.[32] The initial dehydrogenation temperature of alkaline earth metal aluminum hydrogen compound $Mg(AlH_4)_2$ is about 115 °C, and the decomposition dehydrogenation reaction is a weak exothermic reaction.[33] The initial dehydrogenation temperature of $Ca(AlH_4)_2$ is about 170 °C, and its thermal stability is higher than that of $Mg(AlH_4)_2$.[34] In addition, the hydrogen evolution behavior of the above two alkaline earth metal aluminum hydrogen compounds is obviously different from that of alkali metal aluminum hydrogen compounds.

There is no report on the application of metal aluminum hydrogen compounds in propellants, but researchers investigated their effects on the energy properties of propellants by means of chemical equilibrium calculation. The effects of $LiAlH_4$ and $Mg(AlH_4)_2$ on the energy properties of HTPB ternary and quaternary composite propellants were studied,[12] it was found that the contribution of these two metal aluminum hydrogen compounds to the standard theoretical specific impulse was greater than that of Al, and there was an optimal value of energy characteristic parameters. The effect of $LiAlH_4$ is better than that of $Mg(AlH_4)_2$. The effects of the above two aluminum hydrogen compounds on the energy properties of P(BAMO–AMMO) based propellant were calculated.[20] The obtained law is similar to that of HTPB composite propellant. Both of them can improve the energy level of the propellant after replacing Al.

The above results show that the use of metal aluminum hydrogen compounds with high hydrogen storage as hydrogen generators in propellants can significantly improve the energy level of propellants. However, similar to metal borohydride compounds, metal aluminum hydride compounds also have strong reducibility. Therefore, in the research of metal aluminum hydrogen compounds used in propellants, it is also necessary to systematically and deeply understand the interaction between metal aluminum hydrogen compounds and other propellant components. In addition, the thermal stability of metal aluminum hydrogen compounds $LiAlH_4$ and $M(AlH_4)_2$ is low, which may decompose and release hydrogen at about 100 °C, which is poorly matched with the processing and molding process of some existing propellants, which may lead to potential safety hazards in the formation process. It is necessary to improve its thermal stability by composition modification and derivative synthesis to meet the practical requirements.

7.2.3.2 Metal Borohydride Compounds

Metal borohydride compounds include alkali metal borohydride compounds, alkaline earth metal borohydride compounds, transition metal borohydride compounds, lanthanide metal borohydride compounds, actinide metal borohydride compounds, *etc.*[29,35] Among them, alkali metal borohydride compounds such as lithium borohydride ($LiBH_4$, hydrogen storage capacity 18.1%), sodium borohydride ($NaBH_4$, hydrogen storage capacity 11.8%) and potassium borohydride (KBH_4, hydrogen storage capacity 8%), and alkaline earth metal borohydride compounds such as magnesium borohydride ($Mg(BH_4)_2$, hydrogen storage capacity 14.9%), calcium borohydride ($Ca(BH_4)_2$, hydrogen storage capacity 12.9%) and aluminum borohydride ($Al(BH_4)_3$, hydrogen storage capacity 19.7%) contain light metals and are collectively referred to as light metal borohydride compounds, which have high hydrogen storage capacity. Obviously, compared with hydrogen storage alloy hydride, the introduction of light metal borohydride compound into the propellant combustion process can also obtain better results.

Among the above light metal borohydride compounds, $Al(BH_4)_3$ is a liquid at room temperature, with low thermal stability and is easily decomposed to form borane (B_2H_6), which is not suitable for propellant. Except $Al(BH_4)_3$, other light metal borohydride compounds are solid at room temperature and have good thermal stability. Among them, the thermal stability of alkali metal borohydride compounds is better than that of alkaline earth metal borohydride compounds. The main hydrogen evolution temperatures of $LiBH_4$ and $NaBH_4$ are above 400 °C and 450 °C, respectively.[36,37] The dehydrogenation of $Mg(BH_4)_2$ and $Ca(BH_4)_2$ began at 250 °C and 320 °C, respectively.[38,39] From the thermal stability and hydrogen storage capacity of the above light metal borohydride compounds, they have great application potential in propellants.

However, the application of light metal borohydride compounds in propellants is still in its infancy, and there are few reports in the literature. For instance, the effect of $NaBH_4$ on the $BA(NO_3)_2$–RP–Mg combustion pyrotechnic agent (combustion agent) system were investigated,[40] and it was found that adding $NaBH_4$ can significantly improve the calorific value of the combustion agent. When the mass fraction of $NaBH_4$ is 20%, the calorific value of the combustion agent can be increased by 14.3%. In addition, $NaBH_4$ can also reduce the sensitivity of the combustion agent and improve the comprehensive performance of the combustion agent. The interaction between $Mg(BH_4)_2$ and RDX by differential scanning calorimetry (DSC) were studied,[15] and it was found that the addition of $Mg(BH_4)_2$ increased the apparent activation energy and pre-exponential factor of RDX thermal decomposition. The apparent activation energy increased from 159.22 $kJ\,mol^{-1}$ to 180.27 $kJ\,mol^{-1}$, indicating that the stability of RDX containing $Mg(BH_4)_2$ increased, showing a good compatibility between $Mg(BH_4)_2$ and RDX.

The decomposition and dehydrogenation process of light metal borohydride compounds is an endothermic reaction, which is disadvantageous in the field of propellant applications with high requirements for energy release. If the ammonia (NH_3) group is used to complex with a light metal borohydride compound to form a metal borohydride compound ammonia complex, the heat change of the material in the process of hydrogen evolution can be effectively adjusted. At present, the reported ammonia complexes of light metal borohydride compounds mainly include $LiBH_4{\cdot}xNH_3$ ($x = 1,2,3,4$), $Mg(BH_4)_2{\cdot}xNH_3$ ($x = 1,2,3,6$) and $Al(BH_4)_3{\cdot}xNH_3$ ($x = 2,3,4,5,6$).[41] While maintaining the high hydrogen storage capacity of the corresponding metal borohydride compound, the above light metal borohydride compound ammonia complex turns into an exothermic reaction, which can further improve the energy performance of the propellant.

Light metal borohydride compounds and their ammonia complexes have high hydrogen content and are expected to be used as hydrogen generators with excellent properties in propellants. However, these compounds also have strong reducibility, which means that there may be compatibility problems between them and the oxidizing components in the propellant. At present, there are few studies on the compatibility between light metal borohydride compounds and their ammonia complexes and other components of propellant. Therefore, it is vital to study the interaction between light metal borohydride compounds, such as $LiBH_4$, $NaBH_4$, KBH_4, $Mg(BH_4)_2$ and $Ca(BH_4)_2$, and their ammonia complexes with common propellant components.

7.2.4 Ternary Metal Composites

A metal combustor is one of the important components of modern solid propellant. It can not only significantly improve the energy level of propellant, but also affect the combustion performance, sensitivity and characteristic signal of propellant to varying degrees.[42] Al powder is widely used in modern solid propellants because of its high density, low oxygen consumption, high combustion enthalpy, significantly increased specific impulse, rich raw materials and low cost, but its combustion heat and combustion efficiency need to be improved.[43,44] In order to promote the combustion of Al powder, many beneficial attempts have been made, such as using nano Al powder,[45,46] adding metal Mg powder,[47] *etc.* However, due to the high activity and easy oxidation of nano powder, it not only affects the processing, storage and service life of propellant, but also brings safety problems. Mg powder has a low boiling point. Under the extreme conditions of engine operation, it can still maintain good combustion efficiency, promote the full combustion of Al compounds, and improve the supplementary combustion efficiency when the Mg content is high. However, because the combustion heat of Mg powder is low, the content should not be too high, otherwise the calorific value, characteristic velocity and traditional specific impulse of propellant will be significantly reduced.[48,49] Thus, the concept of using alloy as a combustion agent in solid propellants and explosives is proposed, and the research on alloys has also attracted the attention of many researchers all over the world; however these studies often stay in bimetallic alloys, such as Al–Mg, Al–Ce, Al–B, Mg–B, *etc.*

In order to solve the disadvantages of insufficient combustion and low specific impulse of Al powder fuel, flammable metal elements are added to Al powder to form flammable intermetallic compound phase. At the same time, controlling the staggered distribution of the flammable phase and the elemental aluminum phase inside the metal can greatly improve the ignition and combustion performance of particles. For example, the ignition temperature of Al powder can be reduced by adding active flammable metals Li and Mg to Al, which improves the combustion completeness of Al powder and significantly improves the activity of Al powder. However, the density of Li and Mg is small, and their addition reduces the specific impulse of the propellant to a certain extent. Therefore, in order to improve the specific impulse of the alloy fuel, high-density energetic metals such as Zr or Ti can be added to the formulation under the condition of ensuring energy does not decrease. It was experimentally found that compared with Ti, with the same particle size, Zr has shorter combustion time and higher combustion temperature. It was also found that Zr with 9 μm particle size has higher flammability and combustion capacity in solid rocket propellants. Therefore, the application of Zr powder in propellant formulations can improve the combustion of propellant.

Furthermore, Zr is widely used not only in propellants and pyrotechnic explosives, but also in combustion energetic fragments. It was found that the addition of Zr powder has a certain impact on the performance of energetic fragments. Adding Zr to the formula can improve the performance of the energetic fragments and make them more effective in preventing damage. The addition of Zr powder can reduce the friction sensitivity, reduce the moisture absorption and combustion rate of energetic fragments, and improve the combustion temperature. Thus, a new type of ternary highly active metal fuel Al–Mg–Zr was developed to give full play to the synergistic effect of metals Al, Mg and Zr and overcome their shortcomings. It was found that a series of Al–Mg–Zr ternary alloys with Mg content of 5–30% prepared by tight coupling gas atomization has good sphericity; with the increase of Mg mass fraction, the temperature of oxidation reaction decreased significantly, and Al–20Mg–2Zr and Al–25Mg–2Zr powders undergo concentrated and intense oxidation exothermic reactions at 945 °C and 938 °C, respectively. Al–20Mg–2Zr powder has the maximum exothermic enthalpy of oxidation, and the centralized oxidation reaction of the powder is basically complete. The activation energy of the oxidation reaction of Al–20Mg–2Zr powder is about 250 $kJ\,mol^{-1}$, and the influence factor is very large. The small activation energy and large frequency factor determine the high activity and large oxidation reaction rate of the powder at high temperature.[50]

Moreover, in view of excellent thermal reaction performance of Al–Ce binary alloy, Mg–Al–Ce series alloys were prepared, and the thermal performance of Al–Mg–Ce ternary alloy fuels were investigated.[51] It was found that in Al–xMg–1Ce alloys, the alloy powder with Mg mass fraction of 5% and 10% has only one oxidation reaction, and the exothermic peak is relatively flat. When the Mg mass fraction increases to 15% and 20%, the alloy powder has two oxidation reactions. With the increase in Mg mass fraction, the total oxidation exothermic enthalpy of Al–xMg–1Ce alloy powder first increases and then decreases. Al–15Mg–1Ce alloy powder has the highest oxidation exothermic enthalpy and shows better thermal properties.

In order to further improve the oxidation performance of Al_3Zr/Al composite fuel, a combustible metal was introduced into Al_3Zr/Al composite fuel to form a new staggered or isolated structure in Al_3Zr/Al alloy. As an energetic additive, Eu can significantly improve the oxidation and combustion properties of Al based alloy. So, the introduction of Eu into Al_3Zr/Al composite fuel to form a micro nano structure will further improve the oxidation combustion performance of Al_3Zr/Al composite fuel.[52] Compared with Al–30Zr composite fuel, Al–30Zr–Eu composite fuel has higher thermal reaction activity, and the heat release is more concentrated with the increase in Eu content. When the mass fraction of Eu is 3 wt%, it shows more concentrated exothermic and more obvious weight gain, and the weight gain reaches 66.5%, which is twice that of Al–30Zr composite fuel. At the same time, the phase composition and micro morphology of Al–B–3Eu series alloys were analyzed and characterized. The thermal oxidation and combustion characteristics of alloy powders were compared using TG–DTA and oxygen elastic calorimetry.[53] It was found that the alloy has good sphericity, and uniform network grain boundary structure is distributed on the particle surface. When the Eu mass fraction remains unchanged (3 wt%), the combustion heat of the Al–B–Eu alloy first increases and then decreases with the increase in boron mass fraction. The combustion heat release of Al–3.5B–3Eu ternary alloy powder under the oxygen partial pressure of 3.0 MPa is relatively complete, and the combustion heat is 33324.5 J g^{-1}, which is 11% higher than the theoretical combustion heat of elemental aluminum powder. Under the synergistic action of a B atom and a Eu atom, a special micro nano network grain boundary structure (mainly composed of B_6Eu and AlB_2 phase) is formed in the Al–3.5B–3Eu alloy. The thermal oxidation expansion of this grain boundary component will lead to the splitting of particles in the combustion process, thus accelerating the oxidation reaction, making the oxidation combustion of Al–3.5B–3Eu alloy more intense and complete.

For instance, Al–Mg–R (Ga, Zn, Sb, Sn, In) alloys with 80 : 15 : 5 mass ratios were prepared,[54] and through the phase analysis of Al–15% Mg–5% R alloy powder, it can be seen that the metal additives Ga, Sn, In and Sb easily form the Mg–R phase with Mg atoms, while Zn is only a solid dissolved in a Mg lattice without forming intermetallic compounds. The differential thermal analysis (DTA) shows that the addition of Ga, Sn, In and Sb can greatly advance the primary ignition temperature, and the addition of Zn can improve the heat release. Al–Mg–Li ternary alloy with smooth surface and high sphericity was also prepared by the atomization method, and it was found that the addition of Li and Mg can greatly reduce the aluminothermic reaction temperature, improve the reaction intensity and improve the activity of the alloy; the addition of Li promotes the low temperature oxidation of the Al–Mg–Li alloy.[55]

As is known, nano copper powder has a significant catalytic effect on AP, according to the “synergy” between Cu and NiB, the catalysis effect of Cu on AP and AP/HTPB were enhanced. Nano Cu–Ni–B composite with the ratio of Cu to Ni of 1 : 1 can reduce the low-temperature and high-temperature thermal decomposition temperature of AP by 13.8 °C and 146.1 °C, showing a good catalytic effect on the low-temperature and high-temperature thermal decomposition of AP, especially on the high-temperature decomposition of AP. In Cu–Ni–B, there may be some "synergy" between Cu, Ni and B that enhances the catalysis. Nano Cu–Ni–B also significantly increased the apparent decomposition heat of AP by nearly twofold. The activation energy of the AP high temperature decomposition reaction was 147 kJ mol^{-1}, which decreased by 30 kJ mol^{-1} after adding nano Cu–Ni–B powder. Moreover, the bright nano Cu–Ni–B composite powder has a significant catalytic effect on the low-temperature and high-temperature thermal decomposition of AP/HTPB propellant, and the catalytic effect on the low-temperature decomposition is greater than that on the high-temperature thermal decomposition.[56]

Generally speaking, liquid borohydride compounds are suitable for composite propellants. They have a plasticizing effect, but the migration problem should be minimized or avoided from the chemical structure. Of course, it is a good way to hang the borohydride group on the polymer chain of the binder, but it is better not to affect the structure of the main chain too much, so as not to damage its mechanical properties. It is interesting to synthesize perchlorate containing a carborane group, but its sensitivity and thermal stability are affected. It should be mentioned here that hydroborates and their coprecipitates with oxidants are worth exploring because they are intermediates for the preparation of carboranes, which are easy to synthesize, stable in performance and low in toxicity.

Finally, it should be pointed out that there are some problems that need to be further studied for the borohydride burning rate regulator to be widely used. For example, the improvement of the synthesis route and process, in order to further reduce the cost; a toxicity study; avoiding the migration of liquid borohydride; research on the mechanism of burning rate regulation; the development of a new type of high efficiency borohydride burning rate regulator. These studies are being vigorously carried out abroad.

Modern war puts forward higher requirements for anti-tank and anti-ballistic missile weapons, so the development of excellent solid propellant with high burning rate has become key in the current propellant research. In order to obtain high burning rate performance of propellants there are still some methods, such as using ultra-fine or porous perchloric acid and embedding metal or carbon fiber, but they still have some difficulties and limitations in technology. The excellent and different characteristics of the high burning rate regulator with borohydride compound introduced in this paper have attracted worldwide attention. Although the research on borohydride in China has a certain foundation, there is still a considerable gap compared with other countries. It is hoped that the relevant departments will pay attention to it and promote the scientific research work in this field, so that the high burning rate performance of solid propellant in China can catch up with and surpass the advanced level of the world in time.

7.2.5 Metallic-based Energetic Composites (MECs)

Metal fuels have long been of interest for propellants and explosives because of their high energy density and high heat release when burned,[57,58] and this advantage is clearly beneficial for volume-limited propulsion systems.[59–61] Metals, as fuels and fuel additives, have other benefits as well, such as mitigation of combustion instability, improvement of burning rate, reduction of sensitivity to detonation, favorable supply, and more.[62,63] Among conventional solid propellants (SPs) and hybrid fuels for aerospace exploration, boron (B) is an important metalloid, and a promising fuel additive because of its high gravimetric (58.30 MJ kg^{-1}) and volumetric (136.44 kJ cm^{-3}) heats of combustion.[64] Figure 7.1 compares the gravimetric and volumetric heats of combustion of different metals.

7.2.5.1 Boron-based Energetic Composites (BECs)

Given the small size and irregular shape of B powder with large specific surface area (SSA), it is difficult to process significant amounts of B powder in propellants or explosives, which is not conducive to tailoring combustion performance.[65] More importantly, a native oxide layer of B_2O_3 with about 2–5 nm thickness is coated on the surface of amorphous or commercial B (B_0), which worsens the condensation reaction between H_3BO_3 and HTPB molecules with large molecular mass polymer chain formation. Besides, B combustion often leads to HOBO with lower energy release, which potentially causes the heat of combustion of B not to be released completely, as opposed to the desired

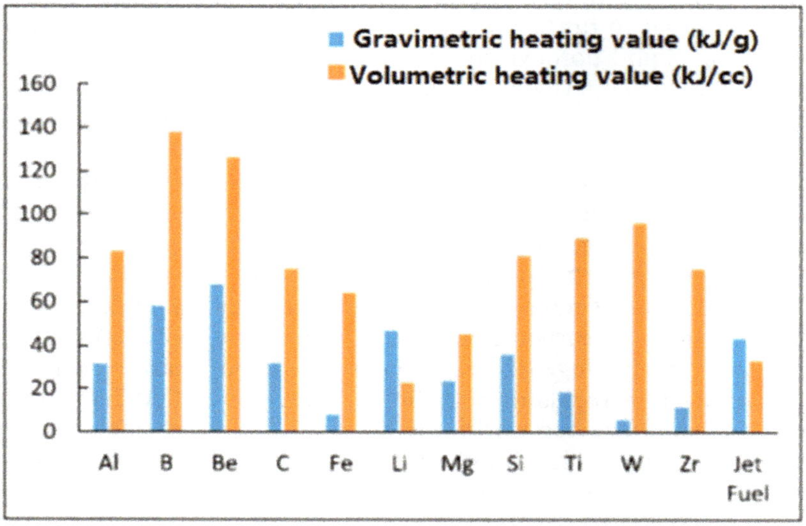

Figure 7.1 Gravimetric and volumetric heats of combustion of different metals. Reproduced from ref. 60 with permission from Elsevier, Copyright 2022.

liquid B_2O_3. Such burning characteristics generally result in relatively long ignition delays and burning times, particularly when compared to metals that vaporize. Hence, for volume limited systems, often only a small fraction of the high theoretical calorific value is released.[66]

Several approaches have been used to diminish the processing difficulties of B powder, solve the ignition and combustion problems, and raise the mass fraction of B and ultrafine oxidizer in propellants.[67] For example, organic fluoropolymers with typically low decomposition temperatures are attractive candidate materials for passivation, as they can burn in many practical oxidizing environments and contribute to the overall energy release.[68] Another advantage of organic coatings and organically coated small particles that have been assembled into larger ones (by agglomeration or other means) is that they result in better particle processability in the propellant formulations.[69] Moreover, it has been proposed that by incorporating fluorine (F) as an added oxidizer with a metal/metal oxide particle, the ignition temperature can be lowered, and reaction rates increased.[70] More recently, B powders have been combined with fluorinated polymer, polytetrafluoroethylene (PTFE), *etc.* to form composites that achieve enhanced combustion performance in practical energetic formulations.[71] Boron particles coated with Mg (or Al, zirconium (Zr), titanium (Ti)), when substituted for conventional Al powder in propellants, have showed significant improvement in terms of the burning rate of SP, and problems related to the formation of agglomerates have been overcome.[72]

As a potential application to SP, B offers significant benefits, such as increased specific impulse of propellants. However, the impurities of B_2O_3 and H_3BO_3 on the B surface can undergo esterification reactions with the terminal hydroxyl groups of hydroxyl-terminated polybutadiene (HTPB) binder, which makes the preparation process of propellant severely difficult,[73] increasing the apparent viscosity and shortening the pot-life of the slurry.[74] Moreover, Boron, because of its high melting and boiling points (2349 K and 4274 K, respectively[57]), oxidizes slowly though heterogeneous reactions on its surface.[75] To accelerate B ignition, B_2O_3 and H_3BO_3 need to be removed or reduced by using methods such as purification,[76] coating,[77] agglomerate formation,[78] prills,[79] incorporation of flammable metal powder,[80] and mechanical milling,[81] which plays an important role in improving the processing performance of B-based propellants.

The ignition difficulty of B can be overcome, and the completeness of oxidation and combustion of B can be increased, by addition of rare earth metals,[82] Mg,[83] metal hydrides,[84] *etc.* The oxidation of B by metal oxides (Bi_2O_3, CuO, Fe_2O_3, Al_2O_3) is another important mechanism that can improve B combustion performance.[85] For example, considering the B_2O_3 shell ($\sim$30% of the mass), 60 nm diameter B particles with Bi_2O_3 were mixed and sonicated in hexane for 30 min, and then dried in a hood overnight allowing hexane to evaporate to prepare B/Bi_2O_3 composite. With pure Bi_2O_3, B could be ignited at a temperature as low as 793 K.[86] Coating B is an effective way to improve the B surface morphology and lower the apparent viscosity and lengthen the pot-life of B-based propellant slurry.[87] Moreover, coating may assist in improving the ignition and combustion performance of B.[88] Agglomeration or granulation of B powder with bigger particle size is an effective way to improve the performance of B and B-based propellant.[89] For example, the agglomerated B with controlled particle size prepared by the mechanical milling method based on preparation parameter optimization,[90] and HTPB (as polymer binder) and AP (as combustion promoter) were added to the composition. After agglomeration, the particle size increases, the particle shape is nearly spherical, and the particle surface is smooth.

It is widely known that the reactivity of metal particles increases when the particle size decreases.[91] Significant improvements in ignition and combustion behaviors of propellants, such as faster burning rates and shorter ignition delay times, have been attributed to the use of nano-sized metal particles, including nano thermite, nB/carbon nanotube, nB/graphene, *etc.*[92] The ignition and combustion processes of B have been explained to occur through two stages, where the first stage involves B reacting while being protected by an oxide layer when the reactants must diffuse across this layer slowing oxidation, and the second stage involves oxide layer removal at higher temperatures and the condensed-phase B reaction occurring directly with oxidizers in the surrounding gas.[93] The combustion efficiency of B can be increased in the case of small particle size or in the case of combustion of B particles embedded in an oxidizer; the combustion mechanism of B has been explained by thermochemical aspects.[94] Moreover, the use of light metal additives in B-CEMs, such as Mg and Al, has been proposed to improve B combustion efficiency.[95]

In summary, an ideal energetic material with chemical energy in its chemical bonds is a substance with high performance, safety and shelf life. Large numbers of investigations have concentrated on the exploration or the development of innovative energetic materials with excellent properties, such as high reactivity, thermal stability, insensitivity, *etc.* The reduction of the particle size of energetic materials from micron to nano is one of the promising approaches for obtaining desirable properties.[96]

7.2.5.2 Al-based Nano-sized Energetic Composites (AECs)

The preparation, storage and application of nano-sized energetic composites (nano-ECs) can be realized, owing to their large specific surface area, high surface energy, and high surface activity.[97] Moreover, nano-ECs can integrate the characteristics of each component, thus the new types of energetic materials with various characteristics can be obtained, which can reduce the size of reactants in composites to nano-scale, and then increase the contact interfacial area effectively between materials, so as to solve the problem of the slow transmission speed of micro-sized energetic composites (micro-ECs) owing to large particles (chemical reaction rate, mass transfer, and heat transfer), and obtain energetic materials with energy density and a high energy release rate.[98] In addition, the energy density of nano-sized energetic composites prepared by the sol–gel method is two times that of mono-molecular energetic material.[99] The mass transfer speed between the oxidant and the aluminum powder in the pore is fast and the reaction is sensitive. The burning speed is more than 1000 times compared with that of the similar ordinary oxides. The energetic performance of nano-ECs depends on the interfacial diffusion and mass transfer during the reacted process. However, the development of the desired structure to significantly enhance the reactivity still remains challenging for researchers. Therefore, nano-ECs as a new type of energetic material have attracted wide interest in the research community. Al is widely used to increase the energy and raise the flame temperature in rocket propellants because of its high combustion enthalpy, easy availability, low toxicity, and good stability in energetic applications, such as propellants, pyrotechnics, and explosives. In particular, nano-sized Al (nAl) is broadly exploited to improve performance incrementing the burning rate and combustion efficiency of energetic systems, leading to shorter ignition delays and shorter agglomerate burning times with respect to propellants containing μAl.[100]

Nano-structured energetic materials have better energy release performance and ignition performance. Nano-structured energetic composite composed of MnO_2 nano sheets with root-embedded Al on silicon wafer could be prepared owing to the close contact area and uniform distribution between metal fuel and oxidant.[101] For example, CuO is a good catalyst for the decomposition and combustion of solid propellant. The functionalized nano-sized Al/CuO particles possess long-term storage stability for as high as 88% of energy left after being exposed in the air for one month, benefitting from enhanced resistibility to a humid environment. The steady combustion process of nano-sized Al/CuO with the flame propagation speed is 100 m s^{-1}. Nonetheless, both qualitative and quantitative characterizations have showed that these materials are indeed energetics, while the materials described here are relatively insensitive to standard impact, electrostatic spark, and friction tests. Qualitatively, it does appear that these nano-sized energetic composites burn faster, which are more sensitive to thermal ignition than their conventional counterparts, and the aerogel materials are more sensitive to ignition than those of xerogels.[102]

At present, most studies use organic solvent as dispersant to disperse nAl powder and oxide. Considering the high activity of nAl and high flammability in the organic solvent, the preparation process is very dangerous. One alternative method is to use water as the dispersant. Because of the high activity of nAl, the nAl powder can still react with water, resulting in a decrease in activity, although the surface is passivated by a layer of oxide.[103] It was found that, based on silane or oleic acid, the organic surface coating can effectively inhibit the reaction between nAl powder and water, ensuring its activity over a long period of time.

The impact ignition energy of nano-sized Al/MoO_3 composites is 0.3–1.7 J, which increases with increased particle size. The maximum reaction rate is related to the initial particle size, and the reaction rate increases with decrease in the particle size. The combustion mechanism and reaction properties of four types of nano-sized Al/MoO_3, Al/WO_3, Al/CuO, and Al/Bi_2O_3 composites were obtained; it was found that the propagation speed depends on the generation of gas and the thermodynamic state of reaction products, and the burning rate decreases with the increase in mixture density, which is contrary to the traditional explosion theory, so the combustion propagation mechanism is different from that of the traditional explosion theory. Nano-sized Al/Cr_2O_3 10–15 nm in diameter was prepared, and the relationship between particle size and combustion performance by means of a CO_2 laser ignition technique was conducted; the performance is significantly improved compared with that of micro-scaled particles. One interesting point is that nano-sized Al/$WO_3 \cdot H_2O$ composites were prepared by means of the wet chemical method, and their performance was compared with that of nano-sized Al/WO_3 composites; it was found that the energy release rate of the reaction of Al/$WO_3 \cdot H_2O$ is significantly improved by the reaction between nAl and H_2O in $WO_3 \cdot H_2O$ to produce hydrogen, which increased the total heat release and energy release rate of the reaction.

Three nanocomposite materials with the same nominal stoichiometric thermite composition of Al/CuO were prepared by three different methods: ultrasonic mixing (USM) of constituent nanopowders, electrospraying (ES), and arrested reactive milling (ARM). The combustion performances of the three prepared types of Al/CuO nanocomposites were conducted.[104] Prepared powders were placed in a 6.7 mm diameter, 0.5 mm deep cavity in a brass substrate and ignited by electro-static discharge. The experiments were performed in air, argon and helium. The mass of powder removed from the sample holder after ignition was measured in each test. Using a multi-anode photo-multiplier tube coupled with a spectrometer, time-resolved light emission traces produced by the ignited samples were recorded in the range of wavelengths of 373–641 nm. Time-resolved temperatures were then determined by fitting the recorded spectra assuming Planck's black body emission. The temporal pressure generated by the ignition events in the enclosed chamber showed that the powder's combustion properties were tied to both their preparation technique as well as the environment in which they were ignited. The agglomeration of nanoparticles hindered the combustion of USM powders, while this was not observed for the ES powders. Lower temperatures and pressures were observed in oxygen-free gas environments for USM and ES powders prepared using starting nanoparticles. For the ES powders, the effect of gas environment was less significant, which was interpreted considering that ES materials included a gasifying nitrocellulose binder, enhanced heat, and mass transfer between individual Al and CuO particles. Higher pressures and temperatures were observed in inert environments for the ARM-prepared powder.

Addition of reactive nanocomposite powders can increase the burn rate of aluminum, and thus the overall reaction rate of the energetic formulation. Replacing only a small fraction of the fuel by a nanocomposite material can enhance the reaction rate with little change to the thermodynamic performance of the formulation. The nanocomposite materials Al/3CuO and Al/MoO_3 prepared by ARM, a scalable "top-down" technique for manufacturing reactive nano-materials, were added to micron-sized aluminum powder and the mixture was aerosolized and burned in a constant volume chamber with varied oxygen, nitrogen, and methane atmospheres.[105] The resulting pressure traces were recorded and processed to compare different types and amounts of modifiers. Additives of nanocomposite powders of Al/MoO_3 to micron-sized aluminum were found to be effective in increasing both the rate of pressure rise and maximum pressure in the respective constant volume explosion experiments. It was observed that 20 wt% of additive resulted in the best combination of the achieved burn rate and pressure.

The potential application of nano-aluminum/nitrocellulose mesoparticles as an ingredient for solid composite rocket propellants was investigated.[106] The basic strategy is to incorporate nano-aluminum in the form of a micrometer scale particle containing a gas generator to enable easier processing and potential benefits resulting from reduced sintering prior to combustion. The mesoparticles were made by electrospray and comprised aluminum nanoparticles (50 nm) and nitrocellulose to form micrometer scale particles. 80% solids loaded composite propellants (AP/HTPB (hydroxil terminated polybutadiene) based) were made with the addition of micrometer sized (2–3 μm) aluminum (10 wt%), and compared directly to propellants made by directly substituting aluminum mesoparticles for traditional micrometer sized particles. The propellant burning rate was relatively insensitive for mesoparticles containing between 5 wt% and 15 wt% nitrocellulose. However, direct comparison between a mesoparticle-based propellant and a propellant containing micrometer scale aluminum particles showed burning rates approximately 35% higher, while having a nearly identical burning rate exponent. High-speed imaging indicates that propellants using mesoparticles have less agglomeration of particles on the propellant surface.

The results show that the burning rate of nano-CEMs is the fastest when the particle size of Al powder and oxide is at the nano-metric level, while the burning rate of CEMs composed of micron Al powder and nano-metric oxide is much faster than that of nanometer Al powder and the micrometric level.

References

1. W. Q. Pang, X. Z. Fan and F. Q. Zhao, Effects of high energetic combustion agent (BHN) on the characteristics of fuel-rich solid propellant, *J. Solid Rocket Technol.*, 2013, **36**(5), 647–653 (in Chinese).
2. W. G. Shan, T. G. Sun and G. D. Zhang, Effects of carbon boron alkyl derivative for the combustion properties of solid propellant, *J. Solid Rocket Technol.*, 1995, **18**(3), 24–26 (in Chinese).

3. Y. Nie, H. Y. Chen and J. L. Miao, The reactivity of boron alkyl $[B_{10}H_{10}]^{2-}$ and its applied development of the derivative, *Org. Chem.*, 2009, **29**(6), 822–834 (in Chinese).
4. Y. Yanjing, Z. Fengqi and Y. Jianhua, *et al.*, Applications of hydrogen-storage materials in high-energy solid rocket propellants, *Chin. J. Explos. Propellants*, 2015, **38**(2), 8–14.
5. X. Guangxian, *Hydrogen storage materials*, Metallurgy Industry Press, Beijing, Lanthanon, 1995, vol. 2, p. 286.
6. P. Weiqiang and F. Xuezhong, Application progress of metal fuels in solid propellants, *Chem. Propellants Polym. Mater.*, 2009, **7**(2), 1–5.
7. L. Leli, L. Fengsheng and T. Linhua, *et al.*, Effect of nano-meter metal and composite metal powders on the thermal decomposition of AP and AP/HTPB composite solid propellant, *Chin. J. Chem. Eng.*, 2004, **12**(4), 595–598.
8. L. Leili, L. Fengsheng and Z. Chunlei, *et al.*, Effect of magnesium-based hydrogen storage materials on the properties of AP/Al/HTPB composite solid propellant, *Chin. J. Energ. Mater.*, 2009, **17**(5), 501–504.
9. L. Jingru and L. Yunjun, Synthesis and properties of Al-based composite applied to solid propellant, *J. Solid Rocket Technol.*, 2010, **33**(5), 545–548.
10. B. Sakintuna, D. F. Lamari and M. Hirscher, Metal hydride materials for solid hydrogen storage: a review, *Int. J. Hydrogen Energy*, 2007, **32**, 1121–1140.
11. G. Sandrock, J. Reilly and J. Graetz, *et al.*, Accelerated thermal decomposition of AlH_3 for hydrogen fueled vehicles, *Appl. Phys. A: Mater. Sci. Process.*, 2005, **80**, 687–690.
12. L. Meng and Z. F. X. Siyu, *et al.*, Energetic characteristics of composite propellant containing different metal hydride, *J. Solid Rocket Technol.*, 2014, **37**(1), 86–90.
13. L. T. DeLuca, L. Galfetti and F. Severini, *et al.*, Physical and ballistic characterization of AlH_3-based space propellants, *Aerosp. Sci. Technol.*, 2007, **11**, 18–25.
14. J. P. Flynn, G. A. Lane and J. J. Plomer, *Nitrocellulose propellant composition containing aluminium hydride*, US3844856, 1965.
15. L. Leili, L. Fengsheng and Z. Chunlei, *et al.*, Synthesis of magnesium hydride and its effect on thermal decomposition of AP, *Rare Met. Mater. Eng.*, 2010, **39**(7), 1289–1292.
16. L. Leili, L. Fengsheng and Z. Chunlei, *et al.*, Effect of magnesium-based hydrogen storage materials on the properties of composite solid propellant, *Chin. J. Energ. Mater.*, 2009, **17**(5), 501–504.
17. Y. Miao, C. Liping and D. Ping, *et al.*, Effect of $Mg(BH_4)_2$ and MgH_2 on thermal decomposition performance of RDX, *China Saf. Sci. J.*, 2013, **23**(1), 115–120.
18. C. Yangfan, M. Honghao and S. Zhaowu, Detonation characteristics of emulsion explosives sensitized by MgH_2, *Chin. J. High Pressure Phys.*, 2013, **27**(1), 45–50.

19. C. Yangfan, M. Honghao and S. Zhaowu, Experimental research on pressure desensitization of emulsion explosive sensitized by MgH_2, *Explos. Shock Waves*, 2014, **34**(4), 427–432.
20. P. Jiangfeng, Z. Fengqi and J. Jianshe, *et al.*, Energetic characteristics of BAMO–AMMO based propellants containing different metal hydride, *Explos. Mater.*, 2014, **43**(4), 11–15.
21. Z. Chunlei, S. Hongchang and L. Fengsheng, *et al.*, Preparation of Mg_2NiH_4 and its catalytic characteristics of thermal decomposition of ammonium perchlorate, *J. Chem. Ind. Eng.*, 2007, **58**(11), 2793–2797.
22. L. Leili, L. Fengsheng and Z. Chunlei, *et al.*, Effect of Mg_2NiH_4 on the thermal decomposition of ammonium perchlorate, *Chem. J. Chin. Univ.*, 2007, **28**(8), 1420–1423.
23. L. Leili, L. Fengsheng and Z. Chunlei, *et al.*, Synthesis of magnesium-copper hydrogen storage alloy and its effect on the thermal decomposition of ammonium perchlorate, *Acta Chim. Sin.*, 2008, **66**(12), 1424–1428.
24. D. Yanmeng, L. Guoping and L. Yunjun, *et al.*, Study on combustion wave structure of hydrogen storage alloy/AP/HTPB propellant, *Acta Armamentarii*, 2013, **34**(2), 162–167.
25. D. Yanmeng, L. Yunjun and L. Guoping, *et al.*, Thermal decomposition properties of hydrogen storage alloy/AP/HTPB propellant, *Chin. J. Explos. Propellants*, 2012, **35**(3), 66–70.
26. D. Yanmeng, L. Guoping and L. Yunjun, *et al.*, Basic performance of hydrogen storage alloy combustion agent, *J. Solid Rocket Technol.*, 2011, **34**(6), 760–771.
27. L. Jingru and L. Yunjun, A study on the compatibility of hydrogen storage alloy incendiary agent with some common energetic components of solid propellants, *Acta Armanentarii*, 2008, **39**(9), 1133–1136.
28. D. Yanmeng, L. Guoping and L. Yunjun, *et al.*, Effects of hydrogen storage alloy on peroperties of GAP binder films, *J. Solid Rocket Technol.*, 2012, **35**(5), 630–634.
29. S. Orimo, Y. Nakamori and J. R. Eliseo, *et al.*, Complex hydrides for hydrogen storage, *Chem. Rev.*, 2007, **107**, 4111–4132.
30. A. Andreasen, Effect of Ti-doping on the dehydrogenation kinetic parameters of lithium aluminum hydride, *J. Alloys Compd.*, 2006, **419**, 40–44.
31. B. Bogdanovic and M. Schwickardi, Ti-doped alkali metal aluminum hydrides as potential novel reversible hydrogen storage materials, *J. Alloys Compd.*, 1997, **253–254**, 1–9.
32. H. Morioka, K. Kakizaki and S. C. Chung, *et al.*, Reversible hydrogen decomposition of $KAlH_4$, *J. Alloys Compd.*, 2003, **353**, 310–314.
33. Y. Kim, E. K. Lee and J. H. Shim, *et al.*, Mechanochemical synthesis and thermal decomposition of $Mg(AlH_4)_2$, *J. Alloys Compd.*, 2006, **422**, 283–287.
34. K. Komiya, N. Morisaku and Y. Shinzato, *et al.*, Synthesis and dehydrogenation of $M(AlH_4)_2$ ($M = Mg$, Ca), *J. Alloys Compd.*, 2007, **446–447**, 237–241.

35. T. J. Marks and J. R. Kolb, Covalent transition metal lanthanide and actinide tetrahydroborate complexes, *Chem. Rev.*, 1977, **77**(2), 263–293.
36. S. Orimo, Y. Nakamori and G. Kitahara, *et al.*, Dehydriding and rehydriding reactions of LiBH4, *J. Alloys Compd.*, 2005, **404–406**, 427–430.
37. P. Ngene, R. Van den Berg and M. H. W. Verkuijlen, *et al.*, Reversibility of the hydrogen desorption from $NaBH_4$ by confinement in nanoporous carbon, *Energy Environ. Sci.*, 2011, **4**, 4108–4115.
38. K. Chopek, C. Frommen and A. Leon, *et al.*, Synthesis and properties of magnesium tetrahydroborate, $Mg(BH_4)_2$, *J. Mater. Chem.*, 2007, **17**, 3496–3503.
39. J. H. Kim, S. A. Jin and J. H. Shim, *et al.*, Thermal decomposition behavior of calcium borohydride $Ca(BH_4)_2$, *J. Alloys Compd.*, 2008, **461**, L20–L22.
40. G. Dayong, L. Jiping and D. Lan, Application of sodium borohydride ($NaBH_4$) in incendiary agent, *Chin. J. Energ. Mater.*, 2007, **15**, 231–234.
41. Y. J. Yang, Y. F. Liu and Y. Li, *et al.*, Synthesis and thermal decomposition behaviors of magnesium borohydride ammoniates with controllable composition as hydrogen storage materials, *Chem. – Asian J.*, 2013, **8**, 476–481.
42. R. Chongzhi, *Modern solid propulsion technology*, Northwestern Polytechnical University Press, Xi'an, 2007.
43. G. P. Sutton and O. Bibraz, *Fundamentals of rocket engine*, ed. Z. Baojiong, *et al.*, Science Press, Beijing, 2003.
44. A. Gany and L. H. Caveny, Agglomeration and ignition mechanism of aluminum particles in solid propellants, *Symp. (Int.) Combust.*, 1979, **17**(1), 1453–1461.
45. K. Jayaraman, K. V. Anand and S. R. Chakravarthy, *et al.*, Effect of nanoaluminum in plateau burning and catalyzed composite solid propellant combustion, *Combust. Flame*, 2009, **156**(8), 1662–1673.
46. P. E. Bocanegra, C. Chauveau and I. Gokalp, Experimental studies on the burning of coated and uncoated micro and nano-sized aluminum particles, *Aerosp. Sci. Technol.*, 2007, **11**(1), 33–38.
47. A. A. Timothy, S. Evgeny and V. Arvind, Ignition mechanism of nickel-coated aluminum particles, *Combust. Flame*, 2007, **150**(1–2), 60–70.
48. X. Xiuyou, L. Baoxuan and W. Yuling, *et al.*, Study on combustion performance of mg Al fuel rich solid propellant at low pressure, *J. Solid Rocket Technol.*, 2005, **28**(1), 49–52.
49. A. Gany and D. W. Nezer, Combustion studies of metallized fuels for solid fuel ram-jet, *J. Propul. Power*, 1986, **2**(5), 423–427.
50. L. Lin-fu, *Preparation and thermal properties of super-reactive Al-based alloy fuels*, PhD thesis, Huazhong University of Science and Technology, 2015.
51. Y. Huan, *Study on Thermal Performance of Al-Ce and Al-Mg-Ce Alloy Fuels*, PhD thesis, Huazhong University of Science & Technology, 2016.
52. S. Wen, *Study on Preparation and Thermal Properties of Al_3Zr/Al Composite Fuels*, PhD thesis, Huazhong University of Science & Technology, 2018.

53. W. Wei, *Research on preparation and thermal properties of boron-containing alloy fuels*, PhD thesis, Huazhong University of Science & Technology, 2019.
54. T. Da-ming, *Preliminary study on combustion performance of Al-based alloy powder and electrochemical study of Al–H_2O reaction*, PhD thesis, Huazhong University of Science and Technology, Wuhan, 2013.
55. X. Changjuan, *Research on the Oxidation Properties of Al-Mg and Al-Mg-Li Alloy Fuels*, PhD thesis, Huazhong University of Science and Technology, Wuhan, 2015.
56. L. Lei-li, *Preparation and catalytic properties of nano metal and composite metal powders*, PhD thesis, Nanjing University of Technology, Nanjing, 2004.
57. W. Q. Pang, T. L. DeLuca, X. Z. Fan, *et al.*, *Boron-based fuel-rich propellant: properties, combustion and technology aspects*, Taylor & Francis Group, CRC Press, 2019.
58. W. Pang and X. Fan, Application progress of metal fuels in solid propellants, *Chem. Propellants Polym. Mater.*, 2009, **7**(2), 1–5.
59. R. C. Brown, C. E. Kolb and S. Y. Cho, *et al.*, Kinetic-model for hydrocarbon-assisted particulate boron combustion, *Int. J. Chem. Kinet.*, 1994, **26**(3), 319–332.
60. P. WeiQiang, R. A. Yetter and L. T. DeLuca, *et al.*, Boron-based composite energetic materials (B-CEMs): preparation, characterization and applications, *Prog. Energy Combust. Sci.*, 2022, **93**, 101038.
61. A. Gany and D. W. Netzer, Combustion studies of metallized fuels for solid-fuel ramjets, *J. Propul. Power*, 1986, **2**, 423–427.
62. D. Sundaram, V. Yang and R. A. Yetter, Metal-based nanoenergetic materials: Synthesis, properties, and applications, *Prog. Energy Combust. Sci.*, 2017, **61**, 293–365.
63. E. L. Dreizin, Metal-based reactive nanomaterials, *Prog. Energy Combust. Sci.*, 2009, **35**, 141–167.
64. K. K. Kuo and R. Pain, *Combustion of boron-based solid propellants and solid fuels*, CRC Press, Boca Raton, FL, 1993.
65. W. Pang, Study of fuel-rich propellants with high mass fraction of boron particles, PhD thesis, Xi'an Modern Chemistry Research Institute, Xi'an, 2011.
66. W. Pang, X. Fan and F. Zhao, *Boron-based fuel-rich solid rocket propellant technology*, National Defense Industry Press, Beijing, 2016.
67. J. Zheng, Study on high-energy boron-based fuel-rich propellant technology, PhD thesis, Beijing Institute of Technology, Beijing, 2004.
68. R. A. Yetter, F. L. Dryer and H. Rabitz, *et al.*, Effect of fluorine on the gasification rate of liquid boron oxide droplets, *Combust. Flame*, 1998, **112**(3), 387–403.
69. H. Xu, X. Feng and X. Tian, *et al.*, Effect of fluorine-binder on the underwater explosion energy of boron-containing explosive, *Initiators Pyrotech.*, 2015, **4**, 38–41.

70. W. Zhou, R. A. Yetter and F. L. Dryer, *et al.*, Effect of fluorine on the combustion of 'clean' surface boron particles, *Combust. Flame*, 1998, **112**(4), 507–521.
71. J. Zhang, Q. Zhang and J. Guo, *et al.*, Surface coating of superfine boron particles with lithium fluoride, *Chin. J. Explos. Propellants*, 2005, **28**(3), 8–11.
72. C. Badiola and E. L. Dreizin, Combustion of micron-sized particles of titanium and zirconium, *Proc. Combust. Inst.*, 2013, **34**, 2237–2243.
73. J. Zheng, A. Wang and A. Pang, Study on process deterioration mechanism of boron based HTPB fuel-rich propellant, *J. Propul. Technol.*, 2003, **24**(3), 282–285.
74. X. Liu, J. Gonzales and M. Schoenitz, *et al.*, Effect of purity and surface modification on stability and oxidation kinetics of boron powders, *Thermochim. Acta*, 2017, **652**, 17–23.
75. K. K. Kuo and M. Summerfield, *Fundamentals of solid propellant combustion*, American Institute of Aeronautics and Astronautics (AIAA), 1984.
76. H. Xu, F. Zhao and X. Li, Purification of amorphous boron powder with solvents, *Chin. J. Explos. Propellants*, 2007, **30**(2), 8–12.
77. B. Chen, J. Liu and D. Liang, *et al.*, Research progress in coating mechanism and technology of boron particles, *Chin. J. Explos. Propellants*, 2016, **39**(5), 13–21.
78. W. Pang, Agglomeration technique of boron and its application to fuel-rich propellants, MS thesis, Northwestern Polytechnical University, Xi'an, 2006.
79. W. Pang, X. Fan and W. Zhang, *et al.*, Application of amorphous boron granulated with hydroxyl-Terminated polybutadiene in fuel-rich solid propellant, *Propellants, Explos., Pyrotech.*, 2011, **36**, 360–366.
80. B. Zhang, C. Huang and S. Yan, *et al.*, Enhanced reactivity of boron, through adding nano-aluminum and wet ball milling, *Appl. Surf. Sci.*, 2013, **286**, 91–98.
81. M. Schoenitz, T. S. Ward and E. L. Dreizin, Fully dense nano-composite powders prepared by arrested reactive milling, *Proc. Combust. Inst.*, 2005, **30**, 2071.
82. S. Karmakar, N. Wang and S. Acharya, *et al.*, Effects of rare-earth oxide catalysts on the ignition and combustion characteristics of boron nanoparticles, *Combust. Flame*, 2013, **160**(12), 3004–3014.
83. W. Pang, X. Fan and H. Xu, *et al.*, Characteristics of B/Mg compounds and the effect on the burning rate properties for fuel-rich propellant, *J. Solid Rocket Technol.*, 2013, **36**(3), 363–367.
84. J. Xi, J. Liu and Y. Wang, *et al.*, Effect of metal hydrides on the burning characteristics of boron, *Thermochim. Acta*, 2014, **597**, 58–64.
85. F. Zhao, Y. Yang, Z. Yuan, *et al.*, *Application of nano metal powder in solid propellants*, National Defense Industry Press, Beijing, 2020.
86. X. Wang, T. Wu and H. Wang, *et al.*, Boron ignition and combustion with doped δ-Bi_2O_3: Bond energy/oxygen vacancy relationships, *Combust. Flame*, 2018, **197**, 127–133.

87. Y. Xu, Q. Cui and C. Zhao, Liquid phase in-situ synthesis of LiF coated boron powder composite and performance study, *Def. Technol.*, 2019, **15**(4), 1–7.
88. S. Li, Optical measurement of size histories of boron particles in ignition and combustion stage, *Combust. Sci. Technol.*, 1991, **77**(1), 149–169.
89. W. Pang, X. Fan and J. Zhang, *et al.*, Agglomerated technology for amorphous boron powder, *Chin. J. Explos. Propellants*, 2008, **31**(2), 46–48.
90. X. Fan, W. Pang and H. Xu, *et al.*, Optimization of preparation process for spherical agglomerated boron particles, *Chin. J. Explos. Propellants*, 2010, **33**(1), 64–67.
91. M. Schoenitz, T. S. Ward and E. L. Dreizin, Fully dense nano-composite powders prepared by arrested reactive milling, *Proc. Combust. Inst.*, 2005, **30**, 2071.
92. V. E. Zarko and A. A. Gromov, *Energetic Nanomaterials: Synthesis, Characterization, and Application*, Elsevier, Amsterdam, 2016.
93. R. A. Yetter and F. L. Dryer, Metal particle combustion and classification, in *Microgravity combustion: fire in free fall*, ed. H. Ross, Academic Press, 2001.
94. N. Kubota, *Propellants and explosives: thermochemical aspects of combustion*, Wiley-VCH, 2002.
95. A. Marinovich, Mg-B dual metal powders as performance enhancer for solid rocket propellants, MS thesis, Politecnico di Milano, Milan, 2010.
96. W. Pang and T. L. DeLuca, *Nano- and micro-scale energetic materials: propellants and explosives*, Wiley, 2023.
97. W. He, P.-J. Liu and G.-Q. He, *et al.*, Highly reactive metastable intermixed composites (MICs): Preparations and characterizations, *Adv. Mater.*, 2018, **30**, 1706293.
98. T. An, F.-Q. Zhao and H.-X. Hao, *et al.*, Effect of thermites on laser ignition characteristics of double base propellants, *Chin. J. Explos. Propellants*, 2011, **34**, 67–72.
99. S. Dilip, Y. Vigor and A. Y. Richard, Metal-based nanoenergetic materials: Synthesis, properties, and applications, *Prog. Energy Combust. Sci.*, 2017, **61**, 293–365.
100. C. Paravan, A. Verga and F. Maggi, *et al.*, Accelerated ageing of micron- and nano-sized aluminum powders: Metal content, composition and non-isothermal oxidation reactivity, *Acta Astronaut.*, 2019, **158**, 397–406.
101. Y. Zhu, X.-X. Ma and K.-L. Zhang, Study on Nano-structured MnO_2/Al Energetic Composite, *Initiators Pyrotech.*, 2018, **1**, 23–27.
102. X. Ke, X. Zhou and H. Gao, *et al.*, Surface functionalized core/shell structured CuO/Al nanothermite with long-term storage stability and steady combustion performance, *Mater. Des.*, 2018, **140**, 179–187.
103. J. A. Puszynski, Processing and characterization of aluminum-based nanothermites, *J. Therm. Anal. Calorim.*, 2009, **96**, 677–685.

104. I. Monk, M. Schoenitz and R. J. Jacob, *et al.*, Combustion characteristics of stoichiometric Al-CuO nanocomposite thermites prepared by different methods, *Combust. Sci. Technol.*, 2017, **189**, 555–574.
105. S. Demitrios, J. Xianjin and B. Ervin, *et al.*, Aluminum burn rate modifiers based on reactive nanocomposite powders, *Propellants, Explos., Pyrotech.*, 2009, **35**, 260–267.
106. G. Young, H. Wang and M. R. Zachariah, Application of nano-aluminum/nitrocellulose mesoparticles in composite solid rocket propellants, *Propellants, Explos., Pyrotech.*, 2015, **43**, 413–418.

Subject Index